T0201306

Information Quality

Information Quality

The Potential of Data and Analytics to Generate Knowledge

Ron S. Kenett

KPA, Israel and University of Turin, Italy

Galit Shmueli

National Tsing Hua University, Taiwan

Library of Congress Cataloging-in-Publication Data

Names: Kenett, Ron. | Shmueli, Galit, 1971–
Title: Information quality : the potential of data and analytics to generate knowledge /
   Ron S. Kenett, Dr. Galit Shmueli.
Description: Chichester, West Sussex : John Wiley & Sons, Inc., 2017. | Includes bibliographical references
   and index.
Identifiers: LCCN 2016022699| ISBN 9781118874448 (cloth) | ISBN 9781118890653 (epub)
Subjects: LCSH: Data mining. | Mathematical statistics.
Classification: LCC QA276 .K4427 2017 | DDC 006.3/12–dc23
LC record available at https://lccn.loc.gov/2016022699

A catalogue record for this book is available from the British Library.

Set in 10/12pt Times by SPi Global, Pondicherry, India
Printed and bound in Malaysia by Vivar Printing Sdn Bhd

10 9 8 7 6 5 4 3 2 1

To Sima; our children Dolav, Ariel, Dror, and Yoed; and their families and especially their children, Yonatan, Alma, Tomer, Yadin, Aviv, Gili, Matan, and Eden, they are my source of pride and motivation.

And to the memory of my dear friend, Roberto Corradetti, who dedicated his career to applied statistics.

RSK

To my family, mentors, colleagues, and students who've sparked and nurtured the creation of new knowledge and innovative thinking

GS

Contents

Foreword

I am often invited to assess research proposals. Included amongst the questions I have to ask myself in such assessments are: Are the goals stated sufficiently clearly? Does the study have a good chance of achieving the stated goals? Will the researchers be able to obtain sufficient quality data for the project? Are the analysis methods adequate to answer the questions? And so on. These questions are fundamental, not merely for research proposals, but for any empirical study – for any study aimed at extracting useful information from evidence or data. And yet they are rarely overtly stated. They tend to lurk in the background, with the capability of springing into the foreground to bite those who failed to think them through.

These questions are precisely the sorts of questions addressed by the InfoQ – Information Quality – framework. Answering such questions allows funding bodies, corporations, national statistical institutes, and other organisations to rank proposals, balance costs against success probability, and also to identify the weaknesses and hence improve proposals and their chance of yielding useful and valuable information. In a context of increasing constraints on financial resources, it is critical that money is well spent, so that maximising the chance that studies will obtain useful information is becoming more and more important. The InfoQ framework provides a structure for maximising these chances.

A glance at the statistics shelves of any technical library will reveal that most books focus narrowly on the details of data analytic methods. The same is true of almost all statistics teaching. This is all very well – it is certainly vital that such material be covered. After all, without an understanding of the basic tools, no analysis, no knowledge extraction would be possible. But such a narrow focus typically fails to place such work in the broader context, without which its chances of success are damaged. This volume will help to rectify that oversight. It will provide readers with insight into and understanding of other key parts of empirical analysis, parts which are vital if studies are to yield valid, accurate, and useful conclusions.

But the book goes beyond merely providing a framework. It also delves into the details of these overlooked aspects of data analysis. It discusses the fact that the same data may be high quality for one purpose and low for another, and that the adequacy of an analysis depends on the data and the goal, as well as depending on other less obvious aspects, such as the accessibility, completeness, and confidentiality of the data. And it illustrates the ideas with a series of illuminating applications.

With computers increasingly taking on the mechanical burden of data analytics the opportunities are becoming greater for us to shift our attention to the higher order aspects of analysis: to precise formulation of the questions, to consideration of data quality to answer those questions, to choice of the best method for the aims, taking account of the entire context of the analysis. In doing so we improve the quality of the conclusions we reach. And this, in turn, leads to improved decisions - for researchers, policy makers, managers, and others. This book will provide an important tool in this process.

David J. Hand
Imperial College London

About the authors

Ron S. Kenett is chairman of the KPA Group; research professor, University of Turin, Italy; visiting professor at the Hebrew University Institute for Drug Research, Jerusalem, Israel and at the Faculty of Economics, Ljubljana University, Slovenia. He is past president of the Israel Statistical Association (ISA) and of the European Network for Business and Industrial Statistics (ENBIS). Ron authored and coauthored over 200 papers and 12 books on topics ranging from industrial statistics, customer surveys, multivariate quality control, risk management, biostatistics and statistical methods in healthcare to performance appraisal systems and integrated management models. The KPA Group he formed in 1990 is a leading Israeli firm focused on generating insights through analytics with international customers such as hp, 3M, Teva, Perrigo, Roche, Intel, Amdocs, Stratasys, Israel Aircraft Industries, the Israel Electricity Corporation, ICL, start-ups, banks, and healthcare providers. He was awarded the 2013 Greenfield Medal by the Royal Statistical Society in recognition for excellence in contributions to the applications of statistics. Among his many activities he is member of the National Public Advisory Council for Statistics Israel; member of the Executive Academic Council, Wingate Academic College; and board member of several pharmaceutical and Internet product companies.

Galit Shmueli is distinguished professor at National Tsing Hua University's Institute of Service Science. She is known for her research and teaching in business analytics, with a focus on statistical and data mining methods in information systems and healthcare. She has authored and coauthored over 70 journal articles, book chapters, books, and textbooks, including *Data Mining for Business Analytics*, *Modeling Online Auctions* and *Getting Started with Business Analytics*. Her research is published in top journals in statistics, management, marketing, information systems, and more. Professor Shmueli has designed and instructed business analytics courses and programs since 2004 at the University of Maryland, the Indian School of Business, Statistics.com, and National Tsing Hua University, Taiwan. She has also taught engineering statistics courses at the Israel Institute of Technology and at Carnegie Mellon University.

Preface

This book is about a strategic and tactical approach to data analysis where providing added value by turning numbers into insights is the main goal of an empirical study. In our long-time experience as applied statisticians and data mining researchers ("data scientists"), we focused on developing methods for data analysis and applying them to real problems. Our experience has been, however, that data analysis is part of a bigger process that begins with problem elicitation that consists of defining unstructured problems and ends with decisions on action items and interventions that reflect on the true impact of a study.

In 2006, the first author published a paper on the statistical education bias where, typically, in courses on statistics and data analytics, only statistical methods are taught, without reference to the statistical analysis process (Kenett and Thyregod, 2006).

In 2010, the second author published a paper showing the differences between statistical modeling aimed at prediction goals versus modeling designed to explain causal effects (Shmueli, 2010), the implication being that the goal of a study should affect the way a study is performed, from data collection to data pre-processing, exploration, modeling, validation, and deployment. A related paper (Shmueli and Koppius, 2011) focused on the role of predictive analytics in theory building and scientific development in the explanatory-dominated social sciences and management research fields.

In 2014, we published "On Information Quality" (Kenett and Shmueli, 2014), a paper designed to lay out the foundation for a holistic approach to data analysis (using statistical modeling, data mining approaches, or any other data analysis methods) by structuring the main ingredients of what turns numbers into information. We called the approach information quality (InfoQ) and identified four InfoQ components and eight InfoQ dimensions.

Our main thesis is that data analysis, and especially the fields of statistics and data science, need to adapt to modern challenges and technologies by developing structured methods that provide a broad life cycle view, that is, from numbers to insights. This life cycle view needs to be focused on generating InfoQ as a key objective (for more on this see Kenett, 2015).

This book, *Information Quality: The Potential of Data and Analytics to Generate Knowledge*, offers an extensive treatment of InfoQ and the InfoQ framework. It is aimed at motivating researchers to further develop InfoQ elements and at students in programs that teach them how to make sure their analytic or statistical work is generating information of high quality.

Addressing this mixed community has been a challenge. On the one hand, we wanted to provide academic considerations, and on the other hand, we wanted to present examples and cases that motivate students and practitioners and give them guidance in their own specific projects.

We try to achieve this mix of objectives by combining Part I, which is mostly methodological, with Part II which is based on examples and case studies.

In Part III, we treat additional topics relevant to InfoQ such as reproducible research, the review of scientific and applied research publications, the incorporation of InfoQ in academic and professional development programs, and how three leading software platforms, R, MINITAB, and JMP support InfoQ implementations.

Researchers interested in applied statistics methods and strategies will most likely start in Part I and then move to Part II to see illustrations of the InfoQ framework applied in different domains. Practitioners and students learning how to turn numbers into information can start in a relevant chapter of Part II and move back to Part I.

A teacher or designer of a course on data analytics, applied statistics, or data science can build on examples in Part II and consolidate the approach by covering Chapter 13 and the chapters in Part I. Chapter 13 on "Integrating InfoQ into data science analytics programs, research methods courses and more" was specially prepared for this audience. We also developed five case studies that can be used by teachers as a rating-based InfoQ assessment exercise (available at http://infoq. galitshmueli.com/class-assignment).

In developing InfoQ, we received generous inputs from many people. In particular, we would like to acknowledge insightful comments by Sir David Cox, Shelley Zacks, Benny Kedem, Shirley Coleman, David Banks, Bill Woodall, Ron Snee, Peter Bruce, Shawndra Hill, Christine Anderson Cook, Ray Chambers, Fritz Sheuren, Ernest Foreman, Philip Stark, and David Steinberg. The motivation to apply InfoQ to the review of papers (Chapter 12) came from a comment by Ross Sparks who wrote to us: "I really like your framework for evaluating information quality and I have started to use it to assess papers that I am asked to review. Particularly applied papers." In preparing the material, we benefited from comprehensive editorial inputs by Raquelle Azran and Noa Shmueli who generously provided us their invaluable expertise—we would like to thank them and recognize their help in improving the text language and style.

The last three chapters were contributed by colleagues. They create a bridge between theory and practice showing how InfoQ is supported by R, MINITAB, and JMP. We thank the authors of these chapters, Silvia Salini, Federica Cugnata, Elena Siletti, Ian Cox, Pere Grima, Lluis Marco-Almagro, and Xavier Tort-Martorell, for their effort, which helped make this work both theoretical and practical.

We are especially thankful to Professor David J. Hand for preparing the foreword of the book. David has been a source of inspiration to us for many years and his contribution highlights the key parts of our work.

In the course of writing this book and developing the InfoQ framework, the first author benefited from numerous discussions with colleagues at the University of Turin, in particular with a great visionary of the role of applied statistics in modern business and industry, the late Professor Roberto Corradetti. Roberto has been a close friend and has greatly influenced this work by continuously emphasizing the need for statistical work to be appreciated by its customers in business and industry. In addition, the financial support of the Diego de Castro Foundation that he managed has provided the time to work in a stimulating academic environment at both the Faculty of Economics and the "Giuseppe Peano" Department of Mathematics of UNITO, the University of Turin. The contributions of Roberto Corradetti cannot be underestimated and are humbly acknowledged. Roberto passed away in June 2015 and left behind a great void. The second author thanks participants of the 2015 Statistical Challenges in eCommerce Research Symposium, where she presented the keynote address on InfoQ, for their feedback and enthusiasm regarding the importance of the InfoQ framework to current social science and management research.

Finally we acknowledge with pleasure the professional help of the Wiley personnel including Heather Kay, Alison Oliver and Adalfin Jayasingh and thank them for their encouragements, comments, and input that were instrumental in improving the form and content of the book.

Ron S. Kenett and Galit Shmueli

References

Kenett, R.S. (2015) Statistics: a life cycle view (with discussion). *Quality Engineering*, 27(1), pp. 111–129.

Kenett, R.S. and Shmueli, G. (2014) On information quality (with discussion). *Journal of the Royal Statistical Society, Series A*, 177(1), pp. 3–38.

Kenett, R.S. and Thyregod, P. (2006) Aspects of statistical consulting not taught by academia. *Statistica Neerlandica*, 60(3), pp. 396–412.

Shmueli, G. (2010) To explain or to predict? *Statistical Science*, 25, pp. 289–310.

Shmueli, G. and Koppius, O.R. (2011) Predictive analytics in information systems research. *MIS Quarterly*, 35(3), pp. 553–572.

Quotes about the book

What experts say about *Information Quality: The Potential of Data and Analytics to Generate Knowledge:*

A glance at the statistics shelves of any technical library will reveal that most books focus narrowly on the details of data analytic methods. The same is true of almost all statistics teaching. This volume will help to rectify that oversight. It will provide readers with insight into and understanding of other key parts of empirical analysis, parts which are vital if studies are to yield valid, accurate, and useful conclusions.

David Hand
Imperial College, London, UK

There is an important distinction between data and information. Data become information only when they serve to inform, but what is the potential of data to inform? With the work Kenett and Shmueli have done, we now have a general framework to answer that question. This framework is relevant to the whole analysis process, showing the potential to achieve higher-quality information at each step.

John Sall
SAS Institute, Cary, NC, USA

The authors have a rare quality: being able to present deep thoughts and sound approaches in a way practitioners can feel comfortable and understand when reading their work and, at the same time, researchers are compelled to think about how they do their work.

Fabrizio Ruggeri
Consiglio Nazionale delle Ricerche
Istituto di Matematica Applicata e Tecnologie Informatiche, Milan, Italy

No amount of technique can make irrelevant data fit for purpose, eliminate unknown biases, or compensate for data paucity. Useful, reliable inferences require balancing real-world and theoretical considerations and recognizing that goals, data, analysis, and costs are necessarily connected. Too often, books on statistics and data analysis put formulae in the limelight at the expense of more important questions about the relevance and limitations of data and the purpose of the analysis. This book elevates these crucial issues to their proper place and provides a systematic structure (and examples) to help practitioners see the larger context of statistical questions and, thus, to do more valuable work.

Phillip Stark
University of California, Berkeley, USA

...the "Q" issue is front and centre for anyone (or any agency) hoping to benefit from the data tsunami that is said to be driving things now ... And so the book will be very timely.

Ray Chambers
University of Wollongong, Australia

Kenett and Shmueli shed light on the biggest contributor to erroneous conclusions in research - poor information quality coming out of a study. This issue - made worse by the advent of Big Data - has received too little attention in the literature and the classroom. Information quality issues can completely undermine the utility and credibility of a study, yet researchers typically deal with it in an ad-hoc, offhand fashion, often when it is too late. Information Quality offers a sensible framework for ensuring that the data going into a study can effectively answer the questions being asked.

Peter Bruce
The Institute for Statistics Education

Policy makers rely on high quality and relevant data to make decisions and it is important that, as more and different types of data become available, we are mindful of all aspects of the quality of the information provided. This includes not only statistical quality, but other dimensions as outlined in this book including, very importantly, whether the data and analyses answer the relevant questions

John Pullinger
National Statistician, UK Statistics Authority, London, UK

This impressive book fills a gap in the teaching of statistical methodology. It deals with a neglected topic in statistical textbooks: the quality of the information provided by the producers of statistical projects and used by the customers of statistical data from surveys , administrative data etc. The emphasis in the book on: defining, discussing, analyzing the goal of the project at a preliminary stage and not less important at the analysis stage and use of the results obtained is of a major importance.

Moshe Sikron
Former Government Statistician of Israel, Jerusalem, Israel

Ron Kenett and Galit Shmueli belong to a class of practitioners who go beyond methodological prowess into questioning what purpose should be served by a data based analysis, and what could be done to gauge the fitness of the analysis to meet its purpose. This kind of insight is all the more urgent given the present climate of controversy surrounding science's own quality control mechanism. In fact science used in support to economic or policy decision – be it natural or social science - has an evident sore point precisely in the sort of statistical and mathematical modelling where the approach they advocate – Information Quality or InfoQ – is more needed. A full chapter is specifically devoted to the contribution InfoQ can make to clarify aspect of reproducibility, repeatability, and replicability of scientific research and publications. InfoQ is an empirical and flexible construct with practically infinite application in data analysis. In a context of policy, one can deploy InfoQ to compare different evidential bases pro or against a policy, or different options in an impact assessment case. InfoQ is a holistic construct encompassing the data, the method and the goal of the analysis. It goes beyond the dimensions of data quality met in official statistics and resemble more holistic concepts of performance such as analysis pedigrees (NUSAP) and sensitivity auditing. Thus InfoQ includes consideration of analysis' Generalizability and Action Operationalization. The latter include both action operationalization (to what extent concrete actions can be derived from the information provided by a study) and construct operationalization (to what extent a construct under analysis is effectively captured by the selected variables for a given goal). A desirable feature of InfoQ is that it demands multidisciplinary skills, which may force statisticians to move out of their comfort zone into the real world. The book illustrates the eight dimensions of InfoQ with a wealth of examples. A recommended read for applied statisticians and econometricians who care about the implications of their work.

Andrea Saltelli
European Centre for Governance in Complexity

Kenett and Shmueli have made a significant contribution to the profession by drawing attention to what is frequently the most important but overlooked aspect of analytics; information quality. For example, statistics textbooks too often assume that data consist of random samples and are measured without error, and data science competitions implicitly assume that massive data sets contain high-quality data and are exactly the data needed for the problem at hand. In reality, of course, random samples are the exception rather than the rule, and many data sets, even very large ones, are not worth the effort required to analyze them. Analytics is akin to mining, not to alchemy; the methods can only extract what is there to begin with. Kenett and Shmueli made clear the point that obtaining good data typically requires significant effort. Fortunately, they present metrics to help analysts understand the limitations of the information in hand, and how to improve it going forward. Kudos to the authors for this important contribution.

Roger Hoerl
Union College, Schenectady, NY USA

About the companion website

Don't forget to visit the companion website for this book:

www.wiley.com/go/information_quality

Here you will find valuable material designed to enhance your learning, including:

1. The JMP add-in presented in Chapter 16

2. Five case studies that can be used as exercises of InfoQ assessment

3. A set of presentations on InfoQ

Scan this QR code to visit the companion website.

Part I

THE INFORMATION QUALITY FRAMEWORK

1

Introduction to information quality

1.1 Introduction

Suppose you are conducting a study on online auctions and consider purchasing a dataset from eBay, the online auction platform, for the purpose of your study. The data vendor offers you four options that are within your budget:

1. Data on all the online auctions that took place in January 2012

2. Data on all the online auctions, for cameras only, that took place in 2012

3. Data on all the online auctions, for cameras only, that will take place in the next year

4. Data on a random sample of online auctions that took place in 2012

Which option would you choose? Perhaps none of these options are of value? Of course, the answer depends on the goal of the study. But it also depends on other considerations such as the analysis methods and tools that you will be using, the quality of the data, and the utility that you are trying to derive from the analysis. In the words of David Hand (2008):

> Statisticians working in a research environment… may well have to explain that the data are inadequate to answer a particular question.

Information Quality: The Potential of Data and Analytics to Generate Knowledge,
First Edition. Ron S. Kenett and Galit Shmueli.
© 2017 John Wiley & Sons, Ltd. Published 2017 by John Wiley & Sons, Ltd.
Companion website: www.wiley.com/go/information_quality

While those experienced with data analysis will find this dilemma familiar, the statistics and related literature do not provide guidance on how to approach this question in a methodical fashion and how to evaluate the value of a dataset in such a scenario.

Statistics, data mining, econometrics, and related areas are disciplines that are focused on extracting knowledge from data. They provide a toolkit for testing hypotheses of interest, predicting new observations, quantifying population effects, and summarizing data efficiently. In these empirical fields, measurable data is used to derive knowledge. Yet, a clean, exact, and complete dataset, which is analyzed professionally, might contain no useful information for the problem under investigation. In contrast, a very "dirty" dataset, with missing values and incomplete coverage, can contain useful information for some goals. In some cases, available data can even be misleading (Patzer, 1995, p. 14):

> Data may be of little or no value, or even negative value, if they misinform.

The focus of this book is on assessing the potential of a particular dataset for achieving a given analysis goal by employing data analysis methods and considering a given utility. We call this concept **information quality** (InfoQ). We propose a formal definition of InfoQ and provide guidelines for its assessment. Our objective is to offer a general framework that applies to empirical research. Such element has not received much attention in the body of knowledge of the statistics profession and can be considered a contribution to both the theory and the practice of applied statistics (Kenett, 2015).

A framework for assessing InfoQ is needed both when designing a study to produce findings of high InfoQ as well as at the postdesign stage, after the data has been collected. Questions regarding the value of data to be collected, or that have already been collected, have important implications both in academic research and in practice. With this motivation in mind, we construct the concept of InfoQ and then operationalize it so that it can be implemented in practice.

In this book, we address and tackle a high-level issue at the core of any data analysis. Rather than concentrate on a specific set of methods or applications, we consider a general concept that underlies any empirical analysis. The InfoQ framework therefore contributes to the literature on statistical strategy, also known as metastatistics (see Hand, 1994).

1.2   Components of InfoQ

Our definition of InfoQ involves four major components that are present in every data analysis: an analysis goal, a dataset, an analysis method, and a utility (Kenett and Shmueli, 2014). The discussion and assessment of InfoQ require examining and considering the complete set of its components as well as the relationships between the components. In such an evaluation we also consider eight dimensions that deconstruct the InfoQ concept. These dimensions are presented in Chapter 3. We start our introduction of InfoQ by defining each of its components.

Before describing each of the four InfoQ components, we introduce the following notation and definitions to help avoid confusion:

- g denotes a specific analysis goal.

- X denotes the available dataset.

- f is an empirical analysis method.

- U is a utility measure.

We use subscript indices to indicate alternatives. For example, to convey K different analysis goals, we use g_1, g_2,..., g_K; J different methods of analysis are denoted $f_1, f_2,..., f_J$.

Following Hand's (2008) definition of statistics as "the technology of extracting meaning from data," we can think of the InfoQ framework as one for evaluating the application of a technology (data analysis) to a resource (data) for a given purpose.

1.2.1   Goal (*g*)

Data analysis is used for a variety of purposes in research and in industry. The term "goal" can refer to two goals: the high-level goal of the study (the "domain goal") and the empirical goal (the "analysis goal"). One starts from the domain goal and then converts it into an analysis goal. A classic example is translating a hypothesis driven by a theory into a set of statistical hypotheses.

There are various classifications of study goals; some classifications span both the domain and analysis goals, while other classification systems focus on describing different analysis goals.

One classification approach divides the domain and analysis goals into three general classes: *causal explanation, empirical prediction,* and *description* (see Shmueli, 2010; Shmueli and Koppius, 2011). Causal explanation is concerned with establishing and quantifying the causal relationship between inputs and outcomes of interest. Lab experiments in the life sciences are often intended to establish causal relationships. Academic research in the social sciences is typically focused on causal explanation. In the social science context, the causality structure is based on a theoretical model that establishes the causal effect of some constructs (abstract concepts) on other constructs. The data collection stage is therefore preceded by a *construct operationalization* stage, where the researcher establishes which measurable variables can represent the constructs of interest. An example is investigating the causal effect of parents' intelligence on their children's intelligence. The construct "intelligence" can be measured in various ways, such as via IQ tests. The goal of empirical prediction differs from causal explanation. Examples include forecasting future values of a time series and predicting the output value for new observations given a set of input variables. Examples include recommendation systems on various websites, which are aimed at predicting services or products that the user is most likely to be interested in. Predictions of the economy are another type of predictive goal, with forecasts of particular economic measures or indices being of interest. Finally, descriptive goals include quantifying and testing for population effects by using data summaries, graphical visualizations, statistical models, and statistical tests.

A different, but related goal classification approach (Deming, 1953) introduces the distinction between *enumerative studies*, aimed at answering the question "how many?," and *analytic studies*, aimed at answering the question "why?"

A third classification (Tukey, 1977) classifies studies into exploratory and confirmatory data analysis.

Our use of the term "goal" includes all these different types of goals and goal classifications. For examples of such goals in the context of customer satisfaction surveys, see Chapter 7 and Kenett and Salini (2012).

1.2.2   Data (X)

Data is a broadly defined term that includes any type of data intended to be used in the empirical analysis. Data can arise from different collection instruments: surveys, laboratory tests, field experiments, computer experiments, simulations, web searches, mobile recordings, observational studies, and more. Data can be primary, collected specifically for the purpose of the study, or secondary, collected for a different reason. Data can be univariate or multivariate, discrete, continuous, or mixed. Data can contain semantic unstructured information in the form of text, images, audio, and video. Data can have various structures, including cross-sectional data, time series, panel data, networked data, geographic data, and more. Data can include information from a single source or from multiple sources. Data can be of any size (from a single observation in case studies to "big data" with zettabytes) and any dimension.

1.2.3   Analysis (f)

We use the general term *data analysis* to encompass any empirical analysis applied to data. This includes statistical models and methods (parametric, semiparametric, nonparametric, Bayesian and classical, etc.), data mining algorithms, econometric models, graphical methods, and operations research methods (such as simplex optimization). Methods can be as simple as summary statistics or complex multilayer models, computationally simple or computationally intensive.

1.2.4   Utility (U)

The extent to which the analysis goal is achieved is typically measured by some performance measure. We call this measure "utility." As with the study goal, utility refers to two dimensions: the utility from the domain point of view and the operationalized measurable utility measure. As with the goal, the linkage between the domain utility and the analysis utility measure should be properly established so that the analysis utility can be used to infer about the domain utility.

In predictive studies, popular utility measures are predictive accuracy, lift, and expected cost per prediction. In descriptive studies, utility is often assessed based on

Figure 1.1   The four InfoQ components.

goodness-of-fit measures. In causal explanatory modeling, statistical significance, statistical power, and strength-of-fit measures (e.g., R^2) are common.

1.3   Definition of information quality

Following Hand's (2008) definition of statistics as "the technology of extracting meaning from data," we consider the utility of applying a technology f to a resource X for a given purpose g. In particular, we focus on the question: What is the potential of a particular dataset to achieve a particular goal using a given data analysis method and utility? To formalize this question, we define the concept of InfoQ as

$$\text{InfoQ}(g, X, f, U) = U\{f(X \mid g)\}$$

The quality of information, InfoQ, is determined by the quality of its components g ("quality of goal definition"), X ("data quality"), f ("analysis quality"), and U ("quality of utility measure") as well as by the relationships between them. (See Figure 1.1 for a visual representation of InfoQ components.)

1.4   Examples from online auction studies

Let us recall the four options of eBay datasets we described at the beginning of the chapter. In order to evaluate the InfoQ of each of these datasets, we would have to specify the study goal, the intended data analysis, and the utility measure.

To better illustrate the role that the different components play, let us examine four studies in the field of online auctions, each using data to address a particular goal.

Case study 1   Determining factors affecting the final price of an auction

Econometricians are interested in determining factors that affect the final price of an online auction. Although game theory provides an underlying theoretical causal model of price in offline auctions, the online environment differs in substantial ways. Online auction platforms such as eBay.com have lowered the entry barrier for sellers and buyers to participate in auctions. Auction rules and settings can differ from classic on-ground auctions, and so can dynamics between bidders.

Let us examine the study "Public versus Secret Reserve Prices in eBay Auctions: Results from a Pokémon Field Experiment" (Katkar and Reiley, 2006) which investigated the effect of two types of reserve prices on the final auction price. A reserve price is a value that is set by the seller at the start of the auction. If the final price does not exceed the reserve price, the auction does not transact. On eBay, sellers can choose to place a public reserve price that is visible to bidders or an invisible secret reserve price, where bidders see only that there is a reserve price but do not know its value.

STUDY GOAL (g)

The researchers' goal is stated as follows:

> We ask, empirically, whether the seller is made better or worse off by setting a secret reserve above a low minimum bid, versus the option of making the reserve public by using it as the minimum bid level.

This question is then converted into the statistical goal (g) of testing a hypothesis "that secret reserve prices actually do produce higher expected revenues."

DATA (X)

The researchers proceed by setting up auctions for Pokémon cards[1] on eBay. com and auctioning off 50 matched pairs of Pokémon cards, half with secret reserves and half with equivalently high public minimum bids. The resulting dataset included information about bids,

---

[1] The Pokémon trading card game was one of the largest collectible toy crazes of 1999 and 2000. Introduced in early 1999, Pokémon game cards appeal both to game players and to collectors. Source: Katkar and Reiley (2006). © National Bureau of Economic Research.

bidders, and the final price in each of the 100 auctions, as well as whether the auction had a secret or public reserve price. The dataset also included information about the sellers' choices, such as the start and close time of each auction, the shipping costs, etc. This dataset constitutes X.

DATA ANALYSIS (f)

The researchers decided to "measure the effects of a secret reserve price (relative to an equivalent public reserve) on three different independent variables: the probability of the auction resulting in a sale, the number of bids received, and the price received for the card in the auction." This was done via linear regression models (f). For example, the sale/no sale outcome was regressed on the type of reserve (public/private) and other control variables, and the statistical significance of the reserve variable was examined.

UTILITY (U)

The authors conclude "The average drop in the probability of sale when using a secret reserve is statistically significant." Using another linear regression model with price as the dependent variable, statistical significance (the p-value) of the regression coefficient was used to test the presence of an effect for a private or public reserve price, and the regression coefficient value was used to quantify the magnitude of the effect, concluding that "a secret-reserve auction will generate a price \$0.63 lower, on average, than will a public-reserve auction." Hence, the utility (U) in this study relies mostly on statistical significance and p-values as well as the practical interpretation of the magnitude of a regression coefficient.

INFOQ COMPONENTS EVALUATION

What is the quality of the information contained in this study's dataset for testing the effect of private versus public reserve price on the final price, using regression models and statistical significance? The authors compare the advantages of their experimental design for answering their question of interest with designs of previous studies using observational data:

> With enough [observational] data and enough identifying econometric assumptions, one could conceivably tease out an empirical measurement of the reserve price effect from eBay field data... Such structural models make strong identifying assumptions in order to recover economic unobservables (such as bidders' private information about the item's value)... In contrast, our research project is much less ambitious, for we focus only on the effect of secret reserve prices relative to public reserve prices (starting bids). Our experiment allows us to carry out this measurement in a manner that is as simple, direct, and assumption-free as possible.

In other words, with a simple two-level experiment, the authors aim to answer a specific research question (g_1) in a robust manner, rather than build an extensive theoretical economic model (g_2) that is based on many assumptions.

Interestingly, when comparing their conclusions against prior literature on the effect of reserve prices in a study that used observational data, the authors mention that they find an opposite effect:

> Our results are somewhat inconsistent with those of Bajari and Hortaçsu.... Perhaps Bajari and Hortaçsu have made an inaccurate modeling assumption, or perhaps there is some important differ-ence between bidding for coin sets and bidding for Pokémon cards.

This discrepancy even leads the researchers to propose a new dataset that can help tackle the original goal with less confounding:

> A new experiment, auctioning one hundred items each in the $100 range, for example could shed some important light on this question.

This means that the InfoQ of the Pokémon card auction dataset is considered lower than that of a more expensive item.

Case study 2   Predicting the final price of an auction at the start of the auction

On any given day, thousands of auctions take place online. Forecasting the price of ongoing auctions is beneficial to buyers, sellers, auction houses, and third parties. For potential bidders, price forecasts can be used for deciding if, when, and how much to bid. For sellers, price forecasts can help decide whether and when to post another item for sale. For auction houses and third parties, services such as seller insurance can be offered with adjustable rates. Hence, there are different possible goals for empirical studies where price is the out-come variable, which translate into different InfoQ of a dataset. We describe in the succeeding text one particular study.

STUDY GOAL (g)

In a study by Ghani and Simmons (2004), the researchers collected historical auction data from eBay and used machine learning algorithms to predict end prices of auction items. Their question (g) was whether end prices of online auctions can be predicted accurately using machine learning methods. This is a predictive forward-looking goal, and the results of the study can improve scientific knowledge about predictability of online auction prices as well as serve as the basis for practical applications.

DATA (X)

The data collected for each closed auction included information about the seller, the item, the auction format, and "temporal features" (price statistics: starting bid, shipping price, and end price) of other auctions that closed recently. Note that all this information is available at the start of an auction of interest and therefore can be used as predictors for its final price. In terms of the outcome variable of interest—price—the data included the numerical end price (in USD). However, the authors considered two versions of this variable: the raw continuous variable and a multiclass categorical price variable where the numerical price is binned into $5 intervals.

DATA ANALYSIS (f)

In this study, several predictive algorithms (f) were used: for the numerical price, they used linear regression (and "polynomial regression with degrees 2 and 3"). For the categorical price, they used classification trees and neural networks.

UTILITY (U)

Because the authors' goal focused on predictive accuracy, their performance measures (U) were computed from a holdout set (RMSE for numerical price and accuracy % for categorical price). This set consisted of 400 auctions that were not used when building ("training") the models. They benchmarked their performance against a naive prediction—the average price (for numerical price) or most common price bin (for categorical price). The authors concluded:

> All of the methods we use[d] are effective at predicting the end-price of auctions. Regression results are not as promising as the ones for classification, mainly because the task is harder since an exact price is being predicted as opposed to a price range. In the future, we plan to narrow the bins for the price range and experiment with using classification algorithms to achieve more fine-grained results.

InfoQ COMPONENTS EVALUATION

For the purpose of their research goal, the dataset proved to be of high InfoQ. Moreover, they were able to assert the difference in InfoQ between two versions of their data (numerical and categorical price). Following their results, the authors proposed two applications where predicting price intervals of an auction might be useful:

> **Price Insurance**: Knowing the end-price before an auction starts provides an opportunity for a third-party to offer price insurance to sellers....
>
> **Listing Optimizer**: The model of the end price based on the input attributes of the auction can also be used to help sellers optimize the selling price of their items.

Case study 3   Predicting the final price of an ongoing auction

We now consider a different study, also related to predicting end prices of online auctions, but in this case predictions will be generated during an ongoing auction. The model used by Ghani and Simmons (2004) for forecasting the price of an auction is a "static model" in the sense that it uses information that is available at the start of the auction, but not later. This must be the case if the price forecasting takes place at the start of the auction. Forecasting the price of an ongoing auction is different: in addition to information available at the start of the auction, we can take into account all the information available at the time of prediction, such as bids that were placed thus far.

Recent literature on online auctions has suggested such models that integrate dynamic information that changes during the auction. Wang et al. (2008) developed a dynamic forecasting model that accounts for the unequal spacing of bids, the changing dynamics of price and bids throughout the auction, as well as static information about the auction, seller, and product. Their model has been used for predicting auction end prices for a variety of products (electronics, contemporary art, etc.) and across different auction websites (see Jank and Shmueli, 2010, Chapter 4). In the following, we briefly describe the Wang et al. (2008) study in terms of the InfoQ components.

STUDY GOAL (*g*)

The goal (*g*) stated by Wang et al. (2008) is to develop a forecasting model that predicts end prices of an ongoing online auction more accurately than traditional models. This is a forward-looking, predictive goal, which aims to benchmark a new modeling approach against existing methods. In addition to the main forecasting goal, the authors also state a secondary goal, to "systematically describe the empirical regularities of auction dynamics."

DATA (X)

The researchers collected data on a set of 190 closed seven-day auctions of Microsoft Xbox gaming systems and *Harry Potter and the Half-Blood Prince* books sold on eBay.com in August–September 2005. For each auction, the data included the bid history (bid amounts, time stamps, and bidder identification) and information on the product characteristics, the auction parameters (e.g., the day of week on which the auction started), and bidder and seller. Bid history information, which includes the timings and amounts of bids placed during the auction, was also used as predictor information.

DATA ANALYSIS (f)

The forecasting model proposed by Wang et al. (2008) is based on representing the sequences of bids from each auction by a smooth curve (using *functional data analysis*). An example for four auctions is shown in Figure 1.2. Then, a regression model for the price at time *t* includes four types of predictors:

  a. Static predictors (such as product characteristics)

  b. Time-varying predictors (such as the number of bids by time *t*)

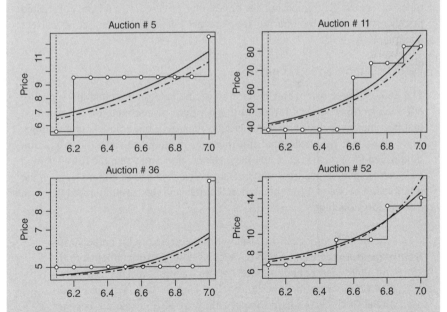

Figure 1.2   Price curves for the last day of four seven-day auctions (x-axis denotes day of auction). Current auction price (line with circles), functional price curve (smooth line) and forecasted price curve (broken line).

c. Price dynamics (estimated from the price curve derivatives)

d. Price lags

Their model for the price at time t is given by

$$y(t) = \alpha + \sum_{i=1}^{Q} \beta_i x_i(t) + \sum_{j=1}^{J} \beta_j D^j y(t) + \sum_{l=1}^{L} \eta_L y(t-l),$$

where $x_1(t),\ldots, x_Q(t)$ is the set of static and time-varying predictors, $D^{(j)}y(t)$ denotes the jth derivative of price at time t, and $y(t-l)$ is the lth price lag. The h-step-ahead forecast, given information up to time T, is given by

$$\bar{y}(T+h|T) = \hat{\alpha} + \sum_{i=1}^{Q} \hat{\beta}_i x_i(T+h|T) + \sum_{j=1}^{J} \hat{\gamma}_j \bar{D}^{(j)} y(T+h|T) + \sum_{l=1}^{L} \hat{\eta}_L \bar{y}(T+h-1|T).$$

UTILITY (U)

As in case study 2, predictive accuracy on a holdout set of auctions was used for evaluating model performance. In this study, the authors looked at two types of errors: (i) comparing the functional price curve and the forecasted price curve and (ii) comparing the forecast curves with the actual current auction prices.

INFOQ COMPONENTS EVALUATION

The authors make use of information in online auction data that are typically not used in other studies forecasting end prices of auctions: the information that becomes available during the auction regarding bid amounts and timings. They show that this additional information, if integrated into the prediction model, can improve forecast accuracy. Hence, they show that the InfoQ is high by generating more accurate forecasts as well as by shedding more light on the relationship between different auction features and the resulting bid dynamics. The authors conclude:

> The model produces forecasts with low errors, and it outperforms standard forecasting methods, such as double exponential smoothing, that severely underpredict the price evolution. This also shows that online auction forecasting is not an easy task. Whereas traditional methods are hard to apply, they are also inaccurate because they do not take into account the dramatic change in auction dynamics. Our model, on the other hand, achieves high forecasting accuracy and accommodates the changing price dynamics well.

Case study 4   Quantifying consumer surplus in eBay auctions

Classic microeconomic theory uses the notion of consumer surplus as the welfare measure that quantifies benefits to a consumer from an exchange. Marshall (1920, p. 124) defined consumer surplus as "the excess of the price which he (a consumer) would be willing to pay rather than go without the thing, over that which he actually does pay …."

Despite the growing research interest in online auctions, little is known about quantifiable consumer surplus levels in such mechanisms. On eBay, the winner is the highest bidder, and she or he pays the second highest bid. Whereas bid histories are publicly available, eBay never reveals the highest bid. Bapna et al. (2008) set out to quantify consumer surplus on eBay by using a unique dataset which revealed the highest bids for a sample of almost 5000 auctions. They found that, under a certain assumption, "eBay's auctions generated at least \$7.05 billion in total consumer surplus in 2003."

STUDY GOAL (g)

The researchers state the goal (g) as estimating the consumer surplus generated in eBay in 2003. This is a descriptive goal, and the purpose is to estimate this quantity with as much accuracy as possible.

DATA (X)

Since eBay does not disclose the highest bid in an auction, the researchers used a large dataset from Cniper.com, a Web-based tool used at the time by many eBay users for placing a "last minute bid." Placing a bid very close to the auction close ("sniping") is a tactic for winning an auction by avoiding the placement of higher bids by competing bidders. The Cniper dataset contained the highest bid for all the winners. The authors then merged the Cniper information with the eBay data for those auctions and obtained a dataset of 4514 auctions that took place between January and April 2003. Their dataset was also unique in that it contained information on auctions in three different currencies and across all eBay product categories.

EMPIRICAL ANALYSIS (f)

The researchers computed the median surplus by using the sample median with a 95% bootstrap confidence interval. They examined various subsets of the data and used regression analysis to correct for possible biases and to evaluate robustness to various assumption violations. For example, they compared their sample with a random sample from eBay in terms of the various variables, to evaluate whether Cniper winners were savvier and hence derived a higher surplus.

UTILITY (*U*)

The precision of the estimated surplus value was measured via a confidence interval. The bias due to nonrepresentative sampling was quantified by calculating an upper bound.

INFOQ COMPONENTS EVALUATION

The unique dataset available to the researchers allowed them to compute a metric that is otherwise unavailable from publicly available information on eBay.com. The researchers conducted special analyses to correct for various biases and arrived at the estimate of interest with conservative bounds. The InfoQ of this dataset is therefore high for the purpose of the study.

1.5   InfoQ and study quality

We defined InfoQ as a framework for answering the question: What is the potential of a particular dataset to achieve a particular goal using a given data analysis method and utility? In each of the four studies in Section 1.4, we examined the four InfoQ components and then evaluated the InfoQ based on examining the components. In Chapter 3 we introduce an InfoQ assessment approach, which is based on eight dimensions of InfoQ. Examining each of the eight dimensions assists researchers and analysts in evaluating the InfoQ of a dataset and its associated study.

In addition to using the InfoQ framework for evaluating the potential of a dataset to generate information of quality, the InfoQ framework can be used for retrospective evaluation of an empirical study. By identifying the four InfoQ components and assessing the eight InfoQ dimensions introduced in Chapter 3, one can determine the usefulness of a study in achieving its stated goal. In part II of the book, we take this approach and examine multiple studies in various domains. Chapter 12 in part III describes how the InfoQ framework can provide a more guided process for authors, reviewers and editors of scientific journals and publications.

1.6   Summary

In this chapter we introduced the concept of InfoQ and its four components. In the following chapters, we discuss how InfoQ differs from the common concepts of data quality and analysis quality. Moving from a concept to a framework that can be applied in practice requires a methodology for assessing InfoQ. In Chapter 3, we break down InfoQ into eight dimensions, to facilitate quantitative assessment of InfoQ. The final chapters (Chapters 4 and 5) in part I examine existing statistical methodology aimed at increasing InfoQ at the study design stage and at the postdata collection stage. Structuring and examining various statistical approaches through the InfoQ lens creates a clearer picture of the role of different statistical approaches and methods, often taught in different courses or used in separate fields. In summary, InfoQ is about assessing and improving the potential of a dataset to achieve a particular goal using a given data analysis method and utility. This book is about structuring and consolidating such an approach.

References

Bapna, R., Jank, W. and Shmueli, G. (2008) Consumer surplus in online auctions. *Information Systems Research*, 19, pp. 400–416.

Deming, W.E. (1953) On the distinction between enumerative and analytic studies. *Journal of the American Statistical Association*, 48, pp. 244–255.

Ghani, R. and Simmons, H. (2004) Predicting the End-Price of Online Auctions. *International Workshop on Data Mining and Adaptive Modelling Methods for Economics and Management*, Pisa, Italy.

Hand, D.J. (1994) Deconstructing statistical questions (with discussion). *Journal of the Royal Statistical Society, Series A*, 157(3), pp. 317–356.

Hand, D.J. (2008) *Statistics: A Very Short Introduction*. Oxford University Press, Oxford.

Jank, W. and Shmueli, G. (2010) *Modeling Online Auctions*. John Wiley & Sons, Inc., Hoboken.

Katkar, R. and Reiley, D.H. (2006) Public versus secret reserve prices in eBay auctions: results from a Pokemon field experiment. *Advances in Economic Analysis and Policy*, 6(2), article 7.

Kenett, R.S. (2015) Statistics: a life cycle view (with discussion). *Quality Engineering*, 27(1), pp. 111–129.

Kenett, R.S. and Salini, S. (2012) Modern analysis of customer surveys: comparison of models and integrated analysis (with discussion). *Applied Stochastic Models in Business and Industry*, 27, pp. 465–475.

Kenett, R.S. and Shmueli, G. (2014) On information quality (with discussion). *Journal of the Royal Statistical Society, Series A*, 177(1), pp. 3–38.

Marshall, A. (1920) *Principles of Economics*, 8th edition. MacMillan, London.

Patzer, G.L. (1995) *Using Secondary Data in Marketing Research*. Praeger, Westport, CT.

Shmueli, G. (2010) To explain or to predict? *Statistical Science*, 25, pp. 289–310.

Shmueli, G. and Koppius, O.R. (2011) Predictive analytics in information systems research. *Management Information Systems Quarterly*, 35, pp. 553–572.

Tukey, J.W. (1977) *Exploratory Data Analysis*. Addison-Wesley, Reading, PA.

Wang, S., Jank, W. and Shmueli, G. (2008) Explaining and forecasting online auction prices and their dynamics using functional data analysis. *Journal of Business and Economics Statistics*, 26, pp. 144–160.

2

Quality of goal, data quality, and analysis quality

2.1 Introduction

> Far better an approximate answer to the **right** question, which is often vague, than an **exact** answer to the wrong question, which can always be made precise.
>
> John Tukey, 1962

At the most basic level, the quality of a goal under investigation depends on whether the stated goal is of interest and relevant either scientifically or practically. At the next level, the quality of a goal is derived from translating a scientific or practical goal into an empirical goal. This challenging step requires knowledge of both the problem domain and data analysis and necessitates close collaboration between the data analyst and the domain expert. A well-defined empirical goal is one that properly reflects the scientific or practical goal. Although a dataset can be useful for one scientific goal g_1, it can be completely useless for a second scientific goal g_2. For example, monthly average temperature data for a city can be utilized to quantify and understand past trends and seasonal patterns, goal g_1, but cannot be used effectively for generating future daily weather forecasts, goal g_2. The challenge is therefore to define the right empirical question under study in order to avoid what Kimball (1957) calls "error of the third kind" or "giving the right answer to the wrong question."

Information Quality: The Potential of Data and Analytics to Generate Knowledge,
First Edition. Ron S. Kenett and Galit Shmueli.
© 2017 John Wiley & Sons, Ltd. Published 2017 by John Wiley & Sons, Ltd.
Companion website: www.wiley.com/go/information_quality

The task of goal definition is often more difficult than any of the other stages in a study. Hand (1994) says:

> It is clear that establishing the mapping from the client's domain to a statistical question is one of the most difficult parts of a statistical analysis.

Moreover, Mackay, and Oldford (2000) note that this important step is rarely mentioned in introductory statistics textbooks:

> Understanding what is to be learned from an investigation is so important that it is surprising that it is rarely, if ever, treated in any introduction to statistics. In a cursory review, we could find no elementary statistics text that provided a structure to understand the problem.

Several authors have indicated that the act of finding and formulating a problem is a key aspect of creative thought and performance, an act that is distinct from, and perhaps more important than, problem solving (see Jay and Perkins, 1997).

Quality issues of goal definition often arise when translating a stakeholder's language into empirical jargon. An example is a marketing manager who requests an analyst to use the company's existing data to "understand what makes customers respond positively or negatively to our advertising." The analyst might translate this statement into an empirical goal of identifying the causal factors that affect customer responsiveness to advertising, which could then lead to designing and conducting a randomized experiment. However, in-depth discussions with the marketing manager may lead the analyst to discover that the analysis results are intended to be used for targeting new customers with ads. While the manager used the English term "understand," his/her goal in empirical language was to "predict future customers' ad responsiveness." Thus, the analyst should develop and evaluate a predictive model rather than an explanatory one. To avoid such miscommunication, a critical step for analysts is to learn how to elicit the required information from the stakeholder and to understand how their goal translates into empirical language.

2.1.1 Goal elicitation

One useful approach for framing the empirical goal is scenario building, where the analyst presents different scenarios to the stakeholder of how the analysis results might be used. The stakeholder's feedback helps narrow the gap between the intended goal and its empirical translation. Another approach, used in developing integrated information technology (IT) systems, is to conduct goal elicitation using organizational maps. A fully developed discipline, sometimes called goal-oriented requirements engineering (GORE), was designed to do just that (Dardenne et al., 1993; Regev and Wegmann, 2005).

2.1.2   From theory to empirical hypotheses

In academic research, different disciplines have different methodologies for translating a scientific question into an empirical goal. In the social sciences, such as economics or psychology, researchers start from a causal theory and then translate it into statistical hypotheses by a step of operationalization. This step, where abstract concepts are mapped into measurable variables, allows the researcher to translate a conceptual theory into an empirical goal. For example, in quantitative linguistics, one translates scientific hypotheses about the human language faculty and its use in the world into statistical hypotheses.

2.1.3   Goal quality, InfoQ, and goal elicitation

Defining the study goal inappropriately, or translating it incorrectly into an empirical goal, will obviously negatively affect information quality (InfoQ). InfoQ relies on, but does not assess the quality of, the goal definition. The InfoQ framework offers an approach that helps assure the alignment of the study goal with the other components of the study. Since goal definition is directly related to the data, data analysis and utility, the InfoQ definition is dependent on the goal, $U(f(X|g))$, thereby requiring a clear goal definition and considering it at every step. By directly considering the goal, using the InfoQ framework raises awareness to the stated goal, thereby presenting opportunities for detecting challenges or issues with the stated goal.

Moreover, the InfoQ framework can be used to enhance the process of *goal elicitation* and *hypothesis generation*. It is often the case that researchers formulate their goals once they see and interact with the data. In his commentary on the paper "On Information Quality" (Kenett and Shmueli, 2013), Schouten (2013) writes about the importance and difficulty of defining the study goal and the role that the InfoQ framework can play in improving the quality of goal definition. He writes:

> A decisive ingredient to information quality is the goal or goals that researchers have set when starting an analysis. From my own experience and looking at analyses done by others, I conclude that research goals may not be that rigorously defined and/or stated beforehand. They should of course be well defined in order to assess the fitness for use of data, but often data exploration and analysis sharpen the mind of the researcher and goals get formed interactively. As such I believe that an assessment of the InfoQ dimensions may actually be helpful in deriving more specific and elaborated analysis goals. Still, I suspect that the framework is only powerful when researchers have well defined goals.

2.2   Data quality

> Raw data, like raw potatoes, usually require cleaning before use.
>
> Thisted, in Hand, 2008

It is rare to meet a data set which does not have quality problems of some kind.

Hand, 2008

Data quality is a critically important subject. Unfortunately, it is one of the least understood subjects in quality management and, far too often, is simply ignored.

Godfrey, 2008

Data quality has long been recognized by statisticians and data analysts as a serious challenge. Almost all data requires some cleaning before it can be further used for analysis. However, the level of cleanliness and the approach to data cleaning depend on the goal. Using the InfoQ notation, data quality typically concerns $U(X|g)$. The same data can contain high-quality information for one purpose and low-quality information for another. This has been recognized and addressed in several fields. Mallows (1998) posed the *zeroth problem*, asking "How do the data relate to the problem, and what other data might be relevant?" In the following we briefly examine several approaches to data quality in different fields and point out how they differ from InfoQ.

2.2.1   MIS-type data quality

In database engineering and management information systems (MIS), the term "data quality" refers to the usefulness of queried data to the person querying it. Wang et al. (1993) gave the following example:

> Acceptable levels of data quality may differ from one user to another. An investor loosely following a stock may consider a ten minute delay for share price sufficiently timely, whereas a trader who needs price quotes in real time may not consider ten minutes timely enough.

Another aspect sometimes attributable to data quality is conformance to specifications or standards. Wang et al. (1993) define data quality as "conformance to requirements." For the purpose of evaluating data quality, they use "data quality indicators." These indicators are based on objective measures such as data source, creation time, collection method and subjective measures such as the level of credibility of a source as assigned by a researcher. In the United Kingdom, for instance, the Department of Health uses an MIS type of definition of data quality with respect to the quality of medical and healthcare patient data in the National Health Service (UK Department of Health, 2004).

Lee et al. (2002) propose a methodology for InfoQ assessment and benchmarking, called assessment of information system methodology and quality (AIMQ). Their focus is on usefulness of organizational data to its users, specifically, data from IT systems. The authors define four categories of InfoQ: intrinsic, contextual, representational and accessibility. While the intrinsic category refers to "information

[that] has quality in its own right," the contextual category takes into account the task at hand (from the perspective of a user), and the last two categories relate to qualities of the information system. Lee et al.'s use of the term "InfoQ" indicates that they consider the data in the context of the user, rather than in isolation (as the term data quality might imply). The AIMQ methodology is used for assessing and benchmarking an organization's own data usage.

A main approach to InfoQ implemented in the context of MIS is the application of entity resolution (ER) analysis. ER is the process of determining whether two references to real-world objects are referring to the same object or two different objects. The degree of completeness, accuracy, timeliness, believability, consistency, accessibility and other aspects of reference data can affect the operation of ER processes and produce better or worse outcomes. This is one of the reasons that ER is so closely related to the MIS field of IQ, an emerging discipline concerned with maximizing the value of an organization's information assets and assuring that the information products it produces meet the expectations of the customers who use them. Improving the quality of reference sources dramatically improves ER process results, and conversely, integrating the references through ER improves the overall quality of the information in the system. ER systems generally use four basic techniques for determining that references are equivalent and should be linked: direct matching, association analysis, asserted equivalence and transitive equivalence. For an introduction to ER, see Talburt (2011). For a case study on open source software conducting ER analysis, in the context of healthcare systems, see Zhou et al. (2010).

Gackowski (2005) reviews popular MIS textbooks and states:

> Current MIS textbooks are deficient with regard to the role of end users and even more so about the information disseminators. The texts are overly technology laden, with oversimplified coverage of the fundamentals on data, information, and particularly the role of informing in business.

The treatment of InfoQ in this book addresses this void and, in some sense, links to data quality communities such as the International Association for Information and Data Quality (IAIDQ).

In a broader context, technology can improve data quality. For example, in manual data entry to an automated system, automatic data validation can provide immediate feedback so that data entry errors can be corrected on the spot. Technological advances in electronic recording, scanners, RFID, electronic entry, electronic data transfer, data verification technologies and robust data storage, as well as more advanced measurement instruments, have produced over time much "cleaner" data (Redman, 2007).

These data quality issues focus on $U(X|g)$, which differs from InfoQ by excluding the data analysis component f. In addition, the MIS reference to utility is typically qualitative and not quantitative. It considers utility as the value of information provided to the receiver in the context of its intended use. In InfoQ, the utility $U(X|g)$ is considered with a quantitative perspective and consists of statistical measures such as prediction error or estimation bias.

2.2.2 Quality of statistical data

A similar concept is the *quality of statistical data* which has been developed and used in official statistics and international organizations that routinely collect data. The concept of *quality of statistical data* refers to the usefulness of summary statistics that are produced by national statistics agencies and other producers of official statistics. This is a special case of InfoQ where f is equivalent to computing summary statistics (although this operation might seem very simple, it is nonetheless considered "analysis," because it in fact involves estimation).

Such organizations have created frameworks for assessing the quality of statistical data. The International Monetary Fund (IMF) and the Organisation for Economic Co-operation and Development (OECD) each developed an assessment framework. Aspects that they assess are *relevance, accuracy, timeliness, accessibility, interpretability, coherence* and *credibility*. These different dimensions are each assessed separately—either subjectively or objectively. For instance, the OECD's definition of *relevance* of statistical data refers to a qualitative assessment of the value contributed by the data. Other aspects are more technical in nature. For example, *accessibility* refers to how readily the data can be located and accessed. See Chapter 3 for further details on the data quality dimensions used by government and international agencies.

In the context of survey quality, official agencies such as Eurostat, the National Center for Science and Engineering Statistics, and Statistics Canada have created quality dimensions for evaluating the quality of a survey for the goal g of obtaining "accurate survey data" as measured by U equivalent to the mean square error (MSE) (see Biemer and Lyberg (2003)). Such agencies have also defined a set of data quality dimensions for the purpose of evaluating data quality. For example, Eurostat's quality dimensions are *relevance of statistical concept, accuracy of estimates, timeliness, and punctuality in disseminating results, accessibility, and clarity of the information, comparability, coherence,* and *completeness* (see www.nsf.gov/statistics for the National Center for Science and Engineering Statistics guidelines and standards).

2.2.3 Data quality in statistics

In the statistics literature, discussions of data quality mostly focus on the cleanliness of the data in terms of data entry errors, missing values, measurement errors and so on. These different aspects of data quality can be classified into different groups using different criteria. For example, Hand (2008) distinguishes between two types of data quality problems: *incomplete data* (including missing values and sampling bias) and *incorrect data.*

In the InfoQ framework, we distinguish between strictly data quality issues and InfoQ issues on the basis of whether they relate only to X or to one or more of the InfoQ components. An issue is a "data quality" issue if it characterizes a technical aspect of the data that can be "cleaned" with adequate technology and without knowledge of the goal. Aspects such as data entry errors, measurement errors and corrupted data are therefore classified as "data quality." Data issues that involve the goal, analysis and/or utility of the study are classified as "InfoQ" issues. These include sampling bias and missing values, which are not simply technical errors: their definition or impact depends on the study goal g. Sampling bias, for example, is relative to the population of interest: the same sample can be biased for one goal and unbiased for another goal. Missing values can add uncertainty for achieving one goal, but reduce uncertainty for achieving another goal (e.g., missing information in financial reports can be harmful for assessing financial performance, but helpful for detecting fraudulent behavior).

Other classifications of data quality issues are also possible. Schouten (2013) distinguishes between data quality and InfoQ, saying "data quality is about the data that one intended to have and InfoQ is about the data that one desired to have." According to his view, different methods are used to improve data quality and InfoQ. "Data processing, editing, imputing and weighting [aim at] reducing the gap between the data at hand and the data that one intended to have. These statistical methods ... aim at improving data quality. Data analysis is about bridging the gap between intended and desired data."

In the remainder of the book, we use the classifications of "data quality" and "InfoQ" based on whether the issue pertains to the data alone (X) or to at least one more InfoQ component.

2.3   Analysis quality

All models are wrong, but some are useful.

Box, 1979

Analysis quality refers to the adequacy of the empirical analysis in light of the data and goal at hand. Analysis quality reflects the adequacy of the modeling with respect to the data and for answering the question of interest. Godfrey (2008) described low analysis quality as "poor models and poor analysis techniques, or even analyzing the data in a totally incorrect way." We add to that the ability of the stakeholder to use the analysis results. Let us consider a few aspects of analysis quality, so that it becomes clear how it differs from InfoQ and how the two are related.

2.3.1   Correctness

Statistics education as well as education in other related fields such as econometrics and data mining is aimed at teaching high-quality data analysis. Techniques for checking analysis quality include graphic and quantitative methods such as residual analysis and cross validation as well as qualitative evaluation such as consideration of endogeneity (reverse causation) in causal studies. Analysis quality depends on the expertise of the analyst and on the empirical methods and software available at the time of analysis.

Analysis quality greatly depends on the goal at hand. The coupling of analysis and goal allows a broader view of analysis adequacy, since "textbook" approaches often consider the suitability of a method for use with specific data types for a specific goal. But usage might fall outside that scope and still be useful. As an example, the textbook use of linear regression models requires data that adheres to the assumption of independent observations. Yet, the use of linear regression for forecasting time series, where the observations are typically auto-correlated, is widely used in practice because it meets the goal of sufficiently accurate forecasts. Naive Bayes classifiers are built on the assumption of conditional independence of predictors, yet despite the violation of the assumption in most applications, naive Bayes provides excellent classification performance.

Analysis quality refers not only to the statistical model used but also to the methodology. For example, comparing a predictive model to a benchmark is a necessary methodological step.

2.3.2   Usability of the analysis

The usability of the analysis method to the stakeholder is another facet of analysis quality. In applications such as credit risk, regulations exist regarding information that must be conveyed to customers whose credit requests are denied. In such cases, using models that are not transparent in terms of the variables used ("blackbox" models) will have low analysis quality, even if they adequately deny/grant credit to prospects.

2.3.3   Culture of data analysis

Finally, there is a subjective aspect of analysis quality. Different academic fields maintain different "cultures" and norms of what is considered acceptable data analysis. For example, regression models are often used for causal inference in the social sciences, while in other fields such a use is deemed unacceptable. Hence, analysis quality also depends on the culture and environment of the researcher or analysis team.

When considering InfoQ, analysis quality is not examined against textbook assumptions or theoretical properties. Instead, it is judged against the specific goal g and the utility U of using the analysis method f with the specific dataset X for that particular goal.

2.4   Quality of utility

Like the study goal, the utility of a study provides the link between the domain and the empirical worlds. The analyst must understand the intended use and purpose of the analysis in order to choose the correct empirical utility or performance measures. The term "utility" refers to the overall usefulness of what the study aims to achieve, as well as to the set of measures and metrics used to assess the empirical results.

2.4.1   Statistical performance

A model may be useful along one dimension and worse than useless along another.

Berk et al., 2013

The field of statistics offers a wealth of utility measures, tests, and charts aimed at gauging the performance of a statistical model. Methods range from classical to Bayesian; loss functions range from L_1-distance to L_2-distance metrics; metrics are based on in-sample or out-of-sample data. They include measures of goodness of fit (e.g., residual analysis) and strength-of-relationship tests and measures (e.g., R^2 and p-values in regression models).

Predictive performance measures include penalized metrics such as the Akaike information criterion (AIC) and Bayes information criterion (BIC) and out-of-sample measures such as root mean square error (RMSE), mean absolute percentage error (MAPE) and other aggregations of prediction errors. One can use symmetric cost functions on prediction errors or asymmetric ones which more heavily penalize over- or underprediction. Even within predictive modeling, there exist a variety of metrics depending on the exact predictive task and data type: for classification (predicting a categorical outcome) one can use a classification matrix, overall error, measures of sensitivity and specificity, recall and precision, receiver operating curves (ROC), and area under curve (AUC) metrics. For predicting numerical records, various aggregations of prediction errors exist that weigh the direction and magnitude of the error differently. For ranking new records, lift charts are most common.

As a side note, Akaike originally called his approach an "entropy maximization principle," because the approach is founded on the concept of entropy in information theory. Minimizing AIC in a statistical model is equivalent to maximizing entropy in a thermodynamic system; in other words, the information-theoretic approach in statistics essentially applies the second law of thermodynamics. As such, AIC generalizes the work of Boltzmann on entropy to model selection in the context of generalized regression (GR). We return to the important dimension of generalization in the context of InfoQ dimensions in the next chapter.

With such a rich set of potential performance measures, utility quality greatly depends on the researcher's ability and knowledge to choose adequate metrics.

2.4.2   Decision-theoretical and economic utility

In practical applications, there are often costs and gains associated with decisions to be made based on the modeling results. For example, in fraud detection, misclassifying a fraudulent case is associated with some costs, while misclassifying a non-fraudulent case is associated with other costs. In studies of this type, it is therefore critical to use cost-based measures for assessing the utility of the model. In the field of statistical process control (SPC), the parameters of classic control charts are based on the sampling distribution of the monitored statistic (typically the sample mean or standard deviation). A different class of control charts is based on "economic design" (see Montgomery, 1980; Kenett et al., 2014), where chart limits, sample size and time between samples are set based on a cost minimization model that takes into account costs due to sampling, investigation, repair and producing defective products (Serel, 2009).

Decision theory provides a rational framework for choosing between alternative courses of action when the consequences resulting from this choice are imperfectly known. In Lindley's foreword to an edited volume by Di Bacco et al. (2004), he treats the question what is meant by statistics by referring to those he considers the founding fathers: Harold Jeffreys, Bruno de Finetti, Frank Ramsey, and Jimmie Savage:

> Both Jeffreys and de Finetti developed probability as the coherent appreciation of uncertainty, but Ramsey and Savage looked at the world rather differently. Their starting point was not the concept of uncertainty but rather decision-making in the face of uncertainty. They thought in terms of action, rather than in the passive contemplation of the uncertain world. Coherence for them was not so much a matter of how your beliefs hung together but of whether your several actions, considered collectively, make sense....

2.4.3   Computational performance

In industrial applications, computation time is often of great importance. Google's search engine must return query results with very little lag time; Amazon and Facebook must choose the product or ad to be presented immediately after a user clicks. Even in academic studies, in some cases research groups cannot wait for months to complete a run, thereby resorting to shortcuts, approximations, parallel computing and other solutions to speed run time. In such cases, the utility also includes computational measures such as run time and computational resources, as well as scalability.

2.4.4   Other types of utility

Interpretability of the analysis results might be considered critical to the utility of a model, where one prefers interpretable models over "blackbox" models, while in other cases the user might be agnostic to interpretation, so that interpretability is not part of the utility function.

For academics, a critical utility of a data analysis is publication! Hence, choices of performance metrics might be geared toward attaining the requirements in their field. These can vary drastically by field. For example, the journal *Basic and Applied Social Psychology* recently announced that it will not publish *p*-values, statistical tests, significance statement or confidence intervals in submitted manuscripts (Trafimow and Marks, 2015).

2.4.5   Connecting empirical utility and study utility

When considering the quality of the utility U, two dangers can lower quality: (i) an absence of a study utility, limiting the study to statistical utility measures, and (ii) a mismatch between the utility measure and the utility of the study.

With respect to focusing solely on statistical utility, we can quote again Lindley (2004) who criticized current publications that use Bayesian methods for neglecting to consider utility. He writes:

> If one looks today at a typical statistical paper that uses the Bayesian method, copious use will be made of probability, but utility, or maximum expected utility (MEU), will rarely get a mention... When I look at statistics today, I am astonished at the almost complete failure to use utility.... Probability is there but not utility. This failure has to be my major criticism of current statistics; we are abandoning our task half-way, producing the inference but declining to explain to others how to act on that inference. The lack of papers that provide discussions on utility is another omission from our publications.

Choosing the right measures depends on correctly identifying the underlying study utility as well as correctly translating the study utility into empirical metrics. This is similar to the case of goal definition and its quality.

Performance measures must depend on the goal at hand, on the nature of the data and on the analysis method. For example, a common error in various fields is using the R^2 statistic for measuring predictive accuracy (see Shmueli, 2010; Shmueli and Koppius, 2011). Recall our earlier example of a marketing manager who tells the analyst the analysis goal is to "understand customer responsiveness to advertising," while effectively the model is to be used for targeting new customers with ads. If the analyst pursues (incorrectly) an explanatory modeling path, then their choice of explanatory performance measures, such as R^2 ("How well does my model explain the effect of customer information on their ad responsiveness?"), would lower the quality of utility. This low-quality utility would typically be discovered in the model deployment stage, where the explanatory model's predictive power would be observed for the first time.

Another error that lowers utility quality is relying solely on p-values for testing hypotheses with very large samples, a common practice in several fields that now use hundreds of thousands or even millions of observations. Because p-values are a function of sample size, with very large samples one can obtain tiny p-values (highly statistically significant) for even miniscule effects. One must therefore examine the effect size and consider its practical relevance (see Lin et al., 2013).

With the proliferation of data mining contests, hosted on public platforms such as kaggle.com, there has been a strong emphasis on finding a model that optimizes a specific performance measure, such as RMSE or lift. However, in real-life studies, it is rarely the case that a model is chosen on the basis of a single utility measure. Instead, the analyst considers several measures and examines the model's utility under various practical considerations, such as adequacy of use by the stakeholder, cost of deployment and robustness under different possible conditions. Similarly, in academic research, model selection is based not on optimizing a single utility measure but rather on additional criteria such as parsimony and robustness and, importantly, on supporting meaningful discoveries.

The quality of the utility therefore directly impacts InfoQ. As with goal quality, the InfoQ framework raises awareness to the relationship between the domain and empirical worlds, thereby helping to avoid disconnects between analysis and reality, as is the case in data mining competitions.

2.5   Summary

This chapter lays the foundation for the remainder of the book by examining each of the four InfoQ components (goal, data, analysis, and utility) from the perspective of quality. We consider the intrinsic quality of these components, thereby differentiating the single components' quality from the overall notion of InfoQ. The next chapter introduces the eight InfoQ dimensions used to deconstruct the general concept of InfoQ. InfoQ combines the four components treated here with the eight dimensions discussed in Chapter 3. The examples in this and other chapters show how InfoQ combines data collection and organization with data analytics and operationalization, designed to achieve specific goals reflecting particular utility functions. In a sense, InfoQ expands on the domain of decision theory by considering modern implications of data availability, advanced and accessible analytics and data-driven systems with operational tasks. Following Chapter 3 we devote specific chapters to the data collection and study design phase and the postdata collection phase, from the perspective of InfoQ. Examples in a wide range of applications are provided in part II of the book.

References

Berk, R.A., Brown, L., George, E., Pitkin, E., Traskin, M., Zhang, K. and Zhao, L. (2013) What You Can Learn from Wrong Causal Models, in *Handbook of Causal Analysis for Social Research*, Morgan, S.L. (editor), Springer, Dordrecht.

Biemer, P. and Lyberg, L. (2003) *Introduction to Survey Quality*. John Wiley & Sons, Inc., Hoboken, NJ.

Box, G.E.P. (1979) Robustness in the Strategy of Scientific Model Building, in *Robustness in Statistics*, Launer, R.L. and Wilkinson, G.N. (editors), Academic Press, New York, pp. 201–236.

Dardenne, A., van Lamsweerde, A. and Fickas, S. (1993) Goal-directed requirements acquisition. *Science of Computer Programming*, 20, pp. 3–50.

Di Bacco, N., d'Amore, G. and Scalfari, F. (2004) *Applied Bayesian Statistical Studies in Biology and Medicine*. Springer, Boston, MA.

Gackowski, Z. (2005) Informing systems in business environments: a purpose-focused view. *Informing Science Journal*, 8, pp. 101–122.

Godfrey, A.B. (2008) Eye on data quality. *Six Sigma Forum Magazine*, 8, pp. 5–6.

Hand, D.J. (1994) Deconstructing statistical questions (with discussion). *Journal of the Royal Statistical Society, Series A*, 157(3), pp. 317–356.

Hand, D.J. (2008) *Statistics: A Very Short Introduction*. Oxford University Press, Oxford.

Jay, E.S. and Perkins, D.N. (1997) Creativity's Compass: A Review of Problem Finding, in *Creativity Research Handbook*, vol. 1, Runco, M.A. (editor), Hampton, Cresskill, NJ, pp. 257–293.

Kenett, R.S. and Shmueli, G. (2013) On information quality. *Journal of the Royal Statistical Society, Series A*, 176(4), pp. 1–25.

Kenett, R., Zacks, S. and Amberti, D. (2014) *Modern Industrial Statistics: With Applications in R, MINITAB and JMP*, 2nd edition. John Wiley & Sons, Chichester, West Sussex, UK.

Kimball, A.W. (1957) Errors of the third kind in statistical consulting. *Journal of the American Statistical Association*, 52, 133–142.

Lee, Y., Strong, D., Kahn, B. and Wang, R. (2002) AIMQ: a methodology for information quality assessment. *Information & Management*, 40, pp. 133–146.

Lin, M., Lucas, H. and Shmueli, G. (2013) Too big to fail: large samples and the p-value problem. *Information Systems Research*, 24(4), pp. 906–917.

Lindley, D.V. (2004) Some Reflections on the Current State of Statistics, in *Applied Bayesian Statistics Studies in Biology and Medicine*, di Bacco, M., d'Amore, G. and Scalfari, F. (editors), Springer, Boston, MA.

Mackay, R.J. and Oldford, R.W. (2000) Scientific method, statistical method, and the speed of light. *Statistical Science*, 15(3), pp. 254–278.

Mallows, C. (1998) The zeroth problem. *The American Statistician*, 52, pp. 1–9.

Montgomery, D.C. (1980) The economic design of control charts: a review and literature survey. *Journal of Quality Technology*, 12, pp. 75–87.

Redman, T. (2007) Statistics in Data and Information Quality, in *Encyclopedia of Statistics in Quality and Reliability*, Ruggeri, F., Kenett, R.S. and Faltin, F. (editors in chief), John Wiley & Sons, Ltd, Chichester, UK.

Regev, G. and Wegmann, W. (2005) Where Do Goals Come from: The Underlying Principles of Goal-Oriented Requirements Engineering. *Proceedings of the 13th IEEE International Requirements Engineering Conference (RE'05)*, Paris, France.

Schouten, B. (2013) Comments on 'on information quality'. *Journal of the Royal Statistical Society, Series A*, 176(4), pp. 27–29.

Serel, D.A. (2009) Economic design of EWMA control charts based on loss function. *Mathematical and Computer Modelling*, 49(3–4), pp. 745–759.

Shmueli, G. (2010) To explain or to predict? *Statistical Science*, 25(3), pp. 289–310.

Shmueli, G. and Koppius, O.R. (2011) Predictive analytics in information systems research. *MIS Quarterly*, 35(3), pp. 553–572.

Talburt, J.R. (2011) *Entity Resolution and Information Quality*. Morgan Kaufmann, Burlington, VT.

Trafimow, D. and Marks, M. (2015) Editorial. *Basic and Applied Social Psychology*, 37(1), pp. 1–2.

Tukey, J.W. (1962) The future of data analysis. *Annals of Mathematical Statistics*, 33(1), pp. 1–67.

UK Department of Health (2004) A Strategy for NHS Information Quality Assurance – Consultation Draft. Department of Health, London. http://webarchive.nationalarchives.gov.uk/20130107105354/http://www.dh.gov.uk/prod_consum_dh/groups/dh_digitalassets/@dh/@en/documents/digitalasset/dh_4087588.pdf (accessed May 2, 2016).

Wang, R.Y., Kon, H.B. and Madnick, S.E. (1993) Data Quality Requirements Analysis and Modeling. *9th International Conference on Data Engineering*, Vienna.

Zhou, Y., Talburt, J., Su, Y. and Yin, L. (2010) OYSTER: A Tool for Entity Resolution in Health Information Exchange. *Proceedings of the Fifth International Conference on the Cooperation and Promotion of Information Resources in Science and Technology (COINFO10)*, pp. 356–362.

3

Dimensions of information quality and InfoQ assessment

3.1 Introduction

Information quality (InfoQ) is a holistic abstraction or a *construct*. To be able to assess such a construct in practice, we operationalize it into measurable variables. Like InfoQ, *data quality* is also a construct that requires operationalization. The issue of assessing data quality has been discussed and implemented in several fields and by several international organizations. We start this chapter by looking at the different approaches to operationalizing data quality. We then take, in Section 3.2, a similar approach for operationalizing InfoQ. Section 3.3 is about methods for assessing InfoQ dimensions and Section 3.4 provides an example of an InfoQ rating-based assessment. Additional in-depth examples are provided in Part II.

3.1.1 Operationalizing "data quality" in marketing research

In marketing research and in the medical literature, *data quality* is assessed by defining the criteria of recency, accuracy, availability, and relevance of a dataset (Patzer, 1995):

1. *Recency* refers to the duration between the time of data collection and the time the study is conducted.

2. *Accuracy* refers to the quality of the data.

Information Quality: The Potential of Data and Analytics to Generate Knowledge,
First Edition. Ron S. Kenett and Galit Shmueli.
© 2017 John Wiley & Sons, Ltd. Published 2017 by John Wiley & Sons, Ltd.
Companion website: www.wiley.com/go/information_quality

3. *Availability* describes the information in the data made available to the analyst.

4. *Relevance* refers to the relevance of the data to the analysis goal: whether the data contains the required variables in the right form and whether they are drawn from the population of interest.

Kaynak and Herbig (2014) mention four criteria to consider for data quality in cross-cultural marketing research:

1. *Compatibility and comparability*—When comparing different sets of data from different countries, are similar units of measurements and definitions used?

2. *Accuracy and reliability* of the data—Has the data been consciously distorted, or was the collection flawed?

3. *Recency*—Is the data infrequently and unpredictably updated?

4. *Availability*

The four criteria of Patzer and of Kaynak and Herbig consider data (X) and goal (g), but do not consider data analysis method (f) and utility (U). Specifically, *recency*, *accuracy*, *reliability*, *availability*, and *comparability* are all characteristics of the dataset and relate implicitly to the analysis goal, while only *relevance* relates directly to the data and the analysis goal.

3.1.2   Operationalizing "data quality" in public health research

Boslaugh (2007) considers three main questions to help assess the quality of secondary data (data collected for purposes other than the study at hand):

1. What was the original purpose for which the data was collected?

2. What kind of data is it, and when and how was the data collected?

3. What cleaning and/or recoding procedures have been applied to the data?

These questions are useful at the prestudy stage, when one must evaluate the usefulness of a dataset for the study at hand. The concepts in the three questions can be summarized into *collection purpose*, *data type*, *data age*, *data collection instrument and process*, and *data preprocessing*. They can be grouped into "source quality" and "data quality" criteria (Kaynak and Herbig, 2014). Obviously, source quality affects data quality:

> It is almost impossible to know too much about the data collection process because it can influence the quality of the data in many ways, some of them not obvious.

Boslaugh (2007, p. 5) further considers availability, completeness, and data format:

> A secondary data set should be examined carefully to confirm that it includes the necessary data, that the data are defined and coded in a manner that allows for the desired analysis, and that the researcher will be allowed to access the data required.

We again note that the questions and criteria mentioned relate to the data and goal, but not to an analysis method or utility; the InfoQ definition, however, requires all four components.

3.1.3 Operationalizing "data quality" in management information systems

In the field of management information systems (MIS), data quality is defined as the level of conformance to specifications or standards. Wang et al. (1993) define data quality as "conformance to requirements." They operationalize this construct by defining quality indicators that are based on objective measures such as data source, creation time, and collection method, as well as subjective measures such as the credibility level of the data at hand, as determined by the researcher.

As mentioned in Chapter 2, Lee et al. (2002) propose a methodology for assessment and benchmarking of InfoQ of IT systems called AIMQ. They collate 15 dimensions from academic papers in MIS: *accessibility, appropriate amount, believability, completeness, concise representation, consistent representation, ease of operation, free of error, interpretability, objectivity, relevancy, reputation, security, timeliness,* and *understandability*. They then group the 15 dimensions into four categories: intrinsic, contextual, representational, and accessibility. While they use the term IQ, it is different from InfoQ. The concept of IQ indicates a consideration of the user of the IT system (and therefore some of the dimensions include relevance, timeliness, etc.). However, IQ does not consider data analysis at all. To operationalize the four categories, Lee et al. (2002) develop a questionnaire with eight items for each of the 15 dimensions. This instrument is then used for scoring an IT system of an organization and for benchmarking it against best practice and other organizations.

3.1.4 Operationalizing "data quality" at government and international organizations

Assessing data quality is one of the core aspects of statistical agencies' work. Government agencies and international organizations that collect data for decision making have developed operationalizations of data quality by considering multiple dimensions. The abstraction *data quality* is usually defined as "fitness for use" in terms of user needs. This construct is operationalized by considering a set of dimensions. We briefly list the dimensions used by several notable organizations.

The concept of *quality of statistical data* has been developed and used in European official statistics as well as organizations such as the International Monetary Fund (IMF),

Statistics Canada, and the Organization for Economic Cooperation and Development (OECD). The OECD operationalizes this construct by defining seven dimensions for quality assessment (see chapter 5 in Giovanni, 2008):

1. *Relevance*—A qualitative assessment of the value contributed by the data

2. *Accuracy*—The degree to which the data correctly estimate or describe the quantities or characteristics that they are designed to measure

3. *Timeliness and punctuality*—The length of elapsed time between data availability and the phenomenon described

4. *Accessibility*—How readily the data can be located and accessed

5. *Interpretability*—The ease with which the data may be understood and analyzed

6. *Coherence*—The degree to which the data is logically connected and mutually consistent

7. *Credibility*—Confidence of users in the data based on their perception of the data producer

The European Commission's Eurostat agency uses seven dimensions for assessing the quality of data from surveys (Ehling and Körner, 2007):

1. *Relevance of statistical concept* refers to whether all statistics that are needed are produced and the extent to which concepts used (definitions, classifications, etc.) reflect user needs.

2. *Accuracy of estimates* denotes the closeness of computations or estimates to the exact or true values.

3. *Timeliness and punctuality in disseminating results*—Timeliness of information reflects the length of time between its availability and the event or phenomenon it describes; punctuality refers to the time lag between the release date of data and the target date when it should have been delivered.

4. *Accessibility and clarity of the information*—Accessibility refers to the physical conditions in which users can obtain data; clarity refers to the data's information environment (whether data are accompanied with appropriate metadata, illustrations such as graphs and maps, etc.).

5. *Comparability* is the extent to which differences between statistics are attributed to differences between the true values of the statistical characteristic.

6. *Coherence of statistics* refers to their adequacy to be reliably combined in different ways and for various uses.

The US Environmental Protection Agency (EPA) has developed the Quality Assurance (QA) Project Plan as a tool for project managers and planners to document the type and quality of data and information needed for making environmental decisions. The program aims to control and enhance data quality in terms of precision, accuracy, representativeness, completeness, and comparability (PARCC) of environmental measurements used in its studies. They define these dimensions as follows:

1. *Precision* is the degree of agreement among repeated measurements of the same characteristic on the same sample or on separate samples collected as close as possible in time and place.

2. *Accuracy* is a measure of confidence in a measurement. The smaller the difference between the measurement of a parameter (the estimate) and its "true" or expected value, the more accurate the measurement.

3. *Representativeness* is the extent to which measurements actually depict the true environmental condition or population being evaluated.

4. *Completeness* is a measure of the number of samples you must take to be able to use the information, as compared to the number of samples you originally planned to take.

5. *Comparability* is the extent to which data from one study can be compared directly to either past data from the current project or data from another study.

The World Health Organization (WHO) established a data quality framework called *Health Metrics Network Framework* (HMN, 2006), based on the IMF Data Quality Assessment Framework (DQAF) and IMF General Data Dissemination System (GDDS). The framework uses six criteria for assessing the quality of health-related data and indicators that are generated from health information systems:

1. *Timeliness*—The period between data collection and its availability to a higher level or its publication

2. *Periodicity*—The frequency with which an indicator is measured

3. *Consistency*—The internal consistency of data within a dataset as well as consistency between datasets and over time and the extent to which revisions follow a regular, well-established, and transparent schedule and process

4. *Representativeness*—The extent to which data adequately represents the population and relevant subpopulations

5. *Disaggregation*—The availability of statistics stratified by sex, age, socioeconomic status, major geographical or administrative region, and ethnicity, as appropriate

6. *Confidentiality, data security, and data accessibility*—The extent to which practices are in accordance with guidelines and other established standards for storage, backup, transport of information (especially over the Internet), and retrieval

These examples provide the background for the assessment of InfoQ. Our goal in presenting the InfoQ dimensions is to propose a generic structure that applies to any empirical analysis and expands on the data quality approaches described above.

3.2   The eight dimensions of InfoQ

Taking an approach that is similar to data quality assessment described in the previous section, we define eight dimensions for assessing InfoQ that consider and affect not only the data and goal, X and g, but also the method of analysis (f) and the utility of the study (U). With this approach, we provide a decomposition of InfoQ that can be used for assessing and improving research initiatives and for evaluating completed studies.

3.2.1   Data resolution

Data resolution refers to the measurement scale and aggregation level of X. The measurement scale of the data should be carefully evaluated in terms of its suitability to the goal, the analysis methods to be used and the required resolution of U. Given the original recorded scale, the researcher should evaluate its adequacy. It is usually easy to produce a more aggregated scale (e.g., two income categories instead of ten), but not a finer scale. Data might be recorded by multiple instruments or by multiple sources. To choose between the multiple measurements, supplemental information about the reliability and precision of the measuring devices or sources of data is useful. A finer measurement scale is often associated with more noise; hence the choice of scale can affect the empirical analysis directly.

The data aggregation level must also be evaluated relative to g. For example, consider daily purchases of over-the-counter medications at a large pharmacy. If the goal of the analysis is to forecast future inventory levels of different medications, when restocking is done weekly, then weekly aggregates are preferable to daily aggregates owing to less data recording errors and noise. However, for the early detection of outbreaks of disease, where alerts that are generated a day or two earlier can make a significant difference in terms of treatment, then weekly aggregates are of low quality. In addition to data frequency, the aggregation level is also important: for purposes of inventory, medication level information is required, whereas for disease outbreak detection medications can be grouped by symptoms, and the symptom-aggregated daily series would be preferable.

Another example relates to the case studies on online auctions in Chapter 1. In many online auction platforms, bid times are typically recorded in seconds and prices in a currency unit. On eBay, for example, bid times are reported at the level of seconds (e.g., August 20, 2010, 03.14.07 Pacific Daylight Time) and prices at the dollar and cent level (e.g., \$23.01). The forecasting model by Wang et al. (2008) uses bid times at second level and cent-level bid amounts until the time of prediction to produce forecasts of price in cents for any second during the auction. In contrast, the forecasting model by Ghani and Simmons (2004) produces forecasts of the final price in terms of \$5 intervals, using only information available at the start of the auction.

The concept of rational subgroup that is used in statistical process control is a special case of aggregation level. The rational subgroup setup determines the level of process variability and the type of signals to detect. If the rational subgroup consists of measurements within a short period of a production process, then statistical process control methods will pick up short-term out-of-control signals, whereas rational subgroups spread over longer periods will support detection of longer-term trends and out-of-control signals (see Kenett et al., 2014). Using our notation, f is the statistical process control method, X is the data, g_1 is the short-term signal, g_2 is the long-term signal, and U is a measure of desirable alerting behavior.

3.2.2 Data structure

Data structure relates to the type(s) of data and data characteristics such as corrupted and missing values due to the study design or data collection mechanism. Data types include structured numerical data in different forms (e.g., cross-sectional, time series, and network data) as well as unstructured nonnumerical data (e.g., text, text with hyperlinks, audio, video, and semantic data). The InfoQ level of a certain data type depends on the goal at hand. Bapna et al. (2006) discussed the value of different "data types" for answering new research questions in electronic commerce research:

> For each research investigation, we seek to identify and utilize the best data type, that is, that data which is most appropriate to help achieve the specific research goals.

An example from the online auctions literature is related to the effect of "seller feedback" on the auction price. Sellers on eBay receive numerical feedback ratings and textual comments. Although most explanatory studies of price determinants use the numerical feedback ratings as a covariate, a study by Pavlou and Dimoka (2006) showed that using the textual comments as a covariate in a model for price leads to much higher R^2 values (U) than using the numerical rating.

Corrupted and missing values require handling by removal, imputation, data recovery, or other methods, depending on g. Wrong values may be treated as missing values when the purpose is to estimate a population parameter, such as in surveys where respondents intentionally enter wrong answers. Yet, for some goals, intentionally submitted wrong values might be informative and therefore should not be discarded or "corrected."

3.2.3 Data integration

Integrating multiple sources and/or types of data often creates new knowledge regarding the goal at hand, thereby increasing InfoQ. An example is the study estimating consumer surplus in online auctions (Bapna et al., 2008a; see Chapter 1), where data from eBay (X_1) that lacked the highest bid values were combined with data from a website called Cniper.com (now no longer active) (X_2) that contained the missing information. Estimating consumer surplus was impossible by using either X_1

or X_2, and only their combination yielded the sufficient InfoQ. In the auction example of Pavlou and Dimoka (2006), textual comments were used as covariates.

New analysis methodologies, such as functional data analysis and text mining, are aimed at increasing InfoQ of new data types and their combination. For example, in the online auction forecasting study by Wang et al. (2008) (see Chapter 1), functional data analysis was used to integrate temporal bid sequences with cross-sectional auction and seller information. The combination allowed more precise forecasts of final prices compared to models based on cross-sectional data alone. The functional approach has also enabled quantifying the effects of different factors on the price process during an auction (Bapna et al., 2008b).

Another aspect of data integration is linking records across databases. Although record linkage algorithms are popular for increasing InfoQ, studies that use record linkage often employ masking techniques that reduce risks of identification and breaches of privacy and confidentiality. Such techniques (e.g., removing identifiers, adding noise, data perturbation, and microaggregation) can obviously decrease InfoQ, even to the degree of making the combined dataset useless for the goal at hand. Solutions, such as "privacy-preserving data mining" and "selective revelation," are aimed at utilizing the linked dataset with high InfoQ without compromising privacy (see, e.g., Fienberg, 2006).

3.2.4   Temporal relevance

The process of deriving knowledge from data can be placed on a timeline that includes the data collection, data analysis, and study deployment periods as well as the temporal gaps between these periods (as depicted in Figure 3.1). These different durations and gaps can each affect InfoQ. The data collection duration can increase or decrease InfoQ, depending on the study goal (e.g., studying longitudinal effects versus a cross-sectional goal). Similarly, uncontrollable transitions during the collection phase can be useful or disruptive, depending on g.

For this reason, online auction studies that collect data on fashionable or popular products (which generate large amounts of data) for estimating an effect try to restrict the data collection period as much as possible. The experiment by Katkar and Reiley (2006) on the effect of reserve prices on online auction prices (see Chapter 1) was conducted over a two-week period in April 2000. The data on auctions for Harry Potter books and Microsoft Xbox consoles in Wang et al. (2008) was collected in the nonholiday months of August and September 2005. In contrast, a study that is

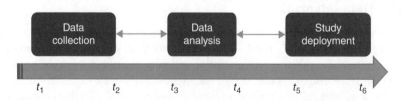

Figure 3.1   Timeline of study, from data collection to study deployment.

interested in comparing preholiday with postholiday bidding or selling behaviors would require collection over a period that includes both preholiday and postholiday times. The gap between data collection and analysis, which coincides with the recency criterion in Section 3.1, is typically larger for secondary data (data not collected for the purpose of the study). In predictive modeling, where the context of prediction should be as close as possible to the data collection context, temporal lags can significantly decrease InfoQ. For instance, a 2010 dataset of online auctions for iPads on eBay will probably be of low InfoQ for forecasting or even estimating current iPad prices because of the fast changing interest in electronic gadgets.

Another aspect affecting temporal relevance is *analysis timeliness*, or the timeliness of $f(X|g)$. Raiffa (1970, p. 264) calls this an "error of the fourth kind: solving the right problem too late." Analysis timeliness is affected by the nature of X, by the complexity of f and ultimately by the application of f to X. The nature of a dataset (size, sparseness, etc.) can affect analysis timeliness and in turn affect its utility for the goal at hand. For example, computing summary statistics for a very large dataset might take several hours, thereby deeming InfoQ low for the purpose of real-time tasks (g_1) but high for retrospective analysis (g_2). The computational complexity of f also determines analysis time: Markov chain Monte Carlo estimation methods and computationally intensive predictive algorithms take longer than estimating linear models or computing summary statistics. In the online auction price forecasting example, the choice of a linear forecasting model was needed for producing timely forecasts of an ongoing auction. Wang et al. (2008) used smoothing splines to estimate price curves for each auction in the dataset—information which is then used in the forecasting model. Although smoothing splines do not necessarily produce monotone curves (as would be expected of a price curve from the start to the end of an eBay-type auction), this method is much faster than fitting monotone smoothing splines, which do produce monotonic curves. Therefore, in this case smoothing splines generated higher InfoQ than monotone splines for real-time forecasting applications. Temporal relevance and analysis timeliness obviously depend on the availability of software and hardware as well as on the efficiency of the researcher or analysis team.

3.2.5   Chronology of data and goal

The choice of variables to collect, the temporal relationship between them and their meaning in the context of g all critically affect InfoQ. We must consider the retrospective versus prospective nature of the goal as well as its type in terms of causal explanation, prediction, or description (Shmueli, 2010). In predictive studies, the input variables must be available at the time of prediction, whereas in explanatory models, causal arguments determine the relationship between dependent and independent variables. The term *endogeneity*, or reverse causation, can occur when a causal input variable is omitted from a model, resulting in biased parameter estimates. Endogeneity therefore yields low InfoQ in explanatory studies, but not necessarily in predictive studies, where omitting input variables can lead to higher predictive accuracy (see Shmueli, 2010). Also related is the Granger causality test (Granger, 1969) aimed at determining whether a lagged time series X contains useful information for predicting future values of another time series Y by using a regression model.

In the online auction context, the level of InfoQ that is contained in the "number of bidders" for models of auction price depends on the study goal. Classic auction theory specifies the number of bidders as an important factor influencing price: the more bidders, the higher the price. Hence, data on number of bidders is of high quality in an explanatory model of price. However, for the purpose of forecasting prices of ongoing online auctions, where the number of bidders is unknown until the end of the auction, the InfoQ of "number of bidders," even if available in a retrospective dataset, is very low. For this reason, the forecasting model by Wang et al. (2008) described in Chapter 1 excludes the number of bidders or number of bids and instead uses the cumulative number of bids until the time of prediction.

3.2.6   Generalizability

The utility of $f(X|g)$ is dependent on the ability to generalize f to the appropriate population. Two types of generalizability are statistical and scientific generalizability. Statistical generalizability refers to inferring from a sample to a target population. Scientific generalizability refers to applying a model based on a particular target population to other populations. This can mean either generalizing an estimated population pattern or model f to other populations or applying f estimated from one population to predict individual observations in other populations.

Determining the level of generalizability requires careful characterization of g. For instance, for inference about a population parameter, statistical generalizability and sampling bias are the focus, and the question of interest is, "What population does the sample represent?" (Rao, 1985). In contrast, for predicting the values of new observations, the question of interest is whether f captures associations in the training data X (the data that are used for model building) that are generalizable to the to-be-predicted data.

Generalizability is a dimension useful for clarifying the concepts of reproducibility, repeatability, and replicability (Kenett and Shmueli, 2015). The three terms are referred to with different and sometimes conflicting meanings, both between and within fields (see Chapter 11). Here we only point out that the distinction between replicating insights and replicating exact identical numerical results is similar and related to the distinction between InfoQ (insights) and data or analysis quality (numerical results).

Another type of generalization, in the context of ability testing, is the concept of specific objectivity (Rasch, 1977). Specific objectivity is achieved if outcomes of questions in a questionnaire that is used to compare levels of students are independent of the specific questions and of other students. In other words, the purpose is to generalize from data on certain students answering a set of questions to the population of outcomes, irrespective of the particular responders or particular questions.

The type of required generalizability affects the choice of f and U. For instance, data-driven methods are more prone to overfitting, which conflicts with scientific generalizability. Statistical generalizability is commonly evaluated by using measures of sampling bias and goodness of fit. In contrast, scientific generalizability for predicting new observations is typically evaluated by the accuracy of predicting a holdout set from the to-be-predicted population, to protect against overfitting.

The online auction studies from Chapter 1 illustrate the different generalizability types. The "effect of reserve price on final price" study (Katkar and Reiley, 2006) is concerned with statistical generalizability. Katkar and Reiley (2006) designed the experiment so that it produces a representative sample. Their focus is on standard errors and statistical significance. The forecasting study by Wang et al. (2008) is concerned with generalizability to new individual auctions. They evaluated predictive accuracy on a holdout set. The third study on "consumer surplus in eBay" is concerned with statistical generalizability from the sample to all eBay auctions in 2003. Because the sample was not drawn randomly from the population, Bapna et al. (2008a) performed a special analysis, comparing their sample with a randomly drawn sample (see appendix B in Bapna et al., 2008a).

3.2.7   Operationalization

Two types of operationalization of the analysis results are considered: construct operationalization and action operationalization.

3.2.7.1   Construct operationalization

Constructs are abstractions that describe a phenomenon of theoretical interest. Measurable data is an operationalization of underlying constructs. For example, psychological stress can be measured via a questionnaire or by physiological measures, such as cortisol levels in saliva (Kirschbaum and Hellhammer, 1989), and economic prosperity can be measured via income or by unemployment rate. The relationship between the underlying construct χ and its operationalization $X = \theta(\chi)$ can vary, and its level relative to g is another important aspect of InfoQ. The role of construct operationalization is dependent on $g(X = \theta(\chi|g))$ and especially on whether the goal is explanatory, predictive, or descriptive. In explanatory models, based on underlying causal theories, multiple operationalizations might be acceptable for representing the construct of interest. As long as X is assumed to measure χ, the variable is considered adequate. Using our earlier example in the preceding text, both questionnaire answers and physiological measurements would be acceptable for measuring psychological stress. In contrast, in a predictive task, where the goal is to create sufficiently accurate predictions of a certain measurable variable, the choice of operationalized variable is critical. Predicting psychological stress as reported in a questionnaire (X_1) is different from predicting levels of a physiological measure (X_2). Hence, the InfoQ in predictive studies relies more heavily on the quality of X and its stability across the periods of model building and deployment, whereas in explanatory studies InfoQ relies more on the adequacy of X for measuring χ.

Returning to the online auction context, the consumer surplus study relies on observable bid amounts, which are considered to reflect an underlying "willingness-to-pay" construct for a bidder. The same construct is operationalized differently in other types of studies. In contrast, in price forecasting studies the measurable variable of interest is auction price, which is always defined very similarly. An example is the work by McShane and Wyner (2011) in the context of climate change, showing that for purposes of predicting temperatures, theoretically based "natural covariates" are inferior to "pseudoproxies" that are lower dimension approximations of the natural covariates. Descriptive tasks are more similar to predictive tasks in the sense of the focus on the observable level. In descriptive studies, the goal is to uncover a signal in a dataset (e.g., to estimate the income distribution or to uncover the temporal patterns in a time series). Because there is no underlying causal theory behind descriptive studies, and because results are reported at the level of the measured variables, InfoQ relies, as in predictive tasks, on the quality of the measured variables rather than on their relationship to an underlying construct.

3.2.7.2   Action operationalization

Action operationalizing is about deriving concrete actions from the information provided by a study. When a report, presenting an analysis of a given dataset in the context of specific goals, leads to clear follow-up actions, we consider such a report of higher InfoQ. The dimension of action operationalization has been discussed in various contexts. In the business and industry settings, an operational definition consists of (i) a criterion to be applied to an object or a group of objects, (ii) a test of compliance for the object or group, and (iii) a decision rule for interpreting the test results as to whether the object or group is, or is not, in compliance. This definition by Deming (2000) closely parallels Shewhart's opening statement in his book *Statistical Method from the Viewpoint of Quality Control* (Shewhart, 1986):

> Broadly speaking there are three steps in a quality control process: the specification of what is wanted, the production of things to satisfy the specification, and the inspection of the things produced to see if they satisfy the specification.

In a broad context of organizational performance, Deming (2000) poses three important questions to help assess the level of action operationalization of a specific organizational study. These are the following:

1. What do you want to accomplish?

2. By what method will you accomplish it?

3. How will you know when you have accomplished it?

In the context of an educational system, the National Education Goals Panel (NEGP) in the United States recommended that states answer four questions on their student reports that are of interest to parents (Goodman and Hambleton, 2004):

1. How did my child do?

2. What types of skills or knowledge does his or her performance reflect?

3. How did my child perform in comparison to other students in the school, district, state, and, if available, the nation?

4. What can I do to help my child improve?

The action operationalization of official statistics has also been discussed extensively by official statistics agencies, internally, and in the literature. Quoting Forbes and Brown (2012):

> An issue that can lead to misconception is that many of the concepts used in official statistics often have specific meanings which are based on, but not identical to, their everyday usage meaning… Official statistics *"need to be used to be useful"* and utility is one of the overarching concepts in official statistics… All staff producing statistics must understand that the conceptual frameworks underlying their work translate the real world into models that interpret reality and make it measurable for statistical purposes… The first step … is to define the issue or question(s) that statistical information is needed to inform. That is, to define the objectives for the framework, and then work through those to create its structure and definitions. An important element … is understanding the relationship between the issues and questions to be informed and the definitions themselves.

3.2.8   Communication

Effective communication of the analysis $f(X|g)$ and its utility U directly affects InfoQ. Common communication media include visual, textual, and verbal presentations and reports. Within research environments, communication focuses on written publications and conference presentations. Research mentoring and the refereeing process are aimed at improving communication (and InfoQ) within the research community. Research results are communicated to the public via articles in the popular media and interviews on television and conferences such as www.ted.com and more recently through blogs and other Internet media. Here the risk of miscommunication is much greater. For example, the "consumer surplus in eBay auctions" study was covered by public media. However, the main results were not always conveyed properly by journalists. For example, the nytimes.com article (http://bits.blogs.nytimes.com/2008/01/28/tracking-consumer-savings-on-ebay/) failed to mention that the study results were evaluated under different assumptions, thereby affecting generalizability. As a result, some readers doubted the study results ("Is the Cniper sample skewed?"). In response, one of the study coauthors posted an online clarification.

In industry, communication is typically done via internal presentations and reports. The failure potential of O-rings at low temperatures that caused the NASA shuttle Challenger disaster was ignored because the engineers failed to communicate the results of their analysis: the 13 charts that were circulated to the teleconferences did not clearly show the relationship between the temperature in 22 previous launches and the 22 recordings of O-ring conditions (see Tufte, 1992). In terms of our notation, the meaning of *f*—in this case risk analysis—and its implications were not properly communicated.

In discussing scientific writing, Gopen and Swan (1990) state that if the reader is to grasp what the writer means, the writer must understand what the reader needs. In general, this is an essential element in effective communication. It is important to emphasize that scientific discourse is not the mere presentation of information, but rather its actual communication. It does not matter how pleased an author might be to have converted all the right data into sentences and paragraphs; it matters only whether a large majority of the reading audience accurately perceives what the author had in mind. Communication is the eighth InfoQ dimension.

3.3   Assessing InfoQ

The eight InfoQ dimensions allow us to evaluate InfoQ for an empirical study (whether implemented or proposed), by evaluating each of the dimensions. In the following, we describe five assessment approaches. The approaches offer different views of the study and one can implement more than a single approach for achieving deeper understanding.

3.3.1   Rating-based evaluation

Similar to the use of "data quality" dimensions by statistical agencies for evaluating data quality, we evaluate each of the eight InfoQ dimensions to assess InfoQ. This evaluation integrates different aspects of a study and assigns an overall InfoQ score based on experts' ratings. The broad perspective of InfoQ dimensions is designed to help researchers enhance the added value of their studies.

Assessing InfoQ using quantitative metrics can be done in several ways. We present a rating-based approach that examines a study report and scores each of the eight InfoQ dimensions. A rough approach is to rate each dimension on a 1–5 scale:

Very low	Low	Acceptable	High	Very high
1	2	3	4	5

The ratings for each of the eight dimensions (Y_i, $i = 1, \ldots, 8$) can then be normalized into a desirability function (see Figini et al., 2010) separately for each dimension ($0 \leq d(Y_i) \leq 1$). The desirability scores are then combined to produce an overall InfoQ score using the geometric mean of the individual desirabilities:

$$\text{InfoQ score} = \left[d_1(Y_1) \times d_2(Y_2) \times \cdots \times d_8(Y_8) \right]^{1/8}$$

The approach using desirability scores produces zero scores when at least one of the elements is rated at the lower values of the scale. In other words, if any of the dimensions is at the lowest rating, InfoQ is considered to be zero. Smoother options consist in averaging the rating scores with an arithmetic mean or geometric mean. In the examples in this book, we used the desirability approach.

We illustrate the use of this rating-based approach for the Katkar and Reiley (2006) study in Section 3.4. We also use this approach for each of the studies described in Parts II and III of the book.

3.3.2   Scenario building

A different approach to assessing InfoQ, especially at the "proof of concept" stage, is to spell out the types of answers that the analysis is expected to yield and then to examine the data in an exploratory fashion, alternatively specifying the ideal data, as if the data analyst has control over the data collection, and then comparing the existing results to the ideal results.

For example, some studies in biosurveillance are aimed at evaluating the usefulness of tracking prediagnostic data for detecting disease outbreaks earlier than traditional diagnostic measures. To evaluate the usefulness of such data (and potential algorithms) in the absence of real outbreak data requires building scenarios of how disease outbreaks manifest themselves in prediagnostic data. Building scenarios can rely on knowledge such as singular historic cases (e.g., Goldenberg et al., 2002) or on integrating epidemiological knowledge into a wide range of data simulations to generate "data with outbreaks" (e.g., Lotze et al., 2010). The wide range of simulations reflects the existing uncertainty in mapping the epidemiological knowledge into data footprints.

3.3.3   Pilot sampling

In many fields it is common practice to begin the analysis with a pilot study based on a small sample. This approach provides initial insights on the dimensions of InfoQ. Following such a pilot, the dataset can be augmented, a new time window for recording the data can be decided, and more in-depth elicitation of the problem at hand and the key stakeholders can be initiated. This strategy is common practice also in survey design, where a pilot with representative responders is conducted to determine the validity and usability of a questionnaire (Kenett and Salini, 2012).

3.3.4   Exploratory data analysis (EDA)

Modern statistical and visualization software provides a range of visualization techniques such as matrix plots, parallel coordinate plots, and dynamic bubble plots and capabilities such as interactive visualization. These techniques support the analyst in exploring and determining, with "freehand format," the level of InfoQ in the data. Exploratory data analysis (EDA) is often conducted iteratively by zooming in on salient features and outliers and triggering further investigations and additional data collection. Other exploratory tools that are useful for assessing InfoQ, termed "exploratory models" by De Veaux (2009), include classification and regression trees, cluster analysis, and data reduction techniques. EDA is therefore another strategy for evaluating and increasing InfoQ.

3.3.5   Sensitivity analysis

Sensitivity analysis is an important type of quantitative assessment applied in a wide range of domains that involve policy making, including economic development, transportation systems, urban planning, and environmental trends. InfoQ provides an efficient approach to sensitivity analysis by changing one of the InfoQ components while holding the other three constant. For example, one might evaluate InfoQ for three different goals, g_1, g_2, g_3, given the same dataset X, a specific analysis method f, and specific utility U. Differences between the InfoQ derived for the different goals can then indicate the boundaries of usefulness of X, f, and U.

For example, consider the use of ensemble models (combining different models from different sources) in predicting climate change. In an incisive review of models used in climate change studies, Saltelli et al. (2015) state that ensembles are not representative of the range of possible (and plausible) models that fit the data generated by the physical model. This implies that the models used represent structural elements with poor generalizability to the physical model. They also claim that the sensitivity analysis performed on these models varies only a subset of the assumptions and only one at a time. Such single-assumption manipulation precludes interactions among the uncertain inputs, which may be highly relevant to climate projections. This also indicates poor generalizability. In terms of operationalization, the authors distinguish policy simulation from policy justification. Policy simulations represent alternative scenarios; policy justification requires establishment of a causal link. The operationalization of the climate models by policy makers requires an ability to justify specific actions. This is the problematic part the authors want to emphasize. An InfoQ assessment of the various studies quoted by the authors can help distinguish between studies providing policy simulations and studies providing policy justifications.

3.4   Example: InfoQ assessment of online auction experimental data

As described in Chapter 1, Katkar and Reiley (2006) investigated the effect of two types of reserve price on the final auction price on eBay. Their data X came from an experiment selling 25 identical pairs of Pokémon cards, where each card was auctioned twice, once with a public reserve price and once with a secret reserve price. The data consists of complete information on all 50 auctions. Katkar and Reiley used linear regression (f) to test for the effect of private or public reserve on the final price and to quantify it. The utility (U) was statistical significance to evaluate the effect of private or public reserve price and the regression coefficient for quantifying the magnitude of the effect. They conclude that

> A secret-reserve auction will generate a price \$0.63 lower, on average, than will a public-reserve auction.

We evaluate the eight InfoQ dimensions on the basis of the paper by Katkar and Reiley (2006). A more thorough evaluation would have required interaction with the authors of the study and access to their data. For demonstration purposes we use a 1–5 scale and generate an InfoQ score based on a desirability function with $d(1)=0$, $d(2)=0.25$, $d(3)=0.5$, $d(4)=0.75$, and $d(5)=1$.

3.4.1   Data resolution

The experiment was conducted over two weeks in April 2000. We therefore have no data on possible seasonal effects during other periods of the year. Data resolution was in USD cents, but individual bids were dropped and only the final price was considered. Other time series (e.g., the cumulative number of bids) were also aggregated to create end-of-auction statistics such as "total number of bids." Given the general goal of quantifying the effect of using a secret versus public reserve price on the final price of an auction, the data appears somewhat restrictive. The two-week data window allows for good control of the experiment but limits data resolution for studying a more general effect. Hence we rate the data resolution as $Y_1=4$ (high).

3.4.2   Data structure

The data included only information on the factor levels that were set by the researchers and the three outcomes: final price, whether the auction transacted, and the number of bids received. The data was either set by the experimenters or collected from the auction website. Although time series data was potentially available for the 50 auctions (e.g., the series of bids and cumulative number of bidders), the researchers aggregated them into auction totals. Textual data was available but not used. For example, bidder usernames can be used to track individual bidders who place multiple bids. With respect to corrupted data, one auction winner unexpectedly rated the sellers, despite the researchers' request to refrain from doing so (to keep the rating constant across the experiment). Luckily, this corruption did not affect the analysis, owing to the study design. Another unexpected source of data corruption was eBay's policy on disallowing bids below a public reserve price. Hence, the total number of bids in auctions with a secret reserve price could not be compared with the same measure in public reserve price auctions. The researchers resorted to deriving a new "total serious bids" variable, which counts the number of bids above the secret reserve price.

Given the level of detailed attention to the experimental conditions, but the lack of use of available time series and textual data, we rate this dimension as $Y_2=4$ (high).

3.4.3   Data integration

The researchers analyzed the two-week data in the context of an experimental design strategy. The integration with the DOE factors was clearly achieved. No textual or other semantic data seems to have been integrated. We rate this dimension as $Y_3=4$ (high).

3.4.4   Temporal relevance

The short duration of the experiment and the experimental design assured that the results would not be confounded with the effect of time. The experimenters tried to avoid confounding the results with a changing seller rating and therefore actively requested winners to avoid rating the seller. Moreover, the choice of Pokémon cards was aligned with timeliness, since at the time such items were in high demand. Finally, because of the retrospective nature of the goal, there is no urgency in conducting the data analysis shortly after data collection. We rate this dimension as $Y_4 = 5$ (very high).

3.4.5   Chronology of data and goal

The causal variable (secret or public reserve) and the blocking variable (week) were determined at the auction design stage and manipulated before the auction started. We rate this dimension as $Y_5 = 5$ (very high).

3.4.6   Generalizability

The study is concerned with statistical generalizability: Do effects that were found in the sample generalize to the larger context of online auctions? One possible bias, which was acknowledged by the authors, is their seller's rating of zero (indicating a new seller) which limits the generalizability of the study to more reputable sellers. In addition, they limited the generality of their results to low value items, which might not generalize to more expensive items. We rate this dimension as $Y_6 = 3$ (acceptable).

3.4.7   Operationalization

In construct operationalization, the researchers considered two theories that explain the effect of a secret versus public reserve price on the final price. One is a psychological explanation: bidders can become "caught up in the bidding" at low bid amounts and end up bidding more than they would have had the bidding started higher. The second theory is a model of rational bidders: "an auction with a low starting bid and a high secret reserve can provide more information to bidders than an auction with a high starting bid." Although these two theories rely on operationalizing constructs such as "information" and "caught up in the bidding," the researchers limited their study to eBay's measurable reserve price options and final prices.

In terms of action operationalization, the study results can be directly used by buyers and sellers in online auction platforms, as well as auction sites (given the restrictions of generalizing beyond eBay and beyond Pokémon cards). Recall that the study examined the effect of a reserve price not only on the final auction price but also on the probability of the auction resulting in a sale. The authors concluded:

> Only 46% of secret-reserve auctions resulted in a sale, compared with 70% of public-reserve auctions for the same goods. Secret-reserve auctions resulted in 0.72 fewer serious bidders per auction, and $0.62 less in final

auction price, than did public-reserve auctions on average. We can therefore recommend that sellers avoid the use of secret reserve prices, particularly for Pokémon cards.

The authors limit their recommendation to low-cost items by quoting from *The Official eBay Guide* (Kaiser and Kaiser, 1999): "If your minimum sale price is below $25, think twice before using a reserve auction. Bidders frequently equate reserve with expensive."

Note that because the study result is applicable to the "average auction," it is most actionable for either an online auction platform which holds many auctions or for sellers who sell many items. The results do not tell us about the predictive accuracy for a single auction.

We rate this dimension as $Y_7 = 4$ (high).

3.4.8 Communication

This research study communicated the analysis via a paper published in a peer-reviewed journal. Analysis results are presented in the form of a scatter plot, a series of estimated regression models (estimated effects and standard errors) and their interpretation in the text. We assume that the researches made additional dissemination efforts (e.g., the paper is publicly available online as a working paper). The paper's abstract is written in nontechnical and clear language and can therefore be easily understood not only by academics and researchers but also by eBay participants. The main communication weakness of the analysis is in terms of visualization, where plots would have conveyed some of the results more clearly. We therefore rate this dimension as $Y_8 = 4$ (high).

3.4.9 Information quality score

The scores that we assigned for each of the dimensions were the following:

1. Data resolution	4
2. Data structure	4
3. Data integration	4
4. Temporal relevance	5
5. Chronology of data and goal	5
6. Generalizability	3
7. Operationalization	4
8. Communication	4

On the basis of these subjective assessments, which represent expert opinions derived from the single publication on the auction experiments, we obtain an InfoQ score based on the geometric mean of desirabilities of 77%, that is, relatively high. The relatively weak dimension is generalizability; the strongest dimensions are temporal relevance and chronology of data and goal. An effort to review the scores with some perspective of time proved these scores to be robust even though expert opinions tend to differ to a certain extent. To derive consensus-based scores, one can ask a number of experts (three to five) to review the case and compare their scores. If the scores are consistent, one can derive a consistent InfoQ score. If they show discrepancies, one would conduct a consensus meeting of the experts where the reasoning behind their score is discussed and some score reconciliation is attempted. If a range of scores remains, then the InfoQ score can be presented as a range of values.

3.5 Summary

In this chapter we break down the InfoQ concept into eight dimensions, each dimension relating to a different aspect of the goal–data–analysis–utility components. Given an empirical study, we can then assess the level of its InfoQ by examining each of the eight dimensions. We present four assessment approaches and illustrate the rating-based approach by applying it to the study by Katkar and Reiley (2006) on the effect of reserve prices in online auctions.

The InfoQ assessment can be done at the planning phase of a study, during a study, or after the study has been reported. In Chapter 13 we discuss the application of InfoQ assessment to research proposals of graduate students. In Chapters 4 and 5, we focus on statistical methods that can be applied, a priori or a posteriori, to enhance InfoQ, and Chapters 6–10 are about InfoQ assessments of completed studies. Such assessments provide opportunities for InfoQ enhancement, at the study design, during or after a study has been completed.

Each of the InfoQ dimensions relates to methods for InfoQ improvement that require multidisciplinary skills. For example, data integration is related to IT capabilities such as extract–transform–load (ETL) technologies, and action operationalization can be related to management processes where action items are defined in order to launch focused interventions. For a comprehensive treatment of data analytic techniques, see Shmueli et al. (2016).

In Part II, we examine a variety of studies from different areas using the rating-based approach for assessing the eight dimensions of InfoQ. The combination of application area and InfoQ assessment provides context-based examples. We suggest starting with a specific domain of interest, reviewing the examples in the respective chapter and then moving on to other domains and chapters. This combination of domain-specific examples and cross-domain case studies was designed to provide in-depth and general perspectives of the added value of InfoQ assessments.

References

Bapna, R., Goes, P., Gopal, R. and Marsden, J.R. (2006) Moving from data-constrained to data-enabled research: experiences and challenges in collecting, validating and analyzing large-scale e-commerce data. *Statistical Science*, 21, pp. 116–130.

Bapna, R., Jank, W. and Shmueli, G. (2008a) Consumer surplus in online auctions. *Information Systems Research*, 19, pp. 400–416.

Bapna, R., Jank, W. and Shmueli, G. (2008b) Price formation and its dynamics in online auctions. *Decision Support Systems*, 44, pp. 641–656.

Boslaugh, S. (2007) *Secondary Data Sources for Public Health: A Practical Guide*. Cambridge University Press, Cambridge, UK.

De Veaux, R.D. (2009) Successful Exploratory Data Mining in Practice. JMP Explorer Series. http://www.williams.edu/Mathematics/rdeveaux/success.pdf (accessed May 24, 2016).

Deming, W.E. (2000) *Out of Crisis*. MIT Press, Cambridge, MA.

Ehling, M. and Körner, T. (2007) Eurostat Handbook on Data Quality Assessment Methods and Tools, Wiesbaden. http://ec.europa.eu/eurostat/web/quality/quality-reporting (accessed April 30, 2016).

Fienberg, S.E. (2006) Privacy and confidentiality in an e-commerce world: data mining, data warehousing, matching and disclosure limitation. *Statistical Science*, 21, pp. 143–154.

Figini, S., Kenett, R.S. and Salini, S. (2010) Integrating operational and financial risk assessments. *Quality and Reliability Engineering International*, 26, pp. 887–897.

Forbes, S. and Brown, D. (2012) Conceptual thinking in national statistics offices. *Statistical Journal of the IAOS*, 28, pp. 89–98.

Ghani, R. and Simmons, H. (2004) Predicting the End-Price of Online Auctions. *International Workshop on Data Mining and Adaptive Modelling Methods for Economics and Management*, Pisa.

Giovanni, E. (2008) *Understanding Economic Statistics*. Organisation for Economic Cooperation and Development Publishing, Geneva.

Goldenberg, A., Shmueli, G., Caruana, R.A. and Fienberg, S.E. (2002) Early statistical detection of anthrax outbreaks by tracking over-the-counter medication sales. *Proceedings of the National Academy of Sciences*, 99(8), pp. 5237–5240.

Goodman, D. and Hambleton, R. (2004) Student test score reports and interpretive guides: review of current practices and suggestions for future research. *Applied Measurement in Education*, 17(2), pp. 145–220.

Gopen, G. and Swan, J. (1990) The science of scientific writing. *American Scientist*, 78, pp. 550–558.

Granger, C.W.J. (1969) Investigating causal relations by econometric models and cross-spectral methods. *Econometrica*, 37, pp. 424–438.

Health Metrics Network Secretariat (2008) *Health Metrics Network Framework and Standards for Country Health Information Systems*, 2nd edition. World Health Organization, Health Metrics Network, Geneva.

Kaiser, L.F. and Kaiser, M. (1999) *The Official eBay Guide to Buying, Selling, and Collecting Just About Anything*. Simon & Schuster, New York.

Katkar, R. and Reiley, D.H. (2006) Public versus secret reserve prices in eBay auctions: results from a Pokemon field experiment. *Advances in Economic Analysis and Policy*, 6(2), article 7.

Kaynak, E. and Herbig, P. (2014) *Handbook of Cross-Cultural Marketing*. Routledge, London.

Kenett, R.S. and Salini, S. (2012) *Modern Analysis of Customer Satisfaction Surveys: With Applications Using R*. John Wiley & Sons, Ltd, Chichester, UK.

Kenett, R.S. and Shmueli, G. (2015) Clarifying the terminology that describes scientific reproducibility. *Nature Methods*, 12, pp. 699.

Kenett, R., Zacks, S. and Amberti, D. (2014) *Modern Industrial Statistics: With Applications in R, MINITAB and JMP*, 2nd edition. John Wiley & Sons, Chichester, West Sussex, UK.

Kirschbaum, C. and Hellhammer, D.H. (1989) Salivary cortisol in psychobiological research: an overview. *Neuropsychobiology*, 22, pp. 150–169.

Lee, Y., Strong, D., Kahn, B. and Wang, R. (2002) AIMQ: a methodology for information quality assessment. *Information & Management*, 40, pp. 133–146.

Lotze, T., Shmueli, G. and Yahav, I. (2010) Simulating and Evaluating Biosurveillance Datasets, in *Biosurveillance: Methods and Case Studies Priority*, Kass-Hout, T. and Zhang, X. (editors), CRC Press, Boca Raton, FL.

McShane, B.B. and Wyner, A.J. (2011) A statistical analysis of multiple temperature proxies: are reconstructions of surface temperatures over the last 1000 years reliable? *Annals of Applied Statistics*, 5, pp. 5–44.

Patzer, G.L. (1995) *Using Secondary Data in Marketing Research*. Praeger, Westport, CT.

Pavlou, P.A. and Dimoka, A. (2006) The nature and role of feedback text comments in online marketplaces: implications for trust building, price premiums, and seller differentiation. *Information Systems Research*, 17(4), pp. 392–414.

Raiffa, H. (1970) *Decision Analysis: Introductory Lectures on Choices under Uncertainty*. Addison-Wesley, Reading, MA.

Rao, C.R. (1985) Weighted Distributions Arising Out of Methods of Ascertainment: What Population Does a Sample Represent?, in *A Celebration of Statistics: The ISI Centenary Volume*, Atkinson, A.C. and Fienberg, S.E. (editors), Springer, New York, pp. 543–569.

Rasch, G. (1977) On specific objectivity: an attempt at formalizing the request for generality and validity of scientific statements. *Danish Yearbook of Philosophy*, 14, pp. 58–93.

Saltelli, A., Stark, P., Becker, W. and Stano, P. (2015) Climate models as economic guides: scientific challenge or quixotic quest? *Issues in Science and Technology*, 31(3). http://issues.org/31-3/climate-models-as-economic-guides-scientific-challenge-or-quixotic-quest (accessed April 30, 2016).

Shewhart, W.A. (1986) *Statistical Method from the Viewpoint of Quality Control*, Deming, W.D. (editor), Dover Publications, New York.

Shmueli, G. (2010) To explain or to predict? *Statistical Science*, 25(3), pp. 289–310.

Shmueli, G., Bruce, P. and Patel, N.R. (2016) *Data Mining for Business Analytics: Concepts, Techniques, and Applications in Microsoft Office Excel with XLMiner*, 3nd edition. John Wiley & Sons, Inc., Hoboken, NJ.

Tufte, R.E. (1992) *The Visual Display of Quantitative Information*. Graphics Press, Cheshire, CT.

Wang, R.Y., Kon, H.B. and Madnick, S.E. (1993) Data Quality Requirements Analysis and Modeling. *9th International Conference on Data Engineering*, Vienna.

Wang, S., Jank, W. and Shmueli, G. (2008) Explaining and forecasting online auction prices and their dynamics using functional data analysis. *Journal of Business and Economic Statistics*, 26, pp. 144–160.

4

InfoQ at the study design stage

4.1 Introduction

Statistical methodology includes study design approaches aimed at generating data with high quality of analysis method, f, and implicitly of high information quality (InfoQ). For example, the field of design of experiments (DoE or DoX) focuses on designing experiments that produce data with sufficient power to detect causal effects of interest, within resource constraints. The domain of clinical trials employs study designs that address ethical and other human subject constraints. And survey methodology offers sampling plans aimed at producing survey data with high InfoQ. In this chapter we review several statistical approaches for increasing InfoQ at the study design stage. In particular, we look at approaches and methodologies for increasing InfoQ prior to the collection of data. Despite the data unavailability at this planning phase, there are various factors that can affect InfoQ even at this stage.

It is useful to distinguish between causes affecting data quality and InfoQ a priori (or ex ante) and a posteriori (or ex post). A priori causes are known during the study design stage and before data collection. They result, for example, from known limitations of resources (e.g., the sample size), ethical, legal, and safety considerations (e.g., an inability to test a certain drug on certain people in a clinical trial) and constraints on factor-level combinations in experimental designs. A posteriori issues (the focus of Chapter 5) result from the actual performance of the mechanism generating or collecting the data and are discovered (or not) after the data has been collected (e.g., data entry errors, measurement error, ex post constraints in experimental conditions and intentional data manipulation). Consider a measured dataset X and a target dataset $X^* \neq X$. We denote data affected by a priori causes by $X = \eta_1(X^*)$, by a posteriori causes by $X = \eta_2(X^*)$ and by both causes by $X = \eta_1\{\eta_2(X^*)\}$. In this chapter we describe existing approaches for increasing InfoQ under different scenarios of a

Information Quality: The Potential of Data and Analytics to Generate Knowledge,
First Edition. Ron S. Kenett and Galit Shmueli.
© 2017 John Wiley & Sons, Ltd. Published 2017 by John Wiley & Sons, Ltd.
Companion website: www.wiley.com/go/information_quality

Table 4.1   Statistical strategies for increasing InfoQ given a priori causes at the design stage.

	Strategies for increasing InfoQ	A priori causes
Design of experiments	Randomization; blocking; replication; linking data collection protocol with appropriate design	Resource constraints; impossible runs
Clinical trials	Randomization; blocking; replication; linking data collection protocol with appropriate design; blinding; placebo	Resource constraints; ethics; safety
Survey sampling	Reducing nonsampling errors (e.g., pretesting questionnaire, reducing nonresponse) and sampling errors (e.g., randomization, stratification, identifying target and sampled populations)	Resource constraints; ethics; safety
Computer experiments	Randomization; blocking; replication; linking data collection protocol with appropriate design; space-filling designs	Impossible or difficult to obtain real data; time and costs associated with computer simulation

priori data issues and related InfoQ-decreasing constraints. Table 4.1 summarizes the strategies and constraints. The next sections expand on each point.

4.2   Primary versus secondary data and experiments versus observational data

Before discussing the four main statistical approaches for increasing InfoQ at the study design stage, we clarify the difference between the two types of data collection methods and their relation to this chapter. The first distinction is between primary and secondary data. The second is between experimental and observational data. Let us examine each of them.

4.2.1   Primary data versus secondary data

The terms *primary data* and *secondary data* are popular in the social sciences, in marketing research and in epidemiology and public health. The distinction between primary and secondary data depends on the relationship between the researcher or team who designed the study and collected the data and the one analyzing it. Hence, the same dataset can be primary data in one analysis and secondary data in another (Boslaugh, 2007).

   Primary data refers to data collected by the researcher for a particular analysis goal. *Secondary data* refers to data collected by someone other than the researcher or collected previously by the researcher for a goal different than the one in the study of interest. The data might have been collected by another researcher or organization for the same purpose of the current analysis or for a completely different purpose. In short, primary data is collected under the control of the researcher with the study purpose in mind, while secondary data is collected independent of the study of interest.

Note that the terms *primary* and *secondary* do not imply an order of importance or usefulness of the data, but only the collection source and purpose relative to the study of interest. The advantages and disadvantages of primary and secondary data are discussed in the marketing research and public health literature (Patzer, 1995, chapter 2; Boslaugh, 2007, pp. 3–4). Considerations for using secondary data over collecting primary data include acquisition costs and time saving, breadth of data, expertise of the data collector, availability of the needed measurements, knowledge of the collection process, and the challenges encountered.

In this chapter we focus on the predata collection design stage of a study. The statistical methodologies we discuss are directed toward a researcher that has control over the data collection. Hence, we are looking at primary data. In Chapter 5 we look at postdata collection methods for increasing InfoQ. We deal with secondary data as well as "primary data with a revised goal"—that is, data collected for one purpose, but then the study goal was revised, thereby making the data somewhat "secondary" in nature.

While the distinction between primary and secondary data might seem clear, there exist some hybrid study designs. One type of study that we discuss in this chapter—a computer experiment—combines primary and secondary data. Specifically, it typically uses secondary data to build a model and then uses the model to simulate "primary data." Another primary–secondary data hybrid is meta-analysis, where "data" refers to results of previous studies studying the same research question. Statistical methodology is then used to combine these prior results for obtaining more precise and reliable results, that is, for increasing InfoQ.

4.2.2   Experimental versus observational data

Another common and important distinction between data collection methods is between *experimental data*, such as data from a randomized controlled trial (RCT), and *observational data*, that is, data collected by observing an existing phenomenon. Experimental data may be collected in a lab (*in vitro*), where the researcher has more control over the environment, or "in the wild" (*in vivo*). Companies like Amazon (www.amazon.com) can run an experiment on a sample of random users and get results within minutes. Web survey companies like SurveyMonkey (www. SurveyMonkey.com) can address a survey to specific groups in a large online panel. For causal study goals, researchers typically prefer to collect experimental data. However, when it is impossible or extremely difficult to conduct an experiment for ethical or legal reasons, or due to prohibitive costs, researchers typically resort to observational data. In analyzing such data they might resort to analytic methods that provide inferential capabilities across groups such as transportability studies (Bareinboim and Pearl, 2013a, 2013b).

Observational data is sometimes the method of choice, especially when the goal of the study is noncausal: descriptive or predictive. The age of big data has seen a huge growth in predictive modeling based on observational data, with applications in many fields. One characteristic of observational data, which is useful for prediction, is that the collected data is typically more similar to the data during the target time of prediction than that in a laboratory experiment.

Note that the primary–secondary data distinction is different from the experimental/observational distinction. Primary data can be experimental or observational. Similarly, secondary data can be obtained from another experiment or from observational data collected for a purpose different from the study of interest. Moreover, observational data can be collected by applying DoE methods. For example, optimal experimental design methods have been used (Steinberg et al., 1995) in deciding where to locate sensors for detecting earthquakes. A general procedure for applying optimal design methods to observational data has been proposed by Berni (2003).

In the context of healthcare, Shavit et al. (2007) propose an approach for cost-effectiveness analysis of studies that are required to evaluate health technologies for reimbursement. They consider bias inherent in study designs as the main factor that differentiates the ability of studies to predict benefits from new healthcare technologies. Their method allows to conduct, at the design stage, an economic evaluation of inherent biases in the study design as an alternative means of acquiring measures of systematic error. This economics-driven analysis is also about enhancing InfoQ at the design stage.

In the following sections in this chapter, we look at both experimental designs and observational designs and the statistical methods aimed at improving the InfoQ of the resulting data.

4.3   Statistical design of experiments

Controlled experiments are considered the gold standard for inferring causality, yet experimentation can be resource intensive. The aim of the field of DoE is to proactively collect data in the most efficient way for answering a causal question of interest. The objective, g, of a statistically designed experiment (DoE) is typically classified into:

- *Screening*—Identifying main factors affecting a response. Typically, a long list of factors with potential effect on the response is reduced to a short list of "active factors" via linear models.

- *Comparing*—Testing the effect of a single factor on a response, often in the presence of other nuisance factors.

- *Optimizing*—Finding a subset of factor space that optimizes the response. Experiments are designed to map out the so-called design space, typically via nonlinear models and optimization methods.

- *Robustifying*—Reducing the sensitivity of a response to noise variables, in the subset of factor space identified as optimal. This is achieved using internal and external experimental arrays reflecting proactive setup of control and noise factors (Phadke, 1989).

Given one of these objectives and typical resource constraints, experimental factor-level combinations are selected and an experimental array is selected. These considerations assume that the final data will be equal to the ideal data: $X = \eta_1(X^*)$. Then, assuming adequate statistical analysis, the data generated by the experiment is expected to yield high InfoQ in terms of answering the question of interest, at the required level of type I and II errors (U).

Note that data collected through a design for achieving goal g_1 should lead to high InfoQ relative to that objective, but it might have low InfoQ for $g_i \neq g_1$. For instance, data from a study that was designed to screen several dozen factors might be of low InfoQ for comparing effects in terms of the chosen factors and their levels. To demonstrate this, consider a 2^{7-3} fractional factorial experiment designed to screen seven factors (labeled M, S, V0, K, P0, T0, and T) affecting the performance of a piston (see Figure 4.1). Such an experiment consists of a design with two levels for each factor and permits to screen factor effects on the basis of linear effects and

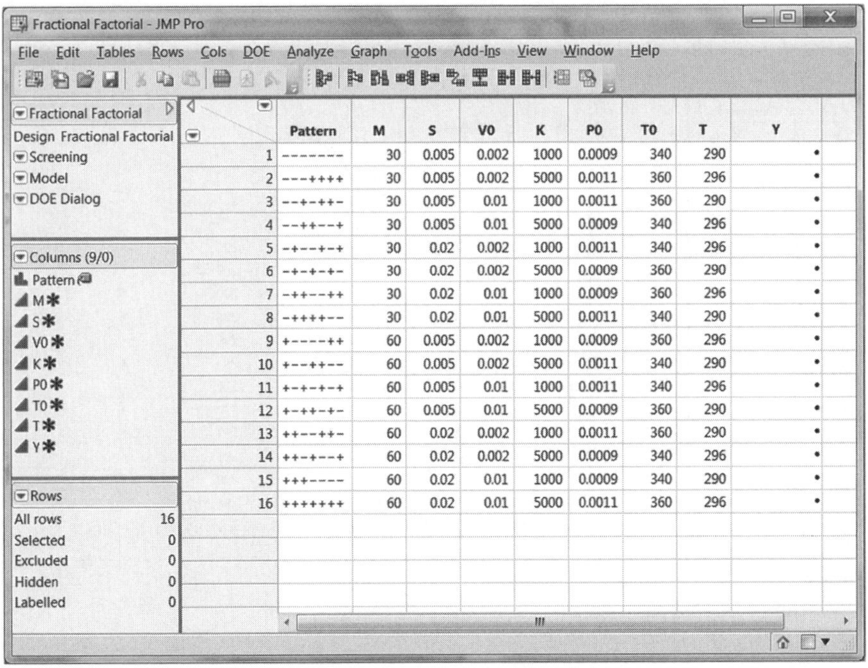

	Pattern	M	S	V0	K	P0	T0	T	Y
1	-------	30	0.005	0.002	1000	0.0009	340	290	•
2	---++++	30	0.005	0.002	5000	0.0011	360	296	•
3	--+-++-	30	0.005	0.01	1000	0.0011	360	290	•
4	--++--+	30	0.005	0.01	5000	0.0009	340	296	•
5	-+--+-+	30	0.02	0.002	1000	0.0011	340	296	•
6	-+-+-+-	30	0.02	0.002	5000	0.0009	360	290	•
7	-++--++	30	0.02	0.01	1000	0.0009	360	296	•
8	-++++--	30	0.02	0.01	5000	0.0011	340	290	•
9	+----++	60	0.005	0.002	1000	0.0009	360	296	•
10	+--++--	60	0.005	0.002	5000	0.0011	340	290	•
11	+-+-+-+	60	0.005	0.01	1000	0.0011	340	296	•
12	+-++-+-	60	0.005	0.01	5000	0.0009	360	290	•
13	++--++-	60	0.02	0.002	1000	0.0011	360	290	•
14	++-+--+	60	0.02	0.002	5000	0.0009	340	296	•
15	+++----	60	0.02	0.01	1000	0.0009	340	290	•
16	+++++++	60	0.02	0.01	5000	0.0011	360	296	•

Figure 4.1   JMP screenshot of a 2^{7-3} fractional factorial experiment with the piston simulator described in Kenett and Zacks (2014). Source: Kenett and Zacks (2014). Reproduced with permission from John Wiley & Sons, Inc.

interactions. This specific experiment is part of a general approach to teaching DoE using simulators, thus providing hands-on interactive experience with DoE methods. The piston simulator is a free add-on to the JMP software used in Kenett and Zacks (2014) to teach statistical process control and DoE. Going from right to left, the experimental array is in standard order and corresponds to a fully balanced orthogonal design.

An alternative design is to use a definite screening design with 17 runs on the same seven factors (see Figure 4.2). The design incorporates three levels for each factor and therefore provides the ability to model quadratic effects with only one more experimental run compared to the previous design.

To compare the designs we can use a fraction of design space plot that characterizes the experimental space in terms of prediction variance (see Figure 4.3). The plots show that the average prediction variance for the 2^{7-3} fractional factorial design is 0.0135 and for the definite screening design, 0.225. In addition we see that the prediction variance in the fractional factorial experiment is more or less uniform across the design region, as opposed to the definite screening design. In that design we see that the prediction variance ranges from 0.1 to 0.4 in the outskirts of the experimental range. The design of choice is clearly dependent on the investigation goals.

	M	S	V0	K	P0	T0	T	Y
1	45	0.02	0.01	5000	0.0011	360	296	•
2	45	0.005	0.002	1000	0.0009	340	290	•
3	60	0.0125	0.01	5000	0.0009	360	290	•
4	30	0.0125	0.002	1000	0.0011	340	296	•
5	60	0.005	0.006	5000	0.0011	340	296	•
6	30	0.02	0.006	1000	0.0009	360	290	•
7	60	0.005	0.002	3000	0.0011	360	290	•
8	30	0.02	0.01	3000	0.0009	340	296	•
9	60	0.02	0.002	1000	0.001	360	296	•
10	30	0.005	0.01	5000	0.001	340	290	•
11	60	0.005	0.01	1000	0.0009	350	296	•
12	30	0.02	0.002	5000	0.0011	350	290	•
13	60	0.02	0.002	5000	0.0009	340	293	•
14	30	0.005	0.01	1000	0.0011	360	293	•
15	60	0.02	0.01	1000	0.0011	340	290	•
16	30	0.005	0.002	5000	0.0009	360	296	•
17	45	0.0125	0.006	3000	0.001	350	293	•

Figure 4.2   JMP screenshot of a definitive screening design experiment with the piston simulator described in Kenett and Zacks (2014). Source: Kenett and Zacks (2014). Reproduced with permission from John Wiley & Sons, Inc.

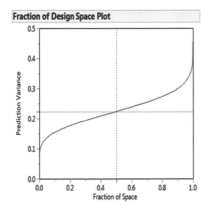
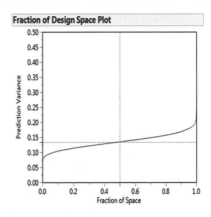

Design Diagnostics

Design Diagnostics

		D Optimal Design	
D Efficiency	84.37606	D Efficiency	100
G Efficiency	91.76629	G Efficiency	100
A Efficiency	84.21053	A Efficiency	100
Average Variance of Prediction	0.22549	Average Variance of Prediction	0.135417
Design Creation Time (seconds)	0	Design Creation Time (seconds)	0

Figure 4.3   JMP screenshot of fraction of design space plots and design diagnostics of fractional (left) and definite screening designs (right).

Screening with the fractional factorial design is better; however the definite screening design allows for better optimization since it picks up quadratic effects. If our objective is to identify how to best operate the piston in order to achieve a target performance, we will choose the definite screening design even though it performs worse than the fractional factorial balanced design. For more on definite screening designs and design optimization under constraints in the levels of experimental factors, see Goos and Jones (2011).

In the online auction experimental study discussed in Chapter 3, we considered the researchers' choice of item to auction (Pokémon cards), the experimental design (selling 25 identical pairs of Pokémon cards, each card auctioned twice: once with a public reserve and once with a secret reserve) and the experimental setting (e.g., all auctions were seven-day auctions and started on Sunday between 7 and 9 p.m. Eastern Standard Time, and the seller rating was kept at zero), which were all directly aimed at achieving high InfoQ for answering the particular comparative question. In contrast, the same data would be of low InfoQ for a screening study to determine the main factors affecting the final price, because many potential factors such as the duration of the auction, start and end day of the week, and seller rating were purposely held constant.

The three main principles of experimental design—*randomization, blocking,* and *replication*—assume that the data collection is under the control of the experimenter and therefore maximize InfoQ via "smart" data generation. However at the postdesign stage, once the data has been collected, all that can be done is to evaluate whether and how these principles have been achieved by auditing the protocol of the experiment execution. We return to this in Chapter 5.

4.4   Clinical trials and experiments with human subjects

A clinical trial is defined as "a prospective study comparing the effect and value of intervention(s) against a control in human beings" (Friedman et al., 1999, p. 2). In clinical trials, key principles are *randomization* and *double blinding* where both doctors and patients do not have knowledge about the treatment assignment; sometimes *triple blinding* is applied, where even the data analyst is not aware of the meaning of the labels of various groups she/he is analyzing. The goal is to generate unbiased data that can then be used to evaluate or compare the intervention effects. However, generating data with high InfoQ via clinical trials is somewhat different from DoE. Although the design of clinical trials and of experiments involving humans has its roots in classic DoE, the human factor involved (e.g., patients and doctors) introduces two important differentiating factors, namely, *ethics* and *safety*, which can limit the level of InfoQ of the generated data $X = \eta_1(X^*)$. Whereas in classic DoE the main constraints are resources, in clinical trials or other experiments where human subjects are involved, an important aspect of the study design is ethical considerations. Ethical considerations constrain the experimental design and can lower InfoQ. For instance, some treatment combinations might not be ethical, some sequences of runs might not be ethical, or even the lack of treatment can be unethical. The early termination of a trial through futility analyses is a reasonable strategy. Clinical trials typically require a control group, yet not providing treatment or providing a placebo might not be ethical in some cases.

Moreover, the strategy of *randomization*, which is at the heart of experiments, is widely ethically debated (Friedman et al., 1999, p. 45). An important principle in the analysis of clinical trials is linking the data collection protocol with data analysis: "as ye randomise so shall ye analyse" (Senn, 2004). For a comprehensive treatment of statistical and ethical issues of first-in-man clinical trials, see Senn et al. (2007).

The Food and Drug Administration (FDA) issued a guidance document titled "Enrichment Strategies for Clinical Trials to Support Approval of Human Drugs and Biological Products" (FDA, 2012). The documents define enrichment as "the prospective use of any patient characteristic to select a study population in which detection of a drug effect (if one is in fact present) is more likely than it would be in an unselected population." The aim of the guidance is to enhance InfoQ by effective design methods.

Enrichment strategies are considered in the context of randomized controlled trials (RCT) and mostly affect patient selection before randomization. They include three main methods:

1. *Strategies to decrease heterogeneity* – These include selecting patients with baseline measurements in a narrow range (decreased interpatient variability) and excluding patients whose disease or symptoms improve spontaneously or whose measurements are highly variable. The decreased variability provided by this approach increases study power.

2. *Prognostic enrichment strategies* – Choosing patients with a greater likelihood of having a disease-related endpoint event or a substantial worsening in condition. This approach increases the absolute effect difference between groups but will not alter relative effect.

3. *Predictive enrichment strategies* – Choosing patients more likely to respond to the drug treatment than other patients with the condition being treated. Such selection can lead to a larger effect size and permit use of a smaller study population.

All these strategies are aiming at increasing the InfoQ of clinical research.

InfoQ is potentially decreased by the human factor of compliance of patients and physicians with the treatment regime mandated by the experimental design. This can limit the types of answerable research questions. In extreme cases, "an investigator may not be able to compare interventions, but only intervention strategies" (Friedman et al., 1999, p. 3). Moreover, human responses can be affected not only by the treatment but also by psychological and other effects. Hence, *placebo* and *blinding* are widely employed.

The FDA guidance document states: "Practices such as encouraging good compliance by making patients aware of the conditions and demands of the trial, avoiding too-rapid titration of drugs that could cause intolerable early side effects, using adherence prompts and alert systems, and counting pills (or using "smart bottles" to monitor drug use) so that noncompliant patients can be encouraged to perform better have become standard. There have also, on occasion, been more specific efforts to identify and enroll good compliers into clinical trials." The last sentence is about increasing InfoQ by proper study design, including the management of patient compliance and effective patient enrolment. Poor compliance by patients with the treatment under study reduces InfoQ. In addition, the FDA provides a warning: "removing poor compliers identified after randomization is generally not acceptable because such patients are not likely to be a random sample of the study population and because compliance itself has been linked to outcome, even compliance in taking a placebo."

Noncompliance that has not been handled by design is therefore negatively affecting the quality of the study and cannot be ignored in the analysis.

Two further factors that decrease InfoQ for comparing interventions are safety issues and the need for informed consent. Safety considerations can affect InfoQ by impacting the study design, for instance, by restricting the dosage to relatively low levels. Another implication of safety is the very small sample size used in phase I studies (for assessing the toxicity level of a drug to humans) and the sequential nature of the drug administration. In fact, multistage and sequential designs are especially popular in the clinical trial context due to ethical and safety considerations. The need to obtain informed consent from participants in clinical trials creates a constraint on the ability to draw "objective" results due to impacting psychological effects and of levels of compliance. Constraints on data that arise from ethical and safety considerations in clinical trials can therefore lead to lower InfoQ.

4.5   Design of observational studies: Survey sampling

The statistical literature on methodology for designing observational studies includes sample surveys. Sampling methodology aims to achieve high precision and low bias estimates, within resource constraints. Designing a survey study consists of determining sampling issues such as sample size, sampling scheme, and sampling allocation in order to reduce sampling errors, as well as addressing nonsampling issues such as nonresponse and questionnaire design to reduce nonsampling errors (measurement bias and selection bias). InfoQ is therefore influenced by both sampling and nonsampling errors, relative to the goal at hand. Survey methodology is aimed at creating a data collection instrument (questionnaire) and survey process that yield reliable and valid data. For example, one should distinguish between *individual surveys*, where the opinions of the respondent are sought, and *household and establishment surveys*, where the respondent is asked to provide an aggregate representative response on behalf of a household or organization. The objectives of the survey determine how the questionnaire is formulated and how the survey is conducted (for the evaluation of establishment survey questionnaires, see Forsyth et al., 1999). An individual survey with high InfoQ can turn out to be of low InfoQ when the opinion sought is that of a company rather than a personal opinion. Another example is of surveys that measure the unemployment rate (such as the Current Population Survey in the United States) but are not designed to produce statistics about the number of jobs held. These two different types of surveys typically produce different unemployment figures.

The first step in designing a survey is generating a clear statistical statement about the desired information as well as a clear definition of the target population. These two factors, which determine g, affect not only the data analysis but also the data collection instrument. As in DoE and clinical trials, the study design must take into account resource constraints as well as ethical, legal, and safety considerations. In all cases, when human subjects are involved (as experimental units or as surveyed participants), a special approval process is usually required in order to carry out a study. In the United States, all federally funded organizations have institutional review boards (IRB), which are in charge of approving, monitoring, and reviewing studies that involve human subjects. In other countries such boards are called Helsinki committees. These review boards' mission is to protect participants' safety and well-being as well as to validate the goals of the proposed study to assure that the study design will achieve sufficient InfoQ. As in clinical trials, InfoQ is potentially limited by such constraints.

Revisiting the online auction example, consider a survey study aimed at comparing behavioral traits of auction winners who placed a bid versus those who paid the "buy-it-now" price (an option popular in many online auctions, which allows individuals to purchase an item at a fixed price prior to the commencement of bidding). Angst et al. (2008) surveyed winners in eBay auctions to test whether competitiveness, impulsiveness, and level of hedonistic need separate bidders from fixed-price buyers. To obtain data with high InfoQ, they tried to reduce nonsampling bias (e.g., by using previously validated scales in the questionnaire and sending multiple follow-ups) as well as sampling bias (by choosing a sample of auctions for a popular product during a limited time period). Several limitations reduced the InfoQ in this study: nonsampling issues include a response rate of 27% (113 usable questionnaires) and a change in eBay's policy during the survey period that lead to a shift from Web surveys to email surveys (thereby introducing a "survey-type" effect). Sampling issues relate to the generalizability from the sample to the larger population, given the small sample size and that a single product was chosen. For more on customer surveys, see Kenett and Salini (2012).

4.6   Computer experiments (simulations)

In computer experiments, a computer runs a computationally intensive stochastic or deterministic model that simulates a scientific phenomenon (e.g., a computational fluid dynamics model). Computer experiments based on a DoE design are then used to collect response values for a set of input values. These are used in turn to build a statistical model (called "emulator" or "metamodel") of the response in terms of the input variables. The models used for this are called Kriging, DACE or Gaussian models, and they account for the lack of experimental errors in the computer runs (see Kenett and Zacks, 2014). Statistical methods such as "space-filling designs" are used to generate data with high InfoQ to help design robust systems and products. Computer experiments range from very basic simulations to computer-intensive complex dynamic systems simulations. This variety is reflected by varying levels of "fidelity" or accuracy and is used in Bayesian-based inference (Huang and Allen, 2005). Validating and calibrating computer simulations are nontrivial tasks, and one major and not uncommon risk is using the wrong simulation model (see Bayarri et al., 2007). A wrong model will obviously lead to low InfoQ. An approach to derive a stochastic emulator from simulation experiments is presented in Bates et al. (2006), the idea being that the emulator modeling the response can be sampled using a space-filling experiment, in order to derive a model of the variability of the responses. The combination of a model for the response levels and a stochastic emulator providing data on the response variability provides an optimal and robust solution. This combination has high InfoQ if the goal is to design a product or system that is on target and robust to variability in the input factors.

4.7   Multiobjective studies

Most of the statistical study design literature, in terms of InfoQ, is focused on a specific goal within the domains of DoE, clinical trials or survey methodology. While an experiment, trial or survey can be aimed at answering several questions, the questions are typically ordered by importance, so that the collected data generates high InfoQ levels for the high-priority questions and lower InfoQ for lower-priority questions.

When it comes to designing studies for achieving multiple objectives of equal importance, a few papers that develop study designs do address multiple objectives. The most popular ubiquitous goal is cost considerations. One example is Ben-Gal and Caramanis (2002), who consider the goal of maximizing both information and economic measures. They propose a sequential strategy to the DoE using a dynamic programming approach that achieves this combined goal. The term *information* used by Ben-Gal and Caramanis is in the sense of Shannon's entropy.

A second example is the data acquisition algorithm developed by Saar-Tsechansky et al. (2009), which is aimed at achieving high predictive accuracy (or some other predictive utility function) while minimizing acquisition cost. The algorithms choose which predictor values (or which missing response labels) should be collected, taking into consideration their cost and their contribution to predictive accuracy.

In another example, Ginsburg and Ben Gal (2006) suggest an experimentation strategy for the robust design of empirically fitted models. The approach is used to design experiments that minimize the variance of the optimal robust solution. This new DoE optimality criterion, termed Vs-optimal, prioritizes the estimation of a model's coefficients, so that the variance of the optimal solution is minimized by the performed experiments. The approach provides for a high InfoQ study focused on achieving robust performance.

As a final example, we mention the work of Engel et al. (2016) that discusses robust design methods when degradation agents affect performance over time. In this context, the authors present a method for specifying targets that account, at the design stage, for a change in requirements that eventually can induce failures. In other words, a system is usually designed to meet certain goals described in requirement documents. However, over time, these requirements might change and typically become more stringent, and performance that was considered acceptable in the past is not so anymore. This change in goal can be anticipated a priori and taken into account in the design specifications. In their paper, Engel et al. (2016) present an example of a heart pacemaker which is affected by such degradation failure agents. This, and the other previous examples, show how a priori considerations can be used to ensure high InfoQ of a study or design.

4.8   Summary

In this chapter, we looked at several statistical approaches applied at the predata collection stage, with the aim of generating data with high InfoQ. We provided examples of primary and secondary data scenarios and of experimental and observational cases. We discussed the main principles of DoE, clinical trials, survey sampling, and computer experiments aimed at maximizing InfoQ while adhering to constraints and requirements that result from factors such as safety, ethical considerations, resource constraints, etc. Such constraints—which we call a priori causes—affect InfoQ even before the data are collected. Table 4.1 summarizes the strategies for increasing InfoQ at the predata collection stage as well as the a priori cases that lower InfoQ. In the next chapter we examine strategies for maximizing InfoQ at the postdata collection stage.

References

Angst, C.M., Agarwal, R. and Kuruzovich, J. (2008) Bid or buy? Individual shopping traits as predictors of strategic exit in on-line auctions. *International Journal of Electronic Commerce*, 13(1), pp. 59–84.

Bareinboim, E. and Pearl, J. (2013a) Meta-Transportability of Causal Effects: A Formal Approach. *Proceedings of the 16th International Conference on Artificial Intelligence and Statistics (AISTATS)*, AIII, Scottsdale, AZ.

Bareinboim, E. and Pearl, J. (2013b) Causal Transportability with Limited Experiments. *Proceedings of the 27th Conference on Artificial Intelligence (AAAI)*, AIII, Bellevue, Washington.

Bates, R., Kenett, R., Steinberg, D. and Wynn, H. (2006) Achieving robust design from computer simulations. *Quality Technology and Quantitative Management*, 3, pp. 161–177.

Bayarri, M., Berger, J., Paulo, R., Sacks, J., Cafeo, J., Cavendish, J., Lin, C.-H. and Tu, J. (2007) A framework for validation of computer models. *Technometrics*, 49, pp. 138–154.

Ben-Gal, I. and Caramanis, M. (2002) Sequential DoE via dynamic programming. *IIE Transactions*, 34, pp. 1087–1100.

Berni, R. (2003) The use of observational data to implement an optimal experimental design. *Quality Reliability Engineering International*, 19, pp. 307–315.

Boslaugh, S. (2007) *Secondary Data Sources for Public Health: A Practical Guide*. Cambridge University Press, Cambridge, UK.

Engel, A., Kenett, R.S., Shahar, S. and Reich, Y. (2016) Optimizing System Design Under Degrading Failure Agents. *Proceedings of the International Symposium on Stochastic Models in Reliability Engineering, Life Sciences and Operations Management (SMRLO16)*, Beer Sheva, Israel.

Food and Drug Administration (2012) Enrichment Strategies for Clinical Trials to Support Approval of Human Drug and Biological Products. http://www.fda.gov/downloads/drugs/guidancecomplianceregulatoryinformation/guidances/ucm332181.pdf (accessed October 20, 2015).

Forsyth, B., Levin, K. and Fisher, S. (1999) Test of an Appraisal Method for Establishment Survey Questionnaires. *Proceedings of the Survey Research Methods Section*, American Statistical Association. www.amstat.org/sections/srms/proceedings/papers/1999_021.pdf (accessed October 20, 2015).

Friedman, L.M., Furberg, C.D. and DeMets, D.L. (1999) *Fundamentals of Clinical Trials*, 3rd edition. Springer, New York.

Ginsburg, H. and Ben Gal, I. (2006) Designing experiments for robust-optimization problems: the Vs-optimality criterion. *IIE Transactions*, 38, pp. 445–461.

Goos, P. and Jones, B. (2011) *Optimal Design of Experiments: A Case Study Approach*. John Wiley & Sons, Inc., Hoboken, NJ.

Huang, D. and Allen, T.T. (2005) Design and analysis of variable fidelity experimentation applied to engine valve heat treatment process design. *Applied Statistics*, 54, pp. 443–463.

Kenett, R.S. and Salini, S. (2012) *Modern Analysis of Customer Satisfaction Surveys: With Applications Using R*. John Wiley & Sons, Ltd, Chichester, UK.

Kenett, R.S. and Zacks, S. (2014), *Modern Industrial Statistics: With Applications Using R, MINITAB and JMP*, 2nd edition. John Wiley & Sons, Inc., Hoboken, NJ.

Patzer, G.L. (1995) *Using Secondary Data in Marketing Research: United States and Worldwide*. Praeger, Westport, CT.

Phadke, M.S. (1989) *Quality Engineering Using Robust Design*. Prentice Hall, Englewood Cliffs, NJ.

Saar-Tsechansky, M., Melville, P. and Provost, F. (2009) Active feature-value acquisition. *Management Science*, 55(4), pp. 664–684.

Senn, S. (2004) Controversies concerning randomization and additivity in clinical trials. *Statistics in Medicine*, 23, pp. 3729–3753.

Senn, S., Amin, D., Bailey, R., Bird, S., Bogacka, B., Colman, P., Garett, A., Grieve, A. and Lachmann, P. (2007) Statistical issues in first-in-man studies. *Journal of the Royal Statistical Society: Series A (Statistics in Society)*, 170(3), pp. 517–579.

Shavit, O., Leshno, M., Goldberger, A., Shmueli, A. and Hoffman, A. (2007) It's time to choose the study design! Net benefit analysis of alternative study designs to acquire information for evaluation of health technologies. *PharmacoEconomics*, 25 (11), pp. 903–911.

Steinberg, D., Rabinowitz, N., Shimshoni, Y. and Mizrachi, D. (1995) Configuring a seismographic network for optimal monitoring of fault lines and multiple sources. *Bulletin of the Seismological Society of America*, 85(6), pp. 1847–1857.

5

InfoQ at the postdata collection stage

5.1 Introduction

In Chapter 4, we examined factors affecting the predata collection study design stage, which yield low InfoQ and dataset X that is related to the target dataset X^*. That chapter presented a range of methods to increase the InfoQ at the predata collection stage.

In this chapter, we turn to the later stage of an empirical study, after the data has been collected. The data may have been collected by the researcher for the purpose of the study (primary data) or otherwise (secondary and semisecondary data). The data may be observational or experimental. Moreover, the study may have revised goals or even revised utility. These changes affect the way the data is analyzed in order to derive high InfoQ of the study.

We begin by laying out key points about primary, secondary, and semisecondary data, as well as revised goals and revised utility. We then move to a discussion of existing methods and approaches designed to increase information quality at the postdata collection stage. The methods range from "fixing" the data to combining data from multiple studies to imputing missing data. In some cases we can directly model the distortion between X and X^*. For the different methods discussed here, we examine the relationship between the target dataset X^* and the actual dataset X as a function of both a priori causes, η_1, and a posteriori causes, η_2, through the relationship $X = \eta_2\{\eta_1(X^*)\}$. Each approach is designed to increase InfoQ of the study by tackling a specific a posteriori cause.

Information Quality: The Potential of Data and Analytics to Generate Knowledge,
First Edition. Ron S. Kenett and Galit Shmueli.
© 2017 John Wiley & Sons, Ltd. Published 2017 by John Wiley & Sons, Ltd.
Companion website: www.wiley.com/go/information_quality

5.2   Postdata collection data

In Chapter 4 we described the terms *primary data* and *secondary data* and the difference between them. Recall that the distinction is based on the relationship between the researcher or team collecting the data and the one analyzing it. Hence, the same dataset can be primary data in one analysis and secondary data in another (Boslaugh, 2007). Primary data refers to data collected by the researcher for a particular analysis goal. Secondary data refers to data collected by someone other than the researcher or collected previously by the researcher for a different goal. Finally, there exist hybrids. In Chapter 4 we looked at computer experiments, which generate primary data (simulations) based on secondary data (from physical models).

In the next sections, we look at existing methods and approaches for increasing InfoQ at the postdata collection stage for data that arise from either primary, secondary, or hybrid sources. While primary data is designed to contain high InfoQ due to the researcher's involvement in the study design, the reality of data collection is that the resulting X is almost always not exactly as intended, due to a posteriori causes. "Unexpected problems may arise when the experiment is carried out. For example, experiments can produce non-quantifiable results or experimental points may generate 'outliers', observations whose values appear quite incompatible with the overall pattern in the data" (Kenett et al., 2006). Therefore, methods for increasing InfoQ at the postdata collection stage can be aimed at secondary data, for example, adjusting for selection bias; at semisecondary data, for example, meta-analysis; and even at primary data, for example, handling missing values.

Primary data can become secondary data if the goal or utility of the study is revised or when secondary goals are addressed. A context which is popular in practice, yet hardly discussed in the statistics or data mining literature from the perspective of information quality, is the case of primary data with revised goals. Cox (2009) notes, "Objectives may be redefined, hopefully improved, and sometimes radically altered as one proceeds."

Similarly, Friedman et al. (2015, p. 182) comment that in clinical trials, "One would like answers to several questions, but the study should be designed with only one major question in mind." Therefore, often multiple questions will be answered using data that was collected through a design for answering a single primary question. In particular, the evaluation of adverse effects is important, yet not the primary goal of a clinical trial. The result is that "clinical trials have inherent methodological limitations in the evaluation of adverse effects. These include inadequate size, duration of follow-up, and restricted participant selection."

In another example discussed in Chapter 4, Engel et al. (2016) consider robust design methods when degradation agents affect performance over time. In this case, specifying targets that do not account for a change in requirements, if realized, will induce failures. Since this change in goal has not been anticipated at the design stage, the a posteriori analysis needs to consider changes in design specifications.

Another common situation where revised goals arise is in the process of journal article reviewing. In the social sciences and in economics, it is common for reviewers to request authors to answer additional research questions. In some cases, it is impossible to collect additional data that would directly have high InfoQ for these new questions, and the authors are forced to use the existing data in answering the new questions. For more on the review process and the information quality it is supposed to ensure, see Chapter 12.

A dramatic practical example of primary data collected through simulations with revised goals is linked to the Columbia shuttle disaster. The Columbia Accident Investigation Board reported that a simulation program called CRATER was used to analyze the impact of foam debris on the shuttle protective tiles. The simulation modeled impact of debris smaller by a factor of 400 than the ones affecting the shuttle at lift-off. The engineers who had developed CRATER had left, and the replacement engineer who used it did not realize the impact of the scale. The analysis he conducted showed that the shuttle was safe. This mistaken information, with obvious low InfoQ, had tragic consequences (see www.nasa.gov/columbia/caib).

The methods and approaches described in the following sections are therefore relevant to a variety of data–goal–utility scenarios. However, their implementation is specific to the goal and utility of interest.

5.3   Data cleaning and preprocessing

Data "cleanliness" has long been recognized by statisticians as a serious challenge. Hand (2008) commented that "it is rare to meet a data set which does not have quality problems of some kind." Godfrey (2008) noted that "Data quality is a critically important subject. Unfortunately, it is one of the least understood subjects in quality management and, far too often, is simply ignored."

Consider a measured dataset X and a target dataset $X^* \neq X$. The data quality literature includes methods for "cleaning" X to achieve X^* and guidelines for data collection, transfer, and storage that reduce the distance between X and X^*.

Denote data quality procedures (cleaning, avoiding errors, etc.) by $h(\cdot)$. We distinguish between two general types of procedures $h(X)$ and $h(X|g)$. The first, $h(X)$, focuses on procedures that generate or clean X to minimize its distance from X^*, without considering anything except the dataset itself.

Advanced data recording devices, such as scanners and radio-frequency identification (RFID) readers, data validation methods, data transfer and verification technologies, and robust data storage, as well as more advanced measurement instruments, have produced "cleaner" data (Redman, 2007) in terms of the distance between X and X^*. Management information systems (MIS)-type data quality (see Chapter 3) focuses on $h(X)$ operations.

In contrast, $h(X|g)$ focuses on quality procedures that generate or clean X conditional on the goal g. One example is classic statistical data imputation (Little and Rubin, 2002; Fuchs and Kenett, 2007), where the type of imputation is based on the assumed missing data generation mechanism and conditional on the purpose of minimizing bias (which is important in explanatory and descriptive studies). Another example is a method for handling missing predictor values in studies with a predictive goal by Saar-Tsechansky and Provost (2007). Their approach builds on multiple predictive models using different subsets of the predictors and then applies, for each new observation, the model that excludes predictors missing for that observation. A third example is a data acquisition algorithm that was developed by Saar-Tsechansky et al. (2009) for data with missing response labels. The algorithm chooses the predictor values or missing response labels to collect, taking into consideration a predictive goal (by considering the cost and contribution to predictive accuracy).

Rounding up recorded values is another type of data cleaning, common in the pharmaceutical industry. Such rounding up is used in order to overcome low-resolution measurements and improve clarity of reporting. We distinguish the difference between *rounding up* and *truncation*, which is treated in Section 5.7. There are different variations on rounding up. For example, *double rounding* is performed when a number is rounded twice, first from n_0 digits to n_1 digits and then from n_1 digits to n_2 digits (where $n_0 > n_1 > n_2$.) Calculations may be conducted on the number between the first and second instances of rounding. Another example is *intermediate rounding*, where values used during a calculation are rounded prior to the derivation of the final result. Rounding of continuous data is performed to obtain a value that is easier to report and communicate than the original. It is also used to avoid reporting measurements or estimates with a number of decimal places that do not reflect the measurement capability or have no practical meaning, a concept known as false precision.

The US Pharmacopeial (USP) Convention state that "Numbers should not be rounded until the final calculations for the reportable value have been completed." Boreman and Chatfield (2015) show with very convincing examples that, from a purely technical point of view, it is always better to work with unrounded data. They recommend that data should only be rounded when required for formal or final reporting purposes, which is usually driven by a specification limit format, that is, the number of decimals quoted in the specification limit. This recommendation implicitly refers to the InfoQ dimension of communication and demonstrates the importance of distinguishing between the need of statistical computations and presentation of outcomes.

Another $h(X|g)$ "data cleaning" strategy is the detection and handling of outliers and influential observations. The choice between removing such observations, including them in the analysis, and otherwise modifying them is goal dependent.

Does data cleaning always increase InfoQ? For $X \neq X^*$ we expect $\text{InfoQ}(f,X,g,U) \neq \text{InfoQ}(f,X^*,g,U)$. In most cases, data quality issues degrade the ability to extract knowledge, thereby leading to $\text{InfoQ}(f,X,g,U) < \text{InfoQ}(f,X^*,g,U)$. Missing values and incorrect values often add noise to our limited sample signal. Yet, sometimes X^* is just as informative as or even more informative than X when conditioning on the goal, and, hence, choosing $h(X) = X$ is optimal. For example, when the goal is to predict the outcome of new observations given a set of predictors, missing predictor values can be a blessing if they are sufficiently informative of the outcome (Ding and Simonoff, 2010). An example is the occurrence of missing data in financial statements, which can be useful for predicting fraudulent reporting. Respondents that refuse to divulge data on their earnings might be more trustworthy (i.e., missing data), so that focusing on covariates in these missing data entries differentiates between reporting types.

5.4   Reweighting and bias adjustment

Selection bias is an a posteriori cause that makes a sample not representative of the population of interest. In surveys, one cause for selection bias is nonresponse. This causes some groups to be over- or underrepresented in the sample.

Another problem is self-selection, where individuals select whether to undergo a treatment or to respond to a survey. Self-selection bias poses a serious challenge for evaluating a treatment effect with observational data. In nonexperimental studies, observations are not randomly assigned to treatment and control groups. Hence, there is always the possibility that people (or animals, firms, or other entities) will select the treatment or control group based on their preference or anticipated outcome. The two main approaches for addressing self-selection bias are the Heckman econometric approach (Heckman, 1979) and the statistical propensity score matching approach (Rosenbaum and Rubin, 1983). Both methods attempt to match the self-selected treatment group with a control group that has the same propensity (or probability) to select the intervention. In propensity score matching, a *propensity score* is computed for each observation, and then the scores are used to create matched samples. Thus $h(X)$ is a subset of the original data which includes a set of matched treated and control observations. Unmatched observations are removed from further analysis. The matched samples are then analyzed using an analysis method f of interest, such as t-test or a linear regression.

Selection bias due to nonresponse or self-selection also poses a challenge in descriptive studies where the goal is to estimate some parameter (e.g., the proportion of voters for a political party or the average household income). A common approach aimed at correcting data for selection bias, especially in survey data, is reweighting or adjustment. Weights are computed based on under- or overrepresentation, so that underrepresented observations in the sample get a weight larger than 1 and overrepresented observations get a weight smaller than 1. Weight calculation requires knowledge of relevant population ratios (or their estimates). For example, if our sample contains 80% men and 20% woman, whereas the population has an equal number of men and women, then each man in the sample gets a weight of $0.5/0.8 = 0.625$ and each woman gets a weight of $0.5/0.2 = 2.5$. Estimating the average income of the population is now done by using the weighted average of the people in the sample. For tests detecting nonresponse bias, see Kenett et al. (2012).

Using weights is aimed at reducing bias at the expense of increased variance, in an effort to maximize the mean squared error (MSE) of the estimator of interest. In other words, $h(X)$ is chosen to maximize $U[f\{h(X|g)\}] = \mathrm{MSE}$. Yet, there is disagreement between survey statisticians regarding the usefulness of reweighting data, because "weighted estimators can do very badly, particularly in small samples" (Little, 2009). When the analysis goal is estimating a population parameter and f is equivalent to estimation, adjusting for estimator bias is common. A comprehensive methodology for handling such issues is called *small area estimation* (Pfeffermann, 2013).

In the eBay consumer surplus example (Section 1.4), Bapna et al. proposed a bias-corrected estimator of consumer surplus in common value auctions (where the auctioned item has the same value to all bidders), which is based on the highest bid.

5.5   Meta-analysis

Meta-analysis is a statistical methodology that has been developed in order to summarize and compare results across studies. It consists of a large battery of tools where the individual study is the experimental unit. In meta-analysis, "data" refers to statistical results of a set of previous studies investigating the same research question. Statistical methodology is then used to combine the results from the disparate studies for obtaining more precise and reliable results, that is, for increasing InfoQ. A posteriori causes that decrease InfoQ include "file drawer" bias, where studies that do not find effects remain unpublished and do not become factored into the meta-analysis; agenda-driven bias, where the researcher intentionally chooses a nonrepresentative set of studies to include in the analysis; and unawareness of Simpson's paradox, which arises due to the aggregation of studies. Meta-analysis consists of identifying all the evidence on a given topic and combining the results of the single studies in order to provide a summary quantitative estimate of the association of interest, which is generally a weighted average of the estimates from individual studies. Quantification and investigation of sources of heterogeneity are also part of the process. Meta-analysis was first developed for the purpose of summarizing results from clinical trials in order to assess the efficacy/effectiveness of a given treatment. Its use has however extended to observational epidemiology and other settings, and meta-analysis of qualitative data has also been proposed (Dixon-Woods et al., 2005).

The choice of the effect measure that represents the results for each individual study depends on which data is available in these studies, the research question investigated, and the properties of the possible measures evaluated in the context of the specific study setting. The methods for obtaining a summary estimate are broadly divided into fixed-effects models and random-effects models. The former assume that all studies measure the same effect, while the latter assume that studies measure different effects and take between-study variation into account. Among the most widely used fixed-effects methods are the inverse variance method and, for binary outcomes, the Mantel–Haenszel and the Peto method. A fundamental component in meta-analysis is quantifying heterogeneity across studies by investigating its sources. This can be accomplished by forming groups of studies according to some given characteristic and comparing the variance within and between groups. Meta-regression investigates whether a linear relationship exists between the outcome measure and one or more covariates (Negri, 2012).

When conducting a meta-analysis, the objective is not merely computing of a combined estimate. In order to achieve information quality, additional aspects of the evidence available for evaluation should be considered, such as the quality of the studies included and hence their adequacy to provide information on the investigated issue, the consistency of results across studies, and the evidence of publication bias.

Once the studies have been identified and retrieved, the data needed to perform the meta-analysis must be extracted from the publications. This may include information on the study design, the study population, number of subjects in categories of exposure/outcome, statistical methods, and so on. Obviously, the data extracted depend on the chosen measure of effect. Other characteristics that are to be used in the analysis of subgroups of studies, as well as indicators of study quality and other variables that may be important to describe the study (e.g., location, response rate) need also to be recorded. The extraction of data from the individual studies is another important step, where unexpected problems often arise. Errors in published articles are quite common, and sometimes a study that meets the inclusion criteria must be excluded because the data in the tables is inconsistent. Data extraction in meta-analysis is an example of a posteriori analysis of secondary data.

5.6   Retrospective experimental design analysis

Designed experiments typically consist of balanced arrays of experimental runs that allow for efficient estimation of factor effects and their interactions (see Chapter 3). However, in running designed experiments, one often meets anticipated and unanticipated problems with the result that the collected data X differs from the target data that was experimental design planned to collect X^*.

In designing experiments, we try to account for anticipated constraints and limitations. For example, the potential impact of raw materials or operating conditions can be accounted for by running the experiment in separate blocks. Practical constraints may dictate that some factors will be "nested" within others or that there will be limitations on the run order. In other examples, there may be some experimental points that we know ahead of time as impossible to execute because of logistical or technological requirements. Nevertheless, unexpected problems may arise when the experiment is carried out. For example, experiments can produce nonquantifiable results, or experimental points may generate "outliers," observations whose values appear quite incompatible with the overall pattern in the data. In analyzing data an underlying model is fitted to the data. For example, two-level factorial experiments are used to estimate parameters of a linear model that, in turn, depends on the estimability properties of the experimental design.

To handle these a posteriori issues and bridge the gap between X and X^*, Kenett et al. (2006) proposed applying bootstrapping methods to handle missing data and validate models used in fitting the data.

When a model is misspecified, an error of the third kind is said to have occurred. Bootstrapping can be used to flag errors of the third kind or, alternatively, validate a specific model. The use of an inadequate model will often lead to an overestimation of the residual variance and to inflated standard errors for the model parameters. Comparison of bootstrap standard errors to those from a regression model analysis is thus a valuable diagnostic. If the bootstrap standard errors are clearly smaller than those from fitting a regression model to the experimental data, it is likely that the model is inadequate.

The general experimentation data analysis strategy with bootstrapping proposed by Kenett et al. (2006) involves six steps:

1. *Evaluation of experimental conditions* including the identification of experimental constraints and a posteriori constraints not planned in the original experimental design. These constraints are reflected by missing or extra experimental runs, constraints on the setting of factor levels or randomization and run-order issues.

2. *Design of bootstrap strategy.* This involves specifying the underlying mathematical model used in the data analysis and the bootstrapping algorithm that matches the experimental setup.

3. *Bootstrap analysis.* This is an iterative step where an initial pilot run of resampled data is evaluated using mostly graphical displays to validate the bootstrapping algorithm accuracy.

4. *Fit of the data* using regression is followed by computation of standard errors from the regression model and the empirically bootstrapped distribution.

5. A *diagnostic check* is performed by comparing standard errors of regression coefficients and bootstrapped standard errors.

6. *Iterative fitting.* Gaps are interpreted through a second iterative cycle until the analysis is completed. The iteration involves sequential adaptation of regression models until a match is achieved with bootstrapping results.

These six steps are an example of how information quality can be enhanced by a posteriori analysis.

5.7   Models that account for data "loss": Censoring and truncation

In fields where the measurement of interest is a duration, a common a posteriori cause is data censoring. Medicine and reliability engineering are two such fields where researchers are interested in survival or time to failure. Telecom providers are interested in customer lifetime (before moving to a different carrier or churn), educators are tracking reasons for student dropout, and risk managers try to identify defaulting loan payment patterns. In all cases, one deals with survival- and censored-type data.

A *censored* observation is one where we observe only part of the duration of interest—for example, if we are measuring time to failure (survival) of a component, then a censored component is one where we did not observe its time of failure. We therefore have partial information X instead of X^*: we only know that the component survived for *at least* the data collection duration. If we observe the "birth" but not the "death" of the observation, it is called *right censoring*, since we do not observe the event of interest (failure) by the end of the data collection period. Right censoring occurs most commonly when the study has a predetermined end-of-collection time, at which point all observations that did not fail are right censored. Another data collection scenario that leads to right-censored data is when the researcher sets a number of "failed" observations to collect and stops collection when the sample size is reached. At that point, any remaining observations are right-censored.

Two other types of censoring are *left censoring* and *interval censoring*. In left censoring, the observation does fail during the data collection period, but the duration of interest starts before the start of the data collection, for example, when we do not know when the component under observation started working. In interval censoring, we do not observe the start or end time of the observation, but we know that during the data collection period, the observation did not fail. This occurs, for example, when we track software system components on a weekly basis with failures aggregated, without information on their failure time. Figure 5.1 illustrates these three types of censoring.

Another, different, type of partial data is the result of *truncation*. Truncation occurs when we cannot observe measurements that exceed or are below a threshold (or interval). For example, one cannot measure body temperature lower or higher than available in a thermometer. In the pharmaceutical industry, one finds many examples of measurements affected by a limit of quantification (LoQ) of the measurement system. Unlike censoring, which represents a form of limit of detection

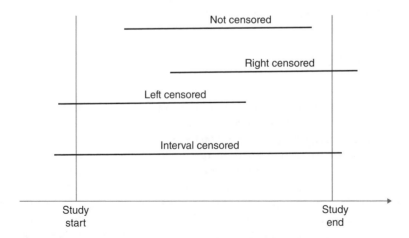

Figure 5.1   Illustration of right, left, and interval censoring. Each line denotes the lifetime of the observation.

(LoD), we have no information about observations that exceed the threshold(s). In other words, while censoring operates on the duration (observations that exceed the data collection duration are censored), truncating affects the magnitude of the measurements (magnitudes that exceed or are before the possible range/value are not observed). In some sense, censoring is a missing data problem, while truncation can potentially introduce bias in statistical estimators. In general, the detection limit of an individual analytical procedure is the lowest amount of analyte in a sample which can be detected but not necessarily quantitated as an exact value so that LoD represents a type of censoring. In contrast, the quantification limit of an individual analytical procedure is the lowest amount of analyte in a sample which can be quantitatively determined with suitable precision and accuracy. The LoQ is a parameter of quantitative assays for low levels of compounds in sample matrices and is used particularly for the determination of impurities and/or degradation products. In practice it is a form of truncation.

A range of statistical models exists for censored and truncated data. Because the a posteriori cause is different in censoring and in truncation, the statistical approaches are different. In models for censored data, both the complete and the censored data are modeled together, using a cumulative distribution function $F(t) = P(T \leq t)$, or the survival function $S(t) = 1 - F(t) = P(T > t)$, where t denotes time. Popular models for censored data are the nonparametric Kaplan–Meier estimator, the Cox semiparametric regression model and the parametric Weibull regression model. For more on this topic, see Mandel (2007).

We note that the choice of censored model (f) must be goal dependent. For example, while the Cox semiparametric model can be useful for a descriptive model, such as estimating the survival rate in a population of interest, it is not useful for predicting the survival of new observations whose survival times are longer than the data collection period from which the model was estimated (Yahav and Shmueli, 2014). In contrast, for truncated data one uses a parametric model that is conditional on the truncation range.

5.8   Summary

In this chapter we describe several common a posteriori causes that potentially deteriorate InfoQ, due to challenges in the data collection stage. We also discuss key statistical approaches for addressing such causes. The approaches range from using the data alone to account for the corruption (MIS-type operations, simple data imputation) to methods that combine information from multiple sources (meta-analysis) or external sources (weights) to incorporating stochastic models (e.g., survival models) to "recover" the original process generating the data. Table 5.1 summarizes the main points. Taking an InfoQ approach helps the researcher or analyst choose the right method among the different possibilities. While reweighting may be useful for some goals and analyses, it might not be for other goals (or analysis methods). Similarly, whether and how to impute data should closely depend on the goal of the study and on the intended method of analysis f.

Table 5.1   Statistical strategies for increasing InfoQ given a posteriori causes at the postdata collection stage and approaches for increasing InfoQ.

	Strategies for increasing InfoQ	A posteriori causes
Missing data	Imputation; observation or measurement deletion; building multiple separate models; other handling of missing values; advanced technologies for data collection, transfer, and storage; detecting and handling outliers and influential observations	Data entry errors, measurement error, and intentional data manipulation; faulty collection instrument; nonresponse
Reweighting	Attach weights to observations; create matched treatment–control samples	Selection bias (self-selection, nonresponse)
Meta-analysis	Reducing nonsampling errors (e.g., pretesting questionnaire, reducing nonresponse) and sampling errors (e.g., randomization, stratification, identifying target and sampled populations)	"File drawer" bias, agenda-driven bias, Simpson's paradox
Retrospective DOE	Randomization; blocking; replication; linking data collection protocol with appropriate design; space filling designs	Nonfeasible experimental runs, hard-to-change factors that do not permit randomization, outlying observations, unexpected constraints
Censoring and truncation	Parametric, semiparametric, and nonparametric models for censored data; parametric models for truncated data	Data collection time constraints; limitations of collection instrument

References

Boreman, P. and Chatfield, M. (2015) Avoid the perils of using rounded data. *Journal of Pharmaceutical and Biomedical Analysis*, 115, pp. 502–508.

Boslaugh, S. (2007) *Secondary Data Sources for Public Health: A Practical Guide*. Cambridge University Press, Cambridge, UK.

Cox, D.R. (2009) Randomization in the design of experiments. *International Statistical Review*, 77, 415–429.

Ding, Y. and Simonoff, J. (2010) An investigation of missing data methods for classification trees applied to binary response data. *Journal of Machine Learning Research*, 11, pp. 131–170.

Dixon-Woods, M., Agarwal, S., Jones, D., Sutton, A., Young, B., Dixon-Woods, M., Agarwal, S., Jones, D. and Young, B. (2005) Synthesising qualitative and quantitative evidence: a review of possible methods. *Journal of Health Services Research & Policy*, 10, pp. 45–53.

Engel, A., Kenett, R.S., Shahar, S. and Reich, Y. (2016) Optimizing System Design Under Degrading Failure Agents. *Proceedings of the International Symposium on Stochastic Models in Reliability Engineering, Life Sciences and Operations Management (SMRLO16)*, Beer Sheva, Israel.

Friedman, L.M., Furberg, C.D., DeMets, D., Reboussin, D.M. and Granger, C.B. (2015) *Fundamentals of Clinical Trials*, 5th edition. Springer International Publishing, Cham.

Fuchs, C. and Kenett, R.S. (2007) Missing Data and Imputation, in *Encyclopedia of Statistics in Quality and Reliability*, Ruggeri, F., Kenett, R.S. and Faltin, F. (editors in chief), John Wiley & Sons, Ltd, Chichester, UK.

Godfrey, A.B. (2008) Eye on data quality. *Six Sigma Forum Magazine*, 8, pp. 5–6.

Hand, D.J. (2008) *Statistics: A Very Short Introduction*. Oxford University Press, Oxford.

Heckman, J.J. (1979) Sample selection bias as a specification error. *Econometrica: Journal of the Econometric Society*, 47, pp. 153–161.

Kenett, R.S., Rahav, E. and Steinberg, D. (2006) Bootstrap analysis of designed experiments. *Quality and Reliability Engineering International*, 22, pp. 659–667.

Kenett, R.S., Deldossi, L. and Zappa, D. (2012) Quality Standards and Control Charts Applied to Customer Surveys, in *Modern Analysis of Customer Satisfaction Surveys*, Kenett, R.S. and Salini, S. (editors), John Wiley & Sons, Ltd, Chichester, UK.

Little, R. (2009) Weighting and Prediction in Sample Surveys. *Working Paper 81*. Department of Biostatistics, University of Michigan, Ann Arbor.

Little, R.J.A. and Rubin, D.B. (2002) *Statistical Analysis with Missing Data*. John Wiley & Sons, Inc., New York.

Mandel, M. (2007) Censoring and truncation – highlighting the differences. *The American Statistician*, 61(4), pp. 321–324.

Negri, E. (2012) Meta-Analysis, in *Statistical Methods in Healthcare*, Faltin, F., Kenett, R.S. and Ruggeri, F. (editors), John Wiley & Sons, Ltd, Chichester, UK.

Pfeffermann, D. (2013). New important developments in small area estimation. *Statistical Science*, 28, pp. 40–68.

Redman, T. (2007) Statistics in Data and Information Quality, in *Encyclopedia of Statistics in Quality and Reliability*, Ruggeri, F., Kenett, R.S. and Faltin, F. (editors in chief), John Wiley & Sons, Ltd, Chichester, UK.

Rosenbaum, P.R., and Rubin, D.B. (1983) The central role of the propensity score in observational studies for causal effects. *Biometrika*, 70 (1), pp. 41–55.

Saar-Tsechansky, M. and Provost, F. (2007) Handling missing features when applying classification models. *Journal of Machine Learning Research*, 8, pp. 1625–1657.

Saar-Tsechansky, M., Melville, P. and Provost, F. (2009) Active feature-value acquisition. *Management Science*, 55, pp. 664–684.

Yahav, I. and Shmueli, G. (2014) Outcomes matter: estimating pre-transplant survival rates of kidney-transplant patients using simulator-based propensity scores. *Annals of Operations Research*, 216(1), pp. 101–128.

Part II

APPLICATIONS OF InfoQ

6

Education

6.1 Introduction

Education is one of the most powerful instruments for reducing poverty and inequality in society and lays the foundation for sustained economic growth. The second millennium development goal of the World Bank has been to achieve universal primary education by 2015 (www.worldbank.org/mdgs/education.html). In this context, the World Bank compiles data on education inputs, participation, efficiency, and outcomes from official responses to surveys and from reports provided by education authorities in each country. The *Key Education Indicators Dashboard* presents a global portrait of education systems, from preprimary to tertiary education. The World Bank *EdStats All Indicator Query* contains around 2500 internationally comparable indicators that describe education access, progression, completion, literacy, teachers, population, and expenditures (http://datatopics.worldbank.org/education). The indicators cover the education cycle from preprimary to vocational and tertiary education. The database also includes learning outcome data from international and regional learning assessments (e.g., PISA, PIACC), equity data from household surveys, and projection/attainment data to 2050. Several quality indicators are tracked and reported including repetition rates, primary completion rates, pupil–teacher ratios, and adult literacy rates. The currently available reports rely on over 2000 quality indicators designed to answer specific questions such as the following:

- How many students complete primary school?

- How many students per teacher are there in primary classrooms?

- Are a few students repeating grades?

Information Quality: The Potential of Data and Analytics to Generate Knowledge,
First Edition. Ron S. Kenett and Galit Shmueli.
© 2017 John Wiley & Sons, Ltd. Published 2017 by John Wiley & Sons, Ltd.
Companion website: www.wiley.com/go/information_quality

- Do females repeat primary grades more than males?

- Which regions have the highest repetition rates?

- Which countries have the highest primary student/teacher ratios?

- Which countries have the highest repetition rates in primary?

- Which countries have the highest repetition rates in secondary?

- Have adult literacy rates improved?

- Which countries have the lowest adult literacy rates?

- Are adult literacy rates equal for men and women?

- Are gender disparities in literacy rates decreasing over time?

The data described earlier, X, is analyzed by methods f to meet goals, g, implied by these questions. The utility function, U, can reflect the needs of a range of stakeholders including parents, teachers, and policy makers. The information provided by the various official reports to address the questions listed earlier is mostly descriptive and relies on a compilation of various data sources with varying levels of quality control and data quality. Assessing the level of information quality (InfoQ) of these reports, with respect to each of the previous questions, would score low on data integration, temporal relevance, and chronology of data and goal. This statement is based on the fact that indicators are considered separately, data is dated, and decision makers interested in forming policy with the support of such data experience a gap between the reported data and their objectives as managers or parliamentarians.

In this chapter, we consider in detail three education-related application areas. The first application is focused on the extensive test reporting industry in the United States. After providing a general context based on work done at the National Assessment of Educational Progress (NAEP), the producer of the nation's report card in the Unites States (http://nces.ed.gov/nationsreportcard), we evaluate the level of InfoQ of the Missouri Assessment Program (MAP) report. The second example interprets the ASA statement on education value-added models (VAMs) using InfoQ dimensions. The third example concerns the assessment of conceptual understanding or "deep understanding" using Meaning Equivalence Reusable Learning Objects (MERLO). The example is based on the application of MERLO in an ongoing assessment program of teachers of mathematics in Italy. Reports based on MERLO assessment are then evaluated using InfoQ dimensions.

6.2   Test scores in schools

In the United States, over 60 000 000 individual reports are sent annually to parents of schoolchildren. Another 6 000 000 reports are generated in Canada. Over 1000 credentialing exams (e.g., securities, accountants, nurses) often exceed 100 000 candidates. The public, educators, policymakers, parents, and examinees want to understand scores and score reports. The types of questions asked by these various stakeholders on the basis of such reports are as follows:

1. Parent questions:

   - Did my child make a year's worth of progress in a year?

   - Is my child growing appropriately toward meeting state standards?

   - Is my child growing as much in math as reading?

   - Did my child grow as much this year as last year?

2. Teacher questions:

   - Did my students make a year's worth of progress in a year?

   - Did my students grow appropriately toward meeting state standards?

   - How close are my students to becoming proficient?

   - Are there students with unusually low growth who need special attention?

3. Administrator questions:

   - Did the students in our district/school make a year's worth of progress in all content areas?

   - Are our students growing appropriately toward meeting state standards?

   - Does this school/program show as much growth as another (specific) one?

   - Can I measure student growth even for students who do not change proficiency categories?

   - Can I pool together results from different grades to draw summary conclusions?

Considerable investments of time and money have been made to address testing programs that produce student reports at various levels of aggregation. The testing field is full of experts working on item response theory (IRT) applications, scoring of performance data, test score comparisons, reliability estimation, and quality control issues such as cheating detection and advancing computer technology. Shortcomings of such student reports are reported in Goodman and Hambleton (2004) and include:

- No stated purpose, no clues about where to start reading.

- Performance categories that are not defined.

- Reports do not indicate that errors of measurement are present.

- Font is often too small to read easily.

- Instructional needs information is not user-friendly—for example, to a parent. Try to interpret the statement: "You need help in extending meaning by drawing conclusions and using critical thinking to connect and synthesize information within and across text, ideas, and concepts."

- Several undefined terms on the displays: percentile, z score, achievement level, and more.

In order to improve test reports, several standards have been developed. For example, the AERA–APA–NCME test standards state:

> When test score information is released….those responsible should provide appropriate interpretations….information is needed about content coverage, meaning of scores, precision of scores, common misinterpretations, and proper use….Score reports should be accompanied by a clear statement of the degree of measurement error associated with each score or classification level and information on how to interpret the scores (http://teststandards.org).

As a concrete example of applying InfoQ to answering a particular question using a school report (data), consider the MAP test report of 8th grader Sara Armstrong presented in Figure 6.1. The score report is not easy to follow. There are multiple scales and the report does not tell a logical story from point A to point D. The report is used as a reference in parent–teacher conferences and for instructional planning, and the quality of the information provided by this report has important consequences. For more information about MAP, see http://dese.mo.gov/college-career-readiness/ assessment/grade-level/map-information-parents. We will review the eight InfoQ dimensions of this report at the end of this section.

Some points to consider in designing test reports include:

1. Number of knowledge/skill areas being reported—too many is problematic, too few is not useful.

2. Either normative or criterion-referenced information (or both) can be provided.

3. If normative, who is in the reference group: all, just passing, all passing, first-time takers?

4. If criterion referenced, what are the cut scores?

5. Report precision of scores.

The related SAT Skills Insight report is available at www.collegeboard.com as a free online tool that helps students put their skills on the map by helping them understand what they know and what they need to know better. Figure 6.2 presents an example of such a report, zooming in on a 500–590 score in critical reading. We present it in contrast to the MAP report of Figure 6.1. As an example, consider the SAT reading and

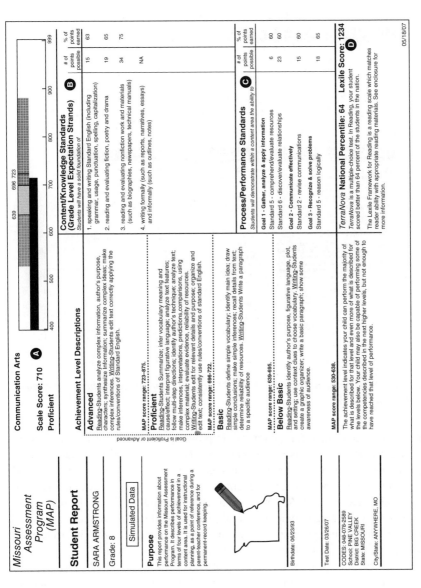

Figure 6.1   The Missouri Assessment Program test report for fictional student Sara Armstrong. Source: http://dese.mo.gov. © Missouri Department of Elementary and Secondary Education.

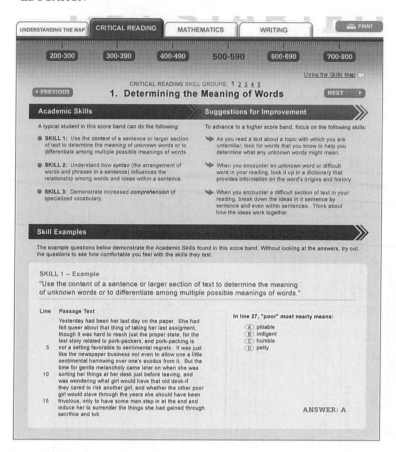

Figure 6.2   SAT Critical Reading skills. Source: https://sat.collegeboard.org/home. © *The College Board.*

writing report diagnostic information: "To improve performance in READING, your child should work on 1) drawing conclusions about the central ideas in a text, 2) understanding the author's techniques and decisions and 3) making, supporting, and extending inferences about contents, events, characters, setting, theme, and style. To improve performance in WRITING, your child should work on 1) organizing the writing around a single topic or central idea, 2) working to avoid errors in conventions of English usage, grammar, spelling, and punctuation that interfere with understanding and 3) supporting the ideas with more specific details."

These instructions provide information of higher InfoQ than the MAP report.

Goodman and Hambleton (2004) point out major problems in score reporting such as providing complex explanations. Consider, for example, the following footnote from a NAEP report: "The between state comparisons take into account sampling and measurement error and that each state is being compared with every other state. Significance is determined by an application of the Bonferroni procedure based on 946 comparisons by comparing the difference between the two means with four times the square root of the sum of the squared standard errors."

Other potential pitfalls listed by Goodman and Hambleton (2004) include small font size, unclear footnotes, acronyms not spelled out, cluttered page, not indicating score precision, not defining key terms, use of jargon, and poorly designed graphs.

With this background on test report design, let us consider the MAP report displayed in Figure 6.1 from an InfoQ lens. We start by identifying the four InfoQ components and then examine each of the eight InfoQ dimensions.

Case study 1   MAP report

INFOQ COMPONENTS

The four InfoQ components in this study are:

Goal (g): As an example, a parent's question: "Is my child growing appropriately toward meeting state standards?"
Data (X): Child's test results in the current year
Analysis (f): MAP report displayed in Figure 6.1
Utility (U): Direct attention to required actions (praise, supplementary instruction, increased follow-up of child's achievements, etc.)

INFOQ DIMENSIONS

The eight InfoQ dimensions in this study are evaluated in the following:

(1) *Data resolution*: Data resolution refers to the measurement scale and aggregation level of the data. The measurement scale of the data should be carefully evaluated in terms of its suitability to the goal. Data might be recorded by multiple instruments or by multiple sources, and, in that case, supplemental information about the reliability and precision of the measuring devices or data sources is useful. The MAP report presents student-specific data for a single subject during a period of assessment. The report uses several measurement scales, some anchored and some continuous, without providing a logic to this complexity.

(2) *Data structure*: Data structure relates to the study design or data collection mechanism. The InfoQ level of a certain data type depends on the goal at hand. The MAP report is based on test results without any comparisons or benchmarks and without considering trends. The data is grouped into content/knowledge standards and process/performance standards.

(3) *Data integration*: With the variety of data source and data types, there is often a need to integrate multiple sources and/or types. Often, the

integration of multiple data types creates new knowledge regarding the goal at hand, thereby increasing InfoQ. The MAP report does not include any data integration.

(4) *Temporal relevance*: The process of deriving knowledge from data can be put on a timeline that includes data collection and data analysis. These different durations and gaps can each affect InfoQ. The data collection duration can increase or decrease InfoQ, depending on the study goal. In the context of test reports, temporal relevance is assured by the necessity to have an updated report during parent–teacher conferences. This practical deadline assures temporal relevance of the MAP report.

(5) *Chronology of data and goal*: The choice of variables to collect, the temporal relationship between them and their meaning in the context of the goal at hand also affect InfoQ. The MAP report relates to test results from tests in the relevant assessment time window. This ensures chronology of data and goal.

(6) *Generalizability*: The utility of an empirical analysis is dependent on the ability to generalize the analysis to the appropriate population. Two types of generalizability are statistical and scientific generalizability. Statistical generalizability refers to inferring from a sample to a target population. Scientific generalizability refers to applying a model based on a particular target population to other populations. The MAP report does not include any report of measurement error and does not provide for statistical generalizability. Nor does it generalize beyond the particular subject or student. Increased generalizability would be achieved by providing a summary statistics of the child's cohort and an assessment of the child's performance relative to this cohort.

(7) *Operationalization*: Operationalization relates to both construct operationalization and action operationalization. Constructs are abstractions that describe a phenomenon of theoretical interest. Measurable data is an operationalization of underlying constructs. The report aims to measure "achievement level" and provides text descriptions of each of five levels. Action operationalization relates to the practical implications of the information provided. The MAP report does not provide support for action operationalization. This is in sharp contrast to the SAT test reports which clearly state which areas need to be further strengthened and which skills need to be reinforced.

(8) *Communication*: Effective communication of the analysis and its utility directly impacts InfoQ. The MAP report clearly is lacking in this respect with a haphazard format, small font, and bad visualization.

We present subjective ratings for each of the dimensions and the InfoQ score in Table 6.1. The overall InfoQ score as a percentage is 33%, which is low. The strongest dimensions are temporal relevance and chronology of data and goal. The information learned from examining each of the eight InfoQ dimensions can be used for improving the MAP presented in Figure 6.1. In fact, it provides a checklist of areas to consider in designing and deploying such improvements.

Table 6.1   InfoQ assessment for MAP report.

InfoQ dimension	Rating
Data resolution	2
Data structure	2
Data integration	2
Temporal relevance	4
Chronology of data and goal	4
Generalizability	2
Operationalization	2
Communication	2
InfoQ score	**33%**

6.3   Value-added models for educational assessment

Spurred in some cases by the federal government's *Race to the Top* initiative, many states and school districts in the United States have included, in their performance evaluations, measures of teacher effectiveness based on student achievement data. States and districts started measuring teacher effectiveness by using test scores and value added models or VAMs. These models provide a measure of teachers' contributions to student achievement that accounts for factors beyond the teacher's control. The basic approach of a VAM is to predict the standardized test score performance that each student would have obtained with the average teacher and then compare the average performance of a given teacher's students to the average of the predicted scores. The difference between the two scores—how the students actually performed with a teacher and how they would have performed with the average teacher—is attributed to the teacher as his or her value added to students' test score performance. VAMs typically use a form of a regression model predicting student scores or growth on standardized tests from background variables (including prior test scores), with terms in the model for the teachers who have taught the student in the past. A percentile is calculated for each student from the model, relating his or her growth to the growth of other students with similar previous test scores. For each teacher, the median or average of the percentiles of his/her students are used to calculate the teacher's VAM score. If a teacher's students have high achievement growth relative to other students with similar prior achievement, then the teacher will have a high VAM score. Some VAMs also include other background variables for the students. The form of the model may lead to biased VAM scores for some teachers. For example, "gifted" students or those with disabilities might exhibit smaller gains in test scores if the model does not accurately account for their status.

Using VAM scores to improve education requires that they provide meaningful information about a teacher's ability to promote student learning. For instance, VAM scores should predict how teachers' students will progress in later grades and how their future students will fare under their tutelage. A VAM score may provide teachers and administrators with information on their students' performance and identify areas where improvement is needed, but it does not provide information on how to improve the teaching. Such improvements need to be targeted at specific goals, and the VAM score should be evaluated in the context of such goals. Without explicitly listing the targeted goal, the InfoQ of the VAM score cannot be assessed.

The models can be used to evaluate the effects of policies or teacher training programs by comparing the average VAM scores of teachers from different programs. In these uses, the VAM scores partially adjust for the differing backgrounds of the students, and averaging the results over different teachers improves the stability of the estimates. For more on statistical properties of VAM, see Ballou et al. (2004), McCaffrey et al. (2003, 2004), Andrabi et al. (2009), Mariano et al. (2010), and Karl et al. (2013, 2014a, 2014b).

In the following, we look at two cases through the InfoQ lens. The first is an empirical study related to VAM, which has important policy implications. The second is a statement issued by the ASA on "Using VAM for Educational Assessment." By examining these two different types of analysis (empirical and written statement), we showcase how the InfoQ framework can help characterize, clarify, and identify good practices as well as challenges in different types of reports.

6.3.1 "Big Study Links Good Teachers to Lasting Gain"

The January 6, 2012 *New York Times* article "Big Study Links Good Teachers to Lasting Gain"[1] covers a research study on "The Long-Term Impacts of Teachers: Teacher Value-Added and Student Outcomes in Adulthood" (Chetty, Friedman, and Rockoff, NBER, www.nber.org/papers/w17699). The authors used econometric models applied to data from test scores of millions of students and their later financial and other demographic information for evaluating the effect of VA teachers on students' future gain. The authors conclude:

> We find that students assigned to higher VA [Value-Added] teachers are more successful in many dimensions. They are more likely to attend college, earn higher salaries, live in better neighborhoods, and save more for retirement. They are also less likely to have children as teenagers.

---

[1] www.nytimes.com/2012/01/06/education/big-study-links-good-teachers-to-lasting-gain.html

Case study 2   Student lifelong earnings

The four InfoQ components in this study are:

Goal (g): Test whether children who get high value-added teachers have better outcomes in adulthood (we focus on this goal, while the study had two goals).

Data (X): Teacher and class assignments from 1991 to 2009 for 2.5 million children, test scores from 1989 to 2009, and selected data from US federal income tax returns from 1996 to 2010 (student outcomes: earnings, college, teenage birth, neighborhood quality, parent characteristics)

Analysis (f): Linear regression. ("Used to predict test scores for students taught by teacher j in year $t+1$ using test score data from previous t years")

Utility: Effect size, minimal prediction error

We now evaluate the study on each of the eight InfoQ dimensions:

(1)   *Data resolution*: The data include one observation per student–subject–year combination. "Research based purely on statistics aggregating over thousands of individuals, not on individual data."

(2)   *Data structure*: Data on teachers' VA scores and socioeconomic and earning variables has been comprehensively considered in the study.

(3)   *Data integration*: Integrated data from US federal income tax returns with school district data ("approximately 90% of student records matched to tax data").

(4)   *Temporal relevance*: The data represents a snapshot relevant for the first decade of the twenty-first century.

(5)   *Chronology of data and goal*: The effect of interest is the presence of a VA teacher on long-term student gains. To assess chronology of data and goal, one has to consider the relevance of the data analysis to the needs of decision makers and policy formulation.

(6)   *Generalizability*: The model aims to generalize to teachers in general and in particular uses statistical inference, thereby indicating that statistical generalization is sought. However, due to the very large sample size, the use of p-values can be misleading for determining meaningful effect sizes.

(7)   *Operationalization*: Operationalizing the information carried out by the report can be translated in the form of teacher hiring and promotion policies and career development.

(8) *Communication*: Interpreting and presenting the results of the statistical analysis are the weakest point of the study. First, the paper reports effects as statistically significant without necessarily reporting the magnitude. Given the one million records sample, statistical significance is achieved with even tiny effects. For example, while the slope in the regression line shown in Figure 6.3 seems dramatic and is statistically significant, earnings fluctuate by less than $1000 per year. To get around this embarrassing magnitude, the authors look at the "lifetime value" of a student. ("On average, having such a [high value-added] teacher for one year raises a child's cumulative lifetime income by $50,000 (equivalent to $9,000 in present value at age 12 with a 5% interest rate).") In other words, there is a large gap between the quantitative analysis results and the broad qualitative statements that the study claims.

Another communications issue relates to visualizing the results. The paper displays the results through several charts and even provides a set of slides and video. However, the charts are misleading because of their choice of scale. For example, the *y*-axis scale does not start from zero in the two charts shown in Figures 6.3 and 6.4 (one measures earnings at age 28 and the other measures average test scores).

Ratings for each of the eight dimensions are given in Table 6.2. The overall InfoQ score for this study is 49%. The strongest dimension is data integration, and the weakest dimensions are generalizability, operationalization and communication.

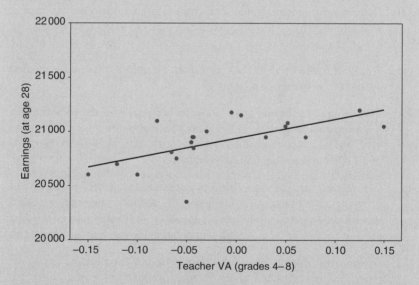

Figure 6.3   Earning per teacher value-added score. Adapted from http:// rajchetty.com/chettyfiles/value_added.htm

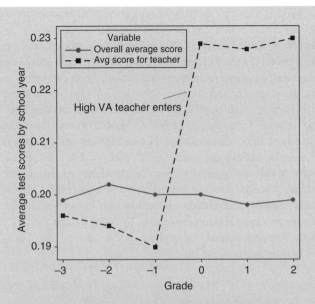

Figure 6.4   Test scores by school by high value-added teacher score. Adapted from http://rajchetty.com/chettyfiles/value_added.htm

Table 6.2   InfoQ assessment for student's lifelong earning study.

InfoQ dimension	Rating
Data resolution	4
Data structure	4
Data integration	5
Temporal relevance	3
Chronology of data and goal	4
Generalizability	2
Operationalization	2
Communication	2
InfoQ score	**49%**

Such conclusions can have critical policy implications. Let us therefore examine the study using the InfoQ framework.

6.3.2   ASA statement on VAM

On April 8, 2014, the ASA issued a statement titled *Using Value-Added Models for Educational Assessment* (ASA, 2014). An excerpt from the executive summary of this document reads as follows: "Many states and school districts have adopted

Value-Added Models (VAMs) as part of educational accountability systems. The goal of these models... is to estimate effects of individual teachers or schools on student achievement while accounting for differences in student background. VAMs are increasingly promoted or mandated as a component in high-stakes decisions such as determining compensation, evaluating and ranking teachers, hiring or dismissing teachers, awarding tenure, and closing schools... VAMs are complex statistical models, and high-level statistical expertise is needed to develop the models and interpret their results. Estimates from VAMs should always be accompanied by measures of precision and a discussion of the assumptions and possible limitations of the model. These limitations are particularly relevant if VAMs are used for high-stakes purposes. VAMs are generally based on standardized test scores, and do not directly measure potential teacher contributions toward other student outcomes. VAMs typically measure correlation, not causation: Effects—positive or negative attributed to a teacher may actually be caused by other factors that are not captured in the model...Ranking teachers by their VAM scores can have unintended consequences that reduce quality."

Case study 3   VAM ASA statement

Let us now evaluate the ASA statement using the InfoQ terminology and framework.

INFOQ COMPONENTS

For an InfoQ assessment, we start by identifying the four InfoQ components:

Goal (g): Evaluate teachers' performance to better manage the educational process, the stakeholders being education administrators and education policy makers.
Data (X): Standardized test results and student background information.
Analysis (f): VAMs based on linear regression.
Utility (U): Minimal prediction error.

The same goals, data, and utility can be considered with an alternative analysis method called student growth percentiles (SGP). We briefly introduce SGP to provide a context for the InfoQ VAM evaluation. As mentioned, SGP is an alternative to VAM that creates a metric of teacher effectiveness by calculating the median or mean conditional percentile rank of student achievement in a given year for students in a teacher's class. For a particular student with current year score A_{ig} and score history $\{A_{i,g-1}, A_{i,g-2}, ..., A_i, 1\}$, one locates the percentile corresponding to the student's actual score, A_{ig}, in the distribution of scores conditional on having a test score history $\{A_{i,g-1}, A_{i,g-2}, ..., A_i, 1\}$. In short, the analyst evaluates how high in the distribution the student achieved, given his or her past scores. Then teachers are evaluated by either the median or the mean conditional percentile rank of their students. Quantile regressions are used to estimate features of the conditional distribution of student achievement. In particular, one estimates the conditional quantiles for all possible test score histories, which are then used for assigning percentile ranks to students. The SGP model does not account for other student background characteristics and excludes other features included in many VAMs used by states and school districts. For more on SGP see Betebenner (2009, 2011). Walsh and Isenberg (2015) find that differences in evaluation scores based on VAM and SGP are not related to the characteristics of students' teachers. Again, our objective here is to review the ASA VAM statement from an InfoQ perspective.

INFOQ DIMENSIONS

With this background on models for teachers' assessment, we consider the ASA VAM statement in terms of the eight InfoQ dimensions:

1. *Data resolution*: VAMs use data on students' scores and backgrounds, by teacher and by class. Data related to the class characteristics such as the level of student engagement or the class social cohesion are not used in VAM. Information on "gifted" students or those with disabilities is also not used.

2. *Data structure*: Data structure is comprehensive in terms of scores but does not include semantic data such as written reports on student's performance. The data used is in fact a type of panel data with information on students and teachers at the individual class level.

3. *Data integration*: Data on scores by students and teachers over time is matched to apply the VAM.

4. *Temporal relevance*: A teacher's added value score is potentially updated at the end of each reporting period.

5. *Chronology of data and goal*: Specific decisions regarding a teacher's assignment or promotion are supported by lagged VAM estimates.

6. *Generalizability*: The VAM analysis reports mostly relate to individual teachers and, as such, provide for statistical generalizability only at the individual teacher level.

7. *Operationalization*: VAM is based on operationalizing the construct "teacher effectiveness" using a function based on test scores. Operationalizing the VAM estimates is exposed to attribution error, as

described in the ASA VAM statement. The statement warns against generalizing a correlation effect to a causation effect. The statement does not provide concrete recommendations (action operationalization is low).

8. *Communication*: The ASA statement about VAM is mostly focused on how the outputs of the models regarding teacher's added value are used and interpreted.

To summarize, the ASA VAM statement is comprehensive in terms of statistical models and their related assumptions. We summarize the ratings for each dimension in Table 6.3. The InfoQ score for the VAM statement is 57%. The caveats and the implication of such assumptions to the operationalization of VAM are the main points of the statement. The data resolution, data structure, data integration, and temporal relevance in VAM are very high. The difficulties lie in the chronology of data and goal, operationalization, generalization, and communication of the VAM outcomes. The ASA statement is designed to reflect this problematic use of VAM. It is however partially ambiguous on these dimensions leaving lots of room for interpretation. Examining and stating these issues through the InfoQ dimensions helps create a clearer and more systematic picture of the VAM approach.

Table 6.3   InfoQ assessment for VAM (based on ASA statement).

InfoQ dimension	Rating
Data resolution	5
Data structure	4
Data integration	5
Temporal relevance	5
Chronology of data and goal	2
Generalizability	3
Operationalization	2
Communication	3
InfoQ score	**57%**

6.4   Assessing understanding of concepts

This section is about the InfoQ of a formative assessment measurement approach used in education. Such assessments are used during training or education sessions to contribute to the learning of students and the improvement of material and delivery style. Before discussing the InfoQ evaluations, we introduce the topic of formative assessment in education with a review of several topics, including concept science and MERLO, with an example on teaching quantitative literacy. In the appendix of this chapter, we also include a MERLO implementation in an introduction to statistics course.

Listening to conversations among content experts reveals a common trend to flexibly reformulate the issue under discussion by introducing alternative points of view, often encoded in alternative representations in different sign systems. For example, a conversation that originated in a strictly spoken exchange may progress to include written statements, images, diagrams, equations, etc., each with its own running—spoken—commentary. The term *meaning equivalence* designates a commonality of meaning across several representations. It signifies the ability to transcode meaning in a polymorphous (one-to-many) transformation of the meaning of a particular conceptual situation through multiple representations within and across sign systems. Listening to conversations among content experts also reveals a common trend to identify patterns of associations among important ideas, relations, and underlying issues. These experts engage in creative discovery and exploration of hidden, but potentially viable, relations that test and extend such patterns of associations that may not be obviously or easily identified. The term "conceptual thinking" is used to describe such ways of considering an issue; it requires the ability, knowledge, and experience to communicate novel ideas through alternative representations of shared meaning and to create lexical labels and practical procedures for their nurturing and further development. This approach was originally developed by Uri Shafrir from the University of Toronto in Canada and Masha Etkind from Ryerson University, also in Toronto (Shafrir and Etkind, 2010). The application of MERLO in education programs of statistics and quantitative literacy was introduced in Etkind et al. (2010). For an application of MERLO and concept mapping to new technologies and e-learning environments including MOOCs, see Shafrir and Kenett (2015).

The pivotal element in conceptual thinking is the application of MERLO assessment and MERLO pedagogy. MERLO items form a multidimensional database that allows the sorting and mapping of important concepts through target statements of particular conceptual situations and relevant statements of shared meaning. Each node of MERLO is an item family, anchored by a target statement that describes a conceptual situation and encodes different features of an important concept and other statements that may—or may not—share equivalence of meaning with the target. Collectively, these item families encode the complete conceptual mapping that covers the full content of a course (a particular content area within a discipline). Figure 6.5 shows a template for constructing an item family anchored in a single target statement.

Statements in the four quadrants of the template—Q1, Q2, Q3, and Q4—are thematically sorted by their relation to the target statement that anchors the particular node (item family). They are classified by two sorting criteria: surface similarity to the target and equivalence of meaning with the target. For example, if the statements contain text in natural language, then by "surface similarity" we mean same/similar words appearing in the same/similar order as in the target statement, and by "meaning

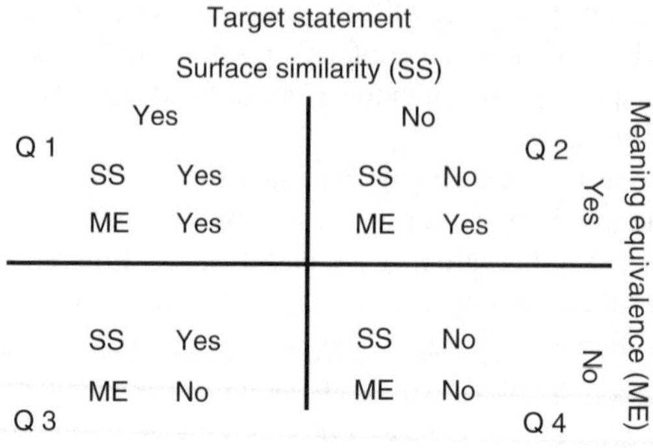

Target statement

Surface similarity (SS)

	Yes	No	
Q 1	SS Yes ME Yes	SS No ME Yes	Q 2
Q 3	SS Yes ME No	SS No ME No	Q 4

Meaning equivalence (ME) — Yes / No (right vertical axis label)

Figure 6.5   Template for constructing an item family in MERLO.

equivalence" we mean that a majority in a community that shares a sublanguage (Cabre, 1998; Kittredge, 1983) with a controlled vocabulary (e.g., statistics) would likely agree that the meaning of the statement being sorted is equivalent to the meaning of the target statement.

MERLO pedagogy guides sequential teaching/learning episodes in a course by focusing learners' attention on meaning. The format of MERLO items allows the instructor to assess deep comprehension of conceptual content by eliciting responses that signal learners' ability to recognize and produce multiple representations that share equivalence of meaning. A typical MERLO item contains five unmarked statements: a target statement plus four additional statements from quadrants Q2, Q3, and, sometimes, also Q4. Task instructions for MERLO test are as follows: "At least two out of these five statements—but possibly more than two—share equivalence-of-meaning: 1) Mark all statements—but only those—that share equivalence-of-meaning and 2) Write down briefly the concept that guided you in making these decisions."

For example, the MERLO item in Figure 6.6 (mathematics/functions) contains five representations (A–E) that include text, equations, tables, and diagrams; at least two of these representations share equivalence of meaning. Thus, the learner is first asked to carry out a recognition task in situations where the particular target statement is not marked, namely, features of the concept to be compared are not made explicit. In order to perform this task, a learner needs to begin by decoding and recognizing the meaning of each statement in the set. This decoding process is carried out, typically, by analyzing concepts that define the "meaning" of each statement. Successful analysis of all the statements in a given five-statement set (item) requires deep understanding of the conceptual content of the specific domain. MERLO item format requires both rule inference and rule application in a similar way to the solution of analogical reasoning items. Once the learner marks those statements that in his/her opinion share equivalence of meaning, he/she formulates and briefly describes the concept/idea/criteria he/she had in mind when making these decisions.

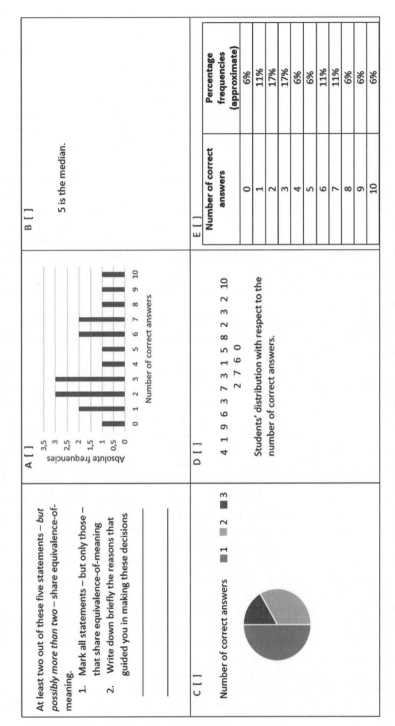

A [ ]

Absolute frequencies

3,5
3
2,5
2
1,5
1
0,5
0

Number of correct answers
0 1 2 3 4 5 6 7 8 9 10

B [ ]

5 is the median.

C [ ]

At least two out of these five statements – *but possibly more than two* – share equivalence-of-meaning.

1. Mark all statements – but only those – that share equivalence-of-meaning
2. Write down briefly the reasons that guided you in making these decisions

Number of correct answers ■ 1 ■ 2 ■ 3

D [ ]

4 1 9 6 3 7 3 1 5 8 2 3 2 10
2 7 6 0

Students' distribution with respect to the number of correct answers.

E [ ]

Number of correct answers	Percentage frequencies (approximate)
0	6%
1	11%
2	17%
3	17%
4	6%
5	6%
6	11%
7	11%
8	6%
9	6%
10	6%

Figure 6.6   Example of MERLO item (mathematics/functions).

A learner's response to a MERLO item combines a multiple choice/multiple response (also called recognition) and a short answer (called production). Subsequently, there are two main scores for each MERLO item: recognition score and production score. Specific comprehension deficits can be traced as low recognition scores on quadrants Q2 and Q3, due to the mismatch between the valence of surface similarity and meaning equivalence (Figure 6.5). Production score of MERLO test items is based on the clarity of the learner's description of the conceptual situation anchoring the item and the explicit inclusion in that description of lexical labels of relevant and important concepts and relations. Classroom implementation of MERLO pedagogy includes interactive MERLO quizzes, as well as inclusion of MERLO items as part of midterm tests and final exams. A MERLO interactive quiz is an in-class procedure that provides learners with opportunities to discuss a PowerPoint display of a MERLO item in small groups and send their individual responses to the instructor's computer via mobile text messaging or by using a clicker (Classroom Response Systems (CRS)). Such a quiz takes 20–30 minutes and includes the following four steps: small group discussion, individual response, feedback on production response, and feedback on recognition response and class discussion. For a live example of such a discussion, see the 1-minute video at https://goo.gl/XENVPn.

The implementation of MERLO has been documented to enhance learning outcomes. Such implementations were carried out at different instructional situations; see Shafrir and Etkind (2006).

To demonstrate the reports derived from MERLO assessments, we refer to results from classes of mathematics in a middle school in Turin, Italy (Arzarello et al., 2015a, 2015b). MERLO assessments were conducted after teaching ten concepts in a middle school. Percentages, powers, transitions, inverse proportions, line, and circumference were assessed in two parallel classes. Fractions, angles, functions, and equations were assessed only in one class. The basic statistics from the MERLO recognition scores are presented in Table 6.4. In Figure 6.7 we display box plots of the recognition scores for ten concepts taught in an Italian middle school in Turin. The conceptual understanding of powers is the lowest, and of angle, the highest. This initial feedback is mostly directed at the teachers and the designers of the material used in the class. The box plots in Figure 6.7 identify specific students with low scores who probably need extra attention. In powers we notice four students with perfect scores; investigating why they understand better than the others might create a learning experience beneficial to the whole group.

In Figure 6.8 we see that powers are less understood than most concepts including percentages and fractions and that angle is better understood than function and equations. Such comparisons provide instructors with useful insights in order to improve pedagogical and teaching strategies.

We see that function, equations, and powers exhibit significantly lower scores than angle, fractions, line, transition, and inverse proportions. These structural differences provide more information to be leveraged by education specialists. The analysis presented in Figures 6.7 and 6.8 and Tables 6.4 and 6.5 was done with Minitab v17.2.

Table 6.4   MERLO recognition scores for ten concepts taught in an Italian middle school.

Variable	N	N*	Mean	Minimum	Maximum
Percentages	42	2	3.500	0.000	5.000
Fractions	29	0	4.172	2.000	5.000
Powers	49	1	2.531	1.000	5.000
Transition	43	1	3.930	2.000	5.000
Line	38	7	4.158	0.000	5.000
Inverse proportions	42	0	3.762	1.000	5.000
Circumference	44	2	3.500	1.000	5.000
Angle	18	1	4.444	2.000	5.000
Function	24	0	3.167	1.000	5.000
Equations	23	1	3.130	2.000	5.000

$N*$ represents missing data.

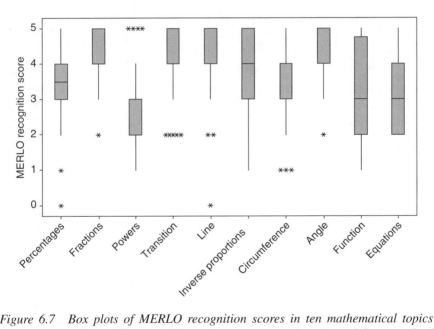

Figure 6.7   Box plots of MERLO recognition scores in ten mathematical topics taught in an Italian middle school. Asterisks represent outliers beyond three standard deviation of mean.

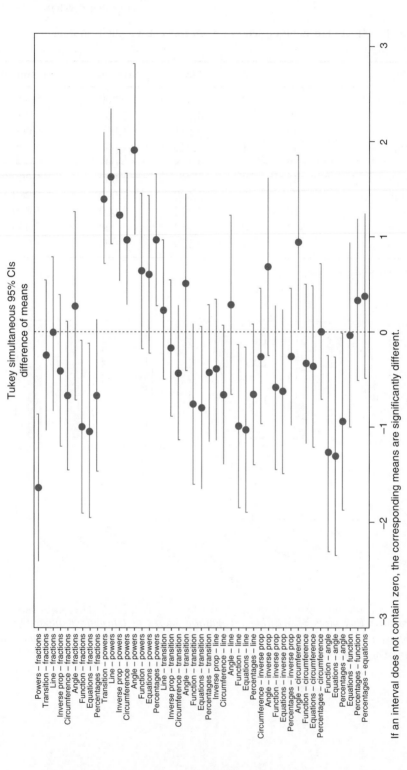

Tukey simultaneous 95% CIs
difference of means

If an interval does not contain zero, the corresponding means are significantly different.

Figure 6.8   Confidence intervals for difference in MERLO recognition scores between topics.

Table 6.5   Grouping of MERLO recognition scores using the Tukey method and 95% confidence.

Factor	N	Mean	Grouping
Angle	18	4.444	A
Fractions	29	4.172	A B
Line	38	4.158	A B
Transition	43	3.930	A B C
Inverse proportions	42	3.762	A B C
Circumference	44	3.500	B C
Percentages	42	3.500	B C
Function	24	3.167	C D
Equations	23	3.130	C D
Powers	49	2.531	D

Means that do not share a letter are significantly different.

Case study 4   MERLO assessment

INFOQ COMPONENTS

Goal (g): Evaluate student understanding of concepts included in the teaching curriculum in order to improve the teaching approach and possibly the teaching material used in class.

Data (X): MERLO recognition scores.

Analysis (f): Descriptive statistics, data visualization, Tukey's simultaneous pairwise comparisons, and grouping of factors.

Utility (U): Minimal standard error in MERLO score mean estimates.

INFOQ DIMENSIONS

We provide below an InfoQ assessment of reports based on MERLO scores in terms of the eight InfoQ dimensions:

1. *Data resolution*: MERLO-derived data combines data from recognition scores of ten target statements. In that sense, data resolution is high.

2. *Data structure*: Data structure of MERLO items is designed to reflect various aspects of concept understanding using self-stated information, as opposed to observed behavioral data. The generated data includes choices of statements (categorical data) as well as descriptions of the concept driving the answer, as reported by the learner (text data).

3. *Data integration*: Data from MERLO scores does not include data on the studied subject matter or the individual whose comprehension level is being assessed.

4. *Temporal relevance*: The process of deriving knowledge on conceptual understanding from data in the education process requires updating in the context of the evolution of the assessed individual.

5. *Chronology of data and goal*: MERLO quiz and interactive education provides for a very high synchronization of chronology of data and goal. Teachers can get quick assessment of students' abilities and difficulties, which can be used for immediately tuning the learning process in specific directions.

6. *Generalizability*: The results from a MERLO quiz or assignment can help the teacher improve the tool for future use. In that sense there is generalization to future offerings of a course on the same topic.

7. *Operationalization*: Concept science provides a comprehensive educational and psychological framework for MERLO assessment. The information derived from MERLO provides focused attention to concepts that need to be better presented or individuals that require special attention.

8. *Communication*: MERLO recognition scores help communicate various aspects of concept understanding, including the differential aspects of low scores to Q2 and Q3 statements.

Ratings for the eight dimensions are shown in Table 6.6. Overall, the InfoQ assessment score of MERLO-derived data is 68%, a relatively high score making it an effective method for conducting formative assessment activities. The impact of MERLO assessments has been proven in a range of educational applications including mathematics and statistics education, architectural design, and healthcare (Shafrir and Kenett, 2015).

Table 6.6   InfoQ assessment for MERLO.

InfoQ dimension	Rating
Data resolution	4
Data structure	3
Data integration	3
Temporal relevance	4
Chronology of data and goal	4
Generalizability	4
Operationalization	4
Communication	4
InfoQ score	**68%**

Table 6.7   Scoring of InfoQ dimensions of examples from education.

InfoQ dimension	(1) MAP report	(2) Students' earnings	(3) VAM statement	(4) MERLO
Data resolution	2	4	5	4
Data structure	2	4	4	3
Data integration	2	5	5	3
Temporal relevance	4	3	5	4
Chronology of data and goal	4	4	2	4
Generalizability	2	2	3	4
Operationalization	2	2	2	4
Communication	2	2	3	4
Use case score	**33**	**49**	**57**	**68**

6.5   Summary

The chapter presents four case studies related to education. Table 6.7 presents the InfoQ assessment from each of the four case studies by qualifying on a scale from 1 ("very poor") to 5 ("very good") the eight InfoQ dimensions of the case studies. This assessment is subjective and is based on discussions we held with colleagues. As a summary measure, we use the InfoQ scores on a 0–100 scale. From Table 6.7 we see that the use cases received an InfoQ score from 33 to 68%. These assessments can also point out to dimensions where focused improvements would increase the level of InfoQ of the related analyses and reports.

Appendix: MERLO implementation for an introduction to statistics course

The motivation for this work is the realization that much of the introduction to statistics classes (generically called "Statistics 101") prove to have very low effectiveness. In some cases, this first exposure of students to statistics generates bias and negative preconceptions and a detrimental effect on lifelong careers with both a personal and professional opportunity loss. These introductory courses typically do not prepare students to apply statistical methods and statistical thinking in their workplace or private life. Here, we apply concept science methodological tools and focus on the quality of the information generated through statistical analysis of MERLO assessment as a remedial intervention reinforcing the constructive and important role of statistics and quantitative literacy in modern life and education.

Teaching statistical methods is a challenging task. Teaching statistical concepts is an even more challenging task that requires skill, experience, and adequate techniques. To demonstrate the use of MERLO in statistical education, we refer to

Example 3.33, page 89, in chapter 3 on probability models and distribution functions from Kenett et al. (2014):

> An insertion machine is designed to insert components into computer printed circuit boards. Every component inserted on a board is scanned optically. An insertion is either error-free or its error is classified to the following two main categories: miss-insertion (broken lead, off pad, etc.) or wrong component. Thus, we have altogether three general categories. Let J_1 = Number of error free components, J_2 = Number of miss-insertions and J3 = Number of wrong components. The probabilities that an insertion belongs to one of these categories are $p_1 = 0.995$, $p_2 = 0.001$, $p_3 = 0.004$. The insertion rate of the machine is fixed at 3500 components per hour.

Question: What is the probability that during one hour of operation there will be no more than 20 insertion errors?

A typical solution i: $Pr(J_2 + J_3 \leq 20) =$ Binomial $(20;3500,0.005) = 0.7699$.

A MERLO target statement for this underlying concept can be stated as *independent Bernoulli event add up as a binomial random variable*, with sample MERLO items being

> Q1: The probability of no more than 20 insertion errors in one hour is derived from a binomial distribution with $n = 3500$ and $p = 0.005$.
>
> Q2: $Pr(J_2 + J_3 \leq 20) =$ binomial $(20;3500,0.005) = 0.7699$.
>
> Q3: To compute the probability of no more than 20 insertion errors in one hour, we assume 3480 insertions and $p = 0.005$.
>
> Q4: To compute the probability of no more than 20 insertion errors in one hour, we assume 3480 insertions and a hypergeometric distribution.

As another example consider the target statement: *The p value is the probability of getting the observed result or more extreme ones, if the null hypothesis is true*, and one could have the following alternative representations:

> Q2: Consider the null hypothesis that the system operates as described earlier, if we reject this hypothesis when we get more than 20 insertion errors, $p = 1 - Pr(J_2 + J_3 \leq 20) = 0.23$.
>
> Q3: The *p* value is the probability that the null hypothesis is true.
>
> Q4: A large *p* value indicates that the alternative hypothesis is true.

As mentioned, preparing MERLO items involves designing Q2–Q4 statements and creating sets consisting of a combination of four such statements in addition to the target statement for evaluation by the students. Test instructions for a MERLO item requires the learner to recognize and mark all but only those statements that share equivalence-of-meaning (at least 2 out of 5 statements in the MERLO item). In addition, students are requested to describe briefly the concept that he/she had in mind while making these decisions. Thus, a learner's response to a MERLO item combines recognition, namely, multiple choice/multiple response, and production, namely, a short answer.

As mentioned, MERLO items are scored by counting the number of correct (marked or unmarked) statements. When a set of MERLO items is given to students, these scores reflect the individual level of understanding. In addition, scores by concept provide feedback to the instructor regarding specific topics that were covered in the course. Specific comprehension deficits can be traced as low recognition scores on quadrants Q2 and Q3, due to the mismatch between the valence of surface similarity and meaning equivalence. A low score on Q2 indicates that the learner fails to include in the "boundary of meaning" of the concept certain statements that share equivalence of meaning (but do not share surface similarity) with the target; such low Q2 score signals an overrestrictive (too exclusive) understanding of the meaning underlying the concept. A low score on Q3 indicates that the learner fails to exclude from the "boundary of meaning" of the concept certain statements that do not share equivalence of meaning (but that do share surface similarity) with the target; this lower Q3 score signals an underrestrictive (too inclusive) understanding of the meaning of the concept. This pedagogical approach is very different from the usual classroom scenario where students are given an exercise (like the one earlier) and are asked to solve it individually.

References

Andrabi, T., Das, J., Khwaja, A. and Zajonc, T. (2009) Do Value-Added Estimates Add Value? Accounting for Learning Dynamics, HKS Faculty Research Working Paper Series RWP09-034, John F. Kennedy School of Government, Harvard University, http://dash. harvard.edu/handle/1/4435671 (accessed April 30, 2016).

Arzarello, F., Kenett, R.S., Robutti, O., Shafrir, U., Prodromou, T. and Carante, P. (2015a) Teaching and Assessing with New Methodological Tools (MERLO): A New Pedagogy? In *IMA International Conference on Barriers and Enablers to Learning Maths: Enhancing Learning and Teaching for All Learners*, Hersh, M.A. and Kotecha, M. (editors), Glasgow, UK.

Arzarello, F., Carante, P., Kenett, R.S., Robutti, O. and Trinchero, G. (2015b) MERLO Project: A New Tool for Education, IES 2015—Statistical Methods for Service Assessment, Bari, Italy.

ASA, American Statistical Association (2014) ASA Statement on Value-Added Models for Educational. https://www.amstat.org/policy/pdfs/ASA_VAM_Statement.pdf (accessed April 30, 2016).

Ballou, D., Sanders, W. and Wright, P. (2004) Controlling for student background in value-added assessment of teachers. *Journal of Educational and Behavioral Statistics*, 29, pp. 37–65.

Betebenner, D.W. (2009) Norm- and criterion-referenced student growth. *Educational Measurement: Issues and Practice*, 28 (4), pp. 42–51.

Betebenner, D.W. (2011) A Technical Overview of the Student Growth Percentile Methodology: Student Growth Percentiles and Percentile Growth Projections/Trajectories. http://www. nj.gov/education/njsmart/performance/SGP_Technical_Overview.pdf (accessed April 30, 2016).

Cabre, M.T. (1998) *Terminology: Theory, Methods, and Applications.* Benjamins, Amsterdam.

Etkind, M., Kenett, R.S. and Shafrir, U. (2010) The Evidence Based Management of Learning: Diagnosis and Development of Conceptual Thinking with Meaning Equivalence Reusable Learning Objects (MERLO). In *The 8th International Conference on Teaching Statistics (ICOTS)*, Ljubljana, Slovenia.

Goodman, D. and Hambleton, R. (2004) Student test score reports and interpretive guides: review of current practices and suggestions for future research. *Applied Measurement in Education*, 17(2), pp. 145–220.

Karl, A., Yang, Y. and Lohr, S. (2013) Efficient maximum likelihood estimation of multiple membership linear mixed models, with an application to educational value-added assessments. *Computational Statistics and Data Analysis*, 59, pp. 13–27.

Karl, A., Yang, Y. and Lohr, S. (2014a) Computation of maximum likelihood estimates for multiresponse generalized linear mixed models with non-nested, correlated random effects. *Computational Statistics and Data Analysis*, 73, pp. 146–162.

Karl, A., Yang, Y. and Lohr, S. (2014b) A correlated random effects model for nonignorable missing data in value-added assessment of teacher effects. *Journal of Educational and Behavioral Statistics*, 38, pp. 577–603.

Kenett, R.S., Zacks, S. and Amberti, D. (2014) *Modern Industrial Statistics: With Applications Using R, MINITAB and JMP*, 2nd edition. John Wiley & Sons, Sussex.

Kittredge, R.I. (1983) Semantic Processing of Texts in Restricted Sublanguages, in *Computational Linguistics*, Cercone, N.J. (editors), Pergamon Press, Oxford, UK, pp. 45–58.

Lohr, S. (2014) Red beads and profound knowledge: deming and quality of education, Deming lecture, Joint Statistical Meetings, Boston, MA.

Mariano, L., McCaffrey, D. and Lockwood, J. (2010) A model for teacher effects from longitudinal data without assuming vertical scaling. *Journal of Educational and Behavioral Statistics*, 35, pp. 253–279.

McCaffrey, D.F., Lockwood, J.R., Koretz, D.M. and Hamiltion, L.S. (2003) *Evaluating Value-Added Models for Teacher Accountability*. The RAND Corporation, Santa Monica.

McCaffrey, D., Lockwood, J.R., Louis, T. and Hamilton, L. (2004) Models for value-added models of teacher effects. *Journal of Educational and Behavioral Statistics*, 29(1), pp. 67–101.

Shafrir, U. and Etkind, M. (2006) eLearning for depth in the semantic web. *British Journal of Educational Technology*, 37(3), pp. 425–444.

Shafrir, U. and Etkind, M. (2010) Concept Science: Content and Structure of Labeled Patterns in Human Experience. Version 31.0.

Shafrir, U. and Kenett, R.S. (2015) Concept Science Evidence-Based MERLO Learning Analytics, in *Handbook of Applied Learning Theory and Design in Modern Education*, IGI Global, Hershey, PA.

Walsh, E. and Isenberg, E. (2015) How does value added compare to student growth percentiles? *Statistics and Public Policy*, 10.1080/2330443X.2015.1034390

7

Customer surveys

7.1 Introduction

Customer satisfaction studies deal with customers, consumers, and user satisfaction from a product or service. The topic was initially developed in marketing theory and applications. The BusinessDictionary (www.businessdictionary.com) defines customer satisfaction as "the degree of satisfaction provided by the goods or services of a company as measured by the number of repeat customers." According to this definition, customer satisfaction seems an objective and easily measured quantity. However, unlike variables such as type of product purchased or geographical location, customer satisfaction is not necessarily observed directly. Typically, in a social science context, analysis of such measures is performed indirectly by employing proxy variables. Unobserved variables are referred to as *latent variables*, while proxy variables are known as *observed variables*. In many cases, the latent variables are very complex and the choice of suitable proxy variables is not immediately obvious. For example, in order to assess customer satisfaction from an airline, it is necessary to identify attributes that characterize this type of service. A general framework for assessing airlines includes attributes such as on board service, timeliness, responsiveness of personnel, airplane seats, and other tangible service characteristics. In general, some attributes are objective, related to the service's technical-specific characteristics, and others are subjective, dealing with behaviors, feelings, and psychological benefits. Eventually, in order to design a survey questionnaire, a set of observed variables must be identified.

In practice, many of the customer satisfaction surveys conducted by companies are analyzed in a very simple way, without using models or statistical methods.

Information Quality: The Potential of Data and Analytics to Generate Knowledge,
First Edition. Ron S. Kenett and Galit Shmueli.
© 2017 John Wiley & Sons, Ltd. Published 2017 by John Wiley & Sons, Ltd.
Companion website: www.wiley.com/go/information_quality

Typical reports include descriptive statistics and basic graphical displays. In this chapter we focus on the information quality of a customer survey. Specifically, we show how the InfoQ of a survey can increase by combining basic analysis with more advanced tools, thereby providing insights on nonobvious patterns and relationships between the survey variables. Specifically, we use the InfoQ framework to compare seven analysis methods (f) popular in customer survey analysis. We assume in all cases that the data (X) is typical survey questionnaire data and that the utility (U) is to inform the administering company or organization of its customer satisfaction for increasing customer satisfaction and/or decreasing customer dissatisfaction. In Section 7.3 we describe and consider a variety of goals (g) that customer surveys aim to achieve.

7.2   Design of customer surveys

Customer surveys are typically based on self-declared completion of questionnaires. Self-administered surveys use structured questioning to map out perceptions and satisfaction levels, into data that can be statistically analyzed. Some surveys target all past and/or current customers; they are in fact a type of census. In event-driven surveys, only customers identified by a specific event, such as a call to a service center or the purchasing of a new system, are included in the surveyed group. In others, a sample is drawn and only customers in the sample receive a questionnaire. In drawing a sample, several sampling schemes can be applied. They range from probability samples, such as cluster, stratified, systematic, or simple random sampling, to non-probability samples, such as quota, convenience, judgment, or snowball sampling. The survey process consists of four main stages: planning, collecting, analyzing, and presenting.

Modern surveys are conducted through a wide variety of techniques including phone interviews, self-reported paper questionnaires, email questionnaires, Internet-based surveys, SMS-based surveys, face-to-face interviews, videoconferencing, and more. In evaluating results of a customer satisfaction survey, three background questions should be checked:

1. Is the questionnaire properly designed?

2. Has the survey been properly conducted?

3. Has the data been properly analyzed?

Responding to these questions requires an understanding of the survey process, the organizational context, and statistical methods. Customer satisfaction surveys can be part of an overall integrated approach. Integrated models are gaining much attention of both researchers and practitioners (Rucci et al., 1998; MacDonald et al., 2003; Godfrey and Kenett, 2007). Kenett (2004) presents a generic integrated model that has been implemented in a variety of industries and businesses. The basic building blocks of the model are datasets representing voice of the customer (VoC), voice of the process (VoP), and voice of the workforce (VoW). The integration, through Bayesian networks (BNs) or other statistical methods, provides links between the variables measured in these three dimensions. These links can show, for example, the extent to which satisfied employees imply happy customers and improved financial performance. The integration at Sears Roebuck has shown that a 5-point increase (out of 100) in employee satisfaction resulted in an increase of 1.5 units (out of 5) in customer satisfaction, which resulted in an estimated 0.5% increase in revenue growth (Rucci et al., 1998).

In handling customer satisfaction, several statements are commonly made regarding the impact of increase in customer loyalty and satisfaction. These are based on practical experience and research (see, e.g., http://tarp.com/home.html). Some of the more popular statements are the following:

1. Growth from retention

   - A very satisfied customer is six times more likely to repurchase your product than a customer who is just satisfied.

   - Loyal customers spend 5–6% more of their spending budget than customers who are not loyal.

2. Profit boost from retention

   - An increase in customer retention of just 5% can boost profits by 25–85%.

   - Loyal customers are not as price sensitive.

3. Reducing the cost of acquisition

   - Acquiring a customer costs five to seven times more than retaining one.

   - Satisfied customers, on average, tell five other people about their good experience.

4. The cost of defection

   - The average customer with a problem eventually tells nine other people about it.

   - 91% of unhappy customers will never buy from you again.

Annual customer satisfaction surveys (ACSS) are conducted on a yearly basis by companies, organizations, and government agencies in order to:

- Identify key drivers of satisfaction and prioritize actions

- Compare data over time to identify patterns in customers' experiences

- Disseminate the results throughout the appropriate audiences within the company to drive change within the organization

Table 7.1   Main deliverables in an Internet-based ACSS project.

Category	Deliverables
Building infrastructure	• Questionnaire evaluation (if relevant) 　○ Last year's questionnaire effectiveness check • Questionnaire design and development 　○ (Re)design of questionnaire 　○ Setting up of a survey website 　○ Testing and validation • Contact list management
Data collection	• Data collection (e-survey and phone) • Open-ended responses
Data analysis	• Data cleanup phase • Reporting and analysis 　○ Full report 　　– Insights, trend analysis 　○ Executive summary 　○ Raw data for drill-down tools
Support and maintenance (personnel)	• Project manager • Technical support for: 　○ Monitoring real time data (during the survey) 　○ Resolving problems operating the questionnaire by customers (via email or phone) • Conducting phone surveys (where relevant) • Quality management—a function that is responsible for KPIs and quality metrics

A typical Internet-based ACSS plan, and its steps/deliverables, is presented in Table 7.1. Typical technical service level agreements (SLA), when conducting Internet-based ACSS, are presented in Table 7.2.

The ACSS is usually part of a larger plan that is designed and approved at the beginning of the financial year. At that point, decisions with strategic and budgetary impact are made.

If the financial year starts in January, the kickoff of the ACSS cycle is usually planned in August. In this context, a general framework for conducting ACSS consists of activities listed in Table 7.3.

To operate this annual cycle, one needs an effective steering committee and improvement methodology. For details on such organizational capabilities in the context of system and software development organizations, see Kenett and Baker (2010).

Tables 7.1, 7.2, and 7.3 sketch an ACSS annual plan and provide the flavor of a typical ACSS, within an overall strategic initiative for achieving operational excellence. When applying an integrated approach, the ACSS initiative is complemented

Table 7.2   Service level agreements for Internet-based customer satisfaction surveys.

Subject	Metric
SLA for maintenance Maintenance includes incidents and problems such as: 1. Customer cannot access survey site 2. Customer cannot enter a specific answer/s 3. Survey is not responsive 4. Response times are poor 5. Progress reports are not accessible	• Mean time to repair (MTTR)—3 hours (working hours, on working days) • Mean time between failures (MTBF)—three days • Mean time between critical failures (MTBCF)—two weeks
SLA for system availability	• % availability—95%
SLA for performance	• Time until webpage is loaded (initially)—four seconds • Time until page is refreshed according to user answers—two seconds.

Table 7.3   A typical ACSS activity plan.

Month	Activity
August	ACSS survey plan and design
September	ACSS survey launch
October	ACSS survey execution
November	ACSS survey data analysis and communication
December	Organization annual budget process
January	Launch of annual strategic initiatives
February	Decision on annual improvement areas and KPIs
March	Launch of improvement initiatives
April	Execution of improvements
May	Execution of improvements
June	Execution of improvements
July	Improvement initiatives progress review

by other initiatives such as employee surveys, dashboards that reflect the VoP and event-driven surveys that are triggered by specific events. Examples of events followed by a satisfaction survey questionnaire include calls to a service center or acquisition of a new product. In the following section, we describe each of the four InfoQ components in customer survey analysis.

7.3   InfoQ components

7.3.1   Goal (g) in customer surveys

There are various goals that companies and organizations aim to achieve with customer satisfaction surveys. Examples of goals of customer satisfaction surveys include:

- Deciding where to launch improvement initiatives

- Identifying the drivers of overall satisfaction

- Detecting positive or negative trends

- Highlighting best practices by comparing products or marketing channels

- Improving the questionnaire

- Setting up improvement goals

- Designing a balanced scorecard using customer inputs

- Determining the meaning of the rating scales

- Effectively communicating the results using graphics or otherwise

7.3.2   Utility (U) in customer surveys

Direct measures of success relate to improvements in customer satisfaction and decrease in customer dissatisfaction. Related utility functions consist of customer loyalty indices such as the willingness to recommend or repurchasing intentions. The bottom-line utility is improved business results.

7.3.3   Data (X) in customer surveys

In customer satisfaction surveys, target individuals are requested to complete questionnaires, which may have between 5 and 100 questions (Salini and Kenett, 2009; Kenett and Salini, 2012). Let us examine an ACSS directed at customers of an electronic product distributed worldwide (we call this company ABC). The survey aims to assess satisfaction levels of customers from different features of the product and related services. The questionnaire is composed of 81 questions including demographics and overall satisfaction from the company. An important output of the survey is to find out which aspects of the product and services influence overall satisfaction, recommendation level and repurchasing intentions. The topics covered by the survey include *equipment, sales support, technical support, training, customer portal, administrative support, terms and conditions,* and *site planning and installation.* Demographic variables that can help profile customer responses include *country, industry type,* and *age of equipment.*

A self-declared "questionnaire" provides an overall assessment of customer satisfaction from a specific product or service. Different measurements of customer satisfaction are used for operationalizing the construct "customer satisfaction." The response (dependent) variables in customer satisfaction models are typically expressed on an anchored scale, with corresponding conventional scores such as a 5-point or a 7-point scale. However, this scale can also be dichotomous or made so by summarizing judgments in two categories. Scheme one splits customers who responded "5" on a 5-point scale from the others. Their percentage yields a satisfaction index labeled TOP5. On the other end of the scale, customers who responded "1" or "2" are aggregated to compose an index labeled BOT12. Some organizations combine the labels "satisfied" and "very satisfied" which produce indices with higher values but much reduced resolution. High frequency of TOP5 from a specific product or service represents a best practice example that needs to be emulated, while high BOT12 frequency presents an opportunity for improvement.

7.3.4   Analysis (f): Models for customer survey data analysis

The analysis of customer surveys is based on a range of models such as regression models, compositional models, and structural models (see Zanella, 2001; Kenett and Salini, 2012; Vives-Mestres et al., 2016). A comprehensive analysis of sampling error in the context of probability sampling is presented by Chambers (2015) within an edited volume on the theory and methods for assessing quality in business surveys. Besides sampling errors, the volume deals with a variety of nonsampling errors, covering coherence and comparability of statistics. The reader is referred to the volume for an example of assessment of quality for an annual and a monthly business survey from Sweden and the United Kingdom. Several models for analyzing customer survey data are presented in the next section.

7.4   Models for customer survey data analysis

7.4.1   Regression models

We present three popular regression-based approaches for modeling customer satisfaction survey data:

a. *Ordinary linear regression model.* Explanatory variables describe dimensions related to specific aspects of a product or service, for example, age of equipment or geographical location. Regression models apply to data that can be expressed on conventional ordered rating scales. Such data can refer to respondents' personal characteristics, such as age, or the number of purchases or total amount spent in a previous period that are measured on continuous scales. The usual statistical analysis techniques for such data apply the least squares criterion for obtaining the estimates of the unknown parameters and related methods for checking goodness of fit.

b. *Regression models and techniques accounting for the ordinal character of the response and of explanatory variables.* In this context, monotonic regression analysis plays an important role (see Kruskal (1965)). In Zanella (1998), a nonlinear regression model with latent variables is presented for deriving a ratio-scale representation of the response.

c. *Logistic regression model.* If one can assume a probability distribution for the response portraying overall satisfaction, the expected value of the response can be presented, with conditioning on the different situations described by the values of the explanatory variables. The logistic regression approach allows us to take into consideration the fact that the values of the response variable are on an ordinal scale, since it refers to the probability distribution of the response in a more direct fashion.

7.4.2   Structural models

An alternative modeling approach is structural equation models (SEM), also known as path models, such as covariance-based models (linear structural models with latent variables (LISREL)) or composite-based models (partial least squares path models (PLSPM)). Such models allow us to establish links between latent variables, which are related to dimensions describing customer satisfaction (Bollen, 1989; Boari and Cantaluppi, 2011). Structural equations models are composed of two equations systems: *structural equations* and *measurement model*. Baumgartner and Homburg (1996) give comments and recommendations on the basis of cases of complete structural model application in marketing. The utility of the model is determined by checking model adequacy, via indicators such as the chi-square statistic, root-mean-square residual, goodness-of-fit index, estimated coefficients for the measurement equations, and so on. The LISREL method is used for calculating the American Customer Satisfaction Index (ACSI) (Anderson and Fornell, 2000; Kenett and Salini, 2012) and likewise the European Customer Satisfaction Index (ECSI). The main problem of the approach described previously is that metric scales are assumed for variables, while in practice they are measured with ordinal scales. Some transformation to obtain metric scales can be used with caution.

7.4.3   SERVQUAL

In the fundamental work by Parasumaran et al. (1985, 1988, 1991), the well-known SERVQUAL model was developed to map gaps between perceived and expected levels of satisfaction (see Figure 7.1). In the model we consider five gaps:

- Gap 1: Customer expectations versus management perceptions of what customers want

- Gap 2: Management perceptions versus service quality specification

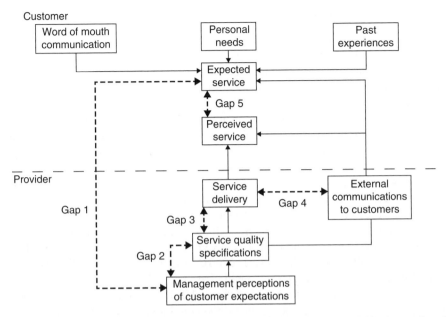

Figure 7.1   SERVQUAL gap model. Source: Parasuraman et al. (1988). Reproduced with permission from Elsevier.

- Gap 3: Service quality specification versus actual service delivery

- Gap 4: Service delivery versus external communications

- Gap 5: Customer expectations versus perceived service

A common strategy for closing Gap 5 consists of first closing Gap 4 and then Gaps 1, 2, and 3.

The model is operationalized with a questionnaire investigating customer expected and perceived performance. Some criticism of the original gap model approach was expressed by Cronin and Taylor (1992), who raised doubts about the SERVQUAL indicator being appropriate to describe service quality. This criticism gave rise to another improved model, SERVPERF. Both models represent structured links between variables representing service components and overall satisfaction.

In general customer satisfaction models such as SERVQUAL or SERPERF, customer satisfaction is considered a "multidimensional attribute," where each component corresponds to a dimension of the conceptual construct, that is, to an aspect of a product or service considered essential in determining customer satisfaction. The synthesis of the evaluations of the single "marginal" satisfaction attributes has a defining and therefore conventional nature. Specifically, these models declare explicitly functional links of the latent variables that correspond to the various dimensions and target one-dimensional variables associated with the concept under investigation such as overall customer satisfaction.

7.4.4   Bayesian networks

BNs implement a graphical model structure known as a directed acyclic graph (DAG) that is popular in statistics, machine learning, and artificial intelligence. BNs are both mathematically rigorous and intuitively understandable. They enable an effective representation and computation of the joint probability distribution (JPD) over a set of random variables (Pearl, 1985, 1988, 2000). The structure of a DAG is defined by two sets: the set of nodes and the set of directed edges. The nodes represent random variables and are drawn as circles labeled by the variable names. The edges represent links among the variables and are represented by arrows between nodes. In particular, an edge from node X_i to node X_j represents a relation between the corresponding variables. Thus, an arrow indicates that a value taken by variable X_j depends on the value taken by variable X_i. This property is used to reduce, sometimes significantly, the number of parameters that are required to characterize the JPD of the variables. This reduction provides an efficient way to compute the posterior probabilities given the evidence present in the data (Jensen, 2001; Ben Gal, 2007; Pearl, 2000). In addition to the DAG structure, which is often considered as the "qualitative" part of the model, one needs to specify the "quantitative" parameters of the model. These parameters are described by applying the Markov property, where the conditional probability distribution (CPD) at each node depends only on its parents. For discrete random variables, this conditional probability is often represented by a table, listing the local probability that a child node takes on each of the feasible values—for each combination of values of its parents. The joint distribution of a collection of variables can be determined uniquely by these local conditional probability tables (CPTs). In learning the network structure, one can include *white lists* of forced causality links imposed by expert opinion and *black lists* of links that are not to be included in the network.

Kenett and Salini (2009) applied a BN to data collected from 266 companies participating in an ACSS. The data includes responses to a questionnaire composed of 81 questions. Figure 7.2 shows the resulting BN, with arrows linking responses to specific questions on a 1–5 scale and customer geographical location. We can see, for example, that the level of satisfaction from supplies affects the level of satisfaction from the equipment, the training, and the web portal. The level of satisfaction from equipment affects overall satisfaction and recommendation levels of the company. The graph presents the structure of the BN, and the CPTs represent the model estimates. On the basis of the network, we can perform various diagnostic checks. For example, we can compute the distribution of responses to various questions for customers who indicated that they are very likely to recommend the product to others. Such an analysis enables profiling of loyal customers and designing early warning indicators that predict customer dissatisfaction. In a sense, the BN provides decision makers with a decision support tool where alternative scenarios can be assessed and operational goals can be established. For more on the application of BNs to customer survey analysis, see Kenett and Baker (2010), Kenett and Salini (2011b), Kenett et al. (2011b), and Cugnata et al. (2014, 2016).

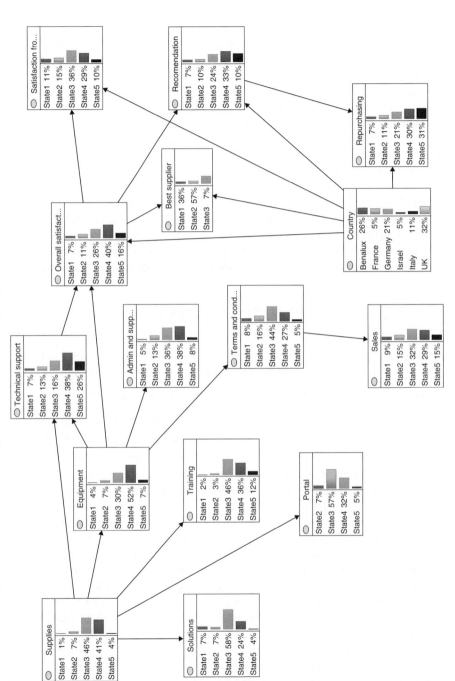

Figure 7.2 Bayesian network of responses to satisfaction questions from various topics, overall satisfaction, repurchasing intentions, recommendation level, and country of respondent.

7.4.5   The Rasch model

The Rasch model (RM) was first proposed in the 1960s to evaluate ability tests (Rasch, 1960). These tests are based on a user assessing a set of items. The assessment of a subject is assumed to depend on two factors: the subject's relative *ability* and the item's intrinsic *difficulty*. Subsequently the RM has been used to evaluate behaviors or attitudes. In the case of customer surveys, the two factors become the subject's *property* and the item's *intensity*, respectively. In recent years the model has been employed in the evaluation of services. In this context, the two factors become an individual customer *satisfaction* and the item (question) an intrinsic level of *quality*. These two factors are measured by the parameters θ_i, referring to the satisfaction of person (customer) i, and β_j, referring to the quality of item (question) j. It is then possible to compare these parameters. Their interaction is expressed by the difference $\theta_i - \beta_j$. A positive difference means that the customer's satisfaction is superior to the item's quality level. The difference $\theta_i - \beta_j$ determines the probability of a specific response to question j. In particular, in the dichotomous case where the question's response is "0" for "not satisfied" and "1" for "satisfied," the probability of a response $x_{ij} = 1$ by customer i with satisfaction level θ_i, when answering question j of quality β_j, is modeled as

$$P\left\{x_{ij} = 1 \mid \theta_i, \beta_j\right\} = \frac{\exp\left(\theta_i - \beta_j\right)}{1 + \exp\left(\theta_i - \beta_j\right)} = p_{ij}$$

In the dichotomous model, data is collected in a *raw score matrix*, with n rows (one for each customer) and J columns (one for each question), whose values are 0 or 1. The sum of each row $r_i = \sum_{j=1}^{J} x_{ij}$ represents the total score of customer i for all the items, while the sum of each column $s_j = \sum_{i=1}^{n} x_{ij}$ represents the score given by all the customers to the question j. The RM possesses several important properties. The first property is that the items measure only one latent feature (*one-dimensionality*). This is a limitation in the applications to customer satisfaction surveys where there are usually several independent dimensions. Another important characteristic of RM is that the answers to an item are independent of the answers to other items (*local independence*). In the customer satisfaction survey context, this is an advantage. For parameters where no assumptions are made, by applying the logit transformation $\log\left(p_{ij} / \left(1 - p_{ij}\right)\right)$, θ_i and β_j can be expressed on the same scale (*parameters linearity*); the estimations of θ_i and β_j are test- and sample-free (*parameter separability*), and the row and column totals on the raw score matrix are sufficient statistics for the estimation of θ_i and β_j (*sufficient statistics*). For more on these properties, see Andrich (2004).

The Rasch dichotomous model has been extended to the case of more than two ordered categories such as a 1–5 Likert scale. This approach assumes that between each category and the next, there is a threshold that qualifies the item's position as a function of the quality level presented by every answer category. A threshold is where two adjacent categories have the same probability to be chosen so that, for example, the probability to choose the first category is the probability not to exceed the first threshold. Thus, the answer to every threshold h of an item j depends on a value

$\beta_j + \tau_h$, where β_j characterizes responses to item j. The second term represents the hth threshold of β_j referring to the item j. The thresholds are ordered $(\tau_{h-1} < \tau_h)$, because they reflect the category order. For more details see De Battisti et al. (2011). This extension allows us to model responses on a 1–5 scale.

The utility of an RM can be evaluated with the Andersen likelihood ratio statistic, which tests the assumption that the estimates of the difficulty parameters are equal. The RM provides many diagnostic tools such as item characteristic curves, goodness-of-fit plots, person–item maps, pathway maps and a wide range of statistical tests (for more details, see chapter 14 in Kenett and Salini, 2011b).

7.4.6 CUB models

Responses to customer satisfaction surveys are governed by specific experiences and psychological considerations. When faced with discrete alternatives, people make choices by pairwise comparison of the items or by sequential removals. Such choices are affected both by uncertainty in the choice and pure randomness. Modeling the distribution of responses is far more precise than considering single summary statistics. Such considerations lead to the development of the combination of uniform and shifted binomial random variables (CUB) model, originally proposed in Piccolo (2003) (see also Iannario and Piccolo (2012)). The CUB model is used in surveys where subjects express a definite opinion selected from an ordered list of categories with m alternatives. The model differentiates between satisfaction level from an item and randomness of the final choice. These unobservable components are defined as *feeling* and *uncertainty*, respectively.

Feeling is the result of several factors related to the respondent such as country of origin, position in the company, and years of experience. This is represented by a sum of random variables which converges to a unimodal continuous distribution. To model this, CUB models feeling by a shifted binomial random variable, characterized by a parameter ξ and a mass b_r for response r where

$$b_r(\xi) = \binom{m-1}{r-1} \xi^{m-r} (1-\xi)^{r-1}, \quad r = 1, 2, \ldots, m.$$

Uncertainty is a result of variables such as the time to answer, the degree of personal involvement of the responder with the topic being surveyed, the availability of information, fatigue, partial understanding of the item, lack of self-confidence, laziness, apathy, boredom, and so on. A basic model for these effects is a discrete uniform random variable:

$$U_r(m) = \frac{1}{m}, \quad r = 1, 2, \ldots, m.$$

The integrated CUB discrete choice model is

$$\Pr(R = r) = \pi b_r(\xi) + (1 - \pi) U_r(m), \quad r = 1, 2, \ldots, m$$

for $0 \le \pi \le 1$, and

$$E(R) = \frac{(m+1)}{2} + \pi(m-1)\left(\frac{1}{2} - \xi\right).$$

7.4.7   Control charts

Perceived quality, satisfaction levels, and customer complaints can be effectively controlled with control charts used in the context of statistical process control (SPC). SPC methods were originally developed in the 1920s to improve the quality of products. Control charts are generally classified into two groups. If the quality characteristic is measured on a continuous scale, we have a *control chart for variables*. When the quality characteristic is classified as conforming or not conforming on the basis of whether or not it possesses certain attributes, then *control charts for attributes* are used. For an introduction to basic and advanced control charts, see Kenett and Zacks (2014). In analyzing customer satisfaction survey data, we can use control charts to identify a shift in satisfaction levels from previous surveys or investigate the achievement of preset targets.

In general, we test the hypothesis

$$\begin{cases} H_0 : \theta = \theta_0 \\ H_1 : \theta \neq \theta_0 \end{cases}$$

where θ can be the mean, the standard error, or a proportion, depending on the particular kind and scope of the control chart (i.e., for variables or for attributes).

All the previous details also hold when we are interested in testing a specific shift of the parameter such as $\theta > \theta_0$ or $\theta < \theta_0$. In these cases, only one control limit, either an upper control limit (UCL) or lower control limit (LCL), is reported on the control chart.

Specifically, the p chart with control limits $= \bar{p} \pm k\sqrt{\bar{p}(1-\bar{p})/n}$ is used to monitor the percentage of respondents who answered "5" (very high) to a question on overall satisfaction. Here n is the number of respondents, and k is a constant multiplier of the binomial standard deviation used to set up the control limits. The value $k=2$ is often applied in applications of control charts to the analysis of customer satisfaction data. For more details, see Kenett et al. (2011a). For an application of multivariate control charts using compositional methods, see Vives-Mestres et al. (2014, 2015, 2016).

7.5   InfoQ evaluation

We now turn to evaluating each of the methods for analyzing customer survey data using the InfoQ framework by considering each of the eight InfoQ dimensions. In this case, we use the InfoQ framework to compare seven analysis methods (f) popular in customer survey analysis. We assume in all cases that the data (X) is typical survey questionnaire data and that the utility (U) is to inform the administering company or organization of its customer satisfaction for increasing customer satisfaction and/or decreasing customer dissatisfaction. While a variety of goals (as described in Section 7.3) do exist, here we focus on identifying the drivers of overall satisfaction, for purpose of illustration.

7.5.1   Regression models

Regression models aim to generate information by providing an explanatory link between covariates and survey responses. The characteristics of the InfoQ dimensions for such models are:

1. *Data resolution*: Regression models can handle any type of data, including ordinal, nominal, and continuous. As such, they provide the ability to adequately handle data resolution.

2. *Data structure*: Through responses to questions and open comments, surveys combine structured and unstructured components. In most cases, regression models do not directly model semantic text. An extra step of text mining is required for that purpose.

3. *Data integration*: Combining data sources and data types can be partially handled by regression models such as data fusion methods.

4. *Temporal relevance*: Time effects can be incorporated in regression models so that temporal relevance, as reflected by the data, can be fully represented.

5. *Chronology of data and goal*: Implementation of regression models in online systems can provide constant updates or retrospective estimates.

6. *Generalizability*: Regression models are based on statistical theory and therefore provide means for statistical inference and generalization from a sample to the population.

7. *Operationalization*: The insights derived from regression models, linking covariates to responses, provide explanations that can prove useful in designing focused action items.

8. *Communication*: The mathematical formulation of regression models provides coefficients that are interpretable in managerial language (i.e., each coefficient expresses the magnitude of the effect of that variable on the outcome). However, the analyst must create explicit statements based on these coefficients to assure they are understood correctly. In many cases, the model fitting is supplemented by graphs representing predictions for specific values of covariates, residual analysis and goodness-of-fit evaluations.

7.5.2   Structural models

Structural models provide information by fitting data to a structural model. The characteristics of the InfoQ dimensions for such models are:

1. *Data resolution*: SEMs are designed to handle questionnaire-based data. As such they typically do not consider continuous or textual covariates.

2. *Data structure*: Structural models explicitly present a data structure combining latent variables. This forced clarification of the data structure is quite unique.

3. *Data integration*: The integration of data from different sources is usually not considered in structural models.

4. *Temporal relevance*: The ACSI, which uses structural models, is based on a rolling series of phone-based surveys. This provides an operational approach to ensure temporal relevance.

5. *Chronology of data and goal*: The use of structural models usually leads to monthly reports without specific considerations for special tailoring of the information from the survey.

6. *Generalizability*: Structural models build on structural equations that represent a generalizable structure presenting links between latent variables. Statistical tests can be used for generalizing the sample relationships to the population.

7. *Operationalization*: Structural models directly treat the unobservable construct of satisfaction (and other constructs) as latent variables and tie them to observable, measurable variables (manifest variables). Hence, it allows the analyst to convey their theoretical model and expert knowledge in terms of both latent and manifest variables.

8. *Communication*: Software programs like Amos (www-03.ibm.com/software/products/en/spss-amos), SmartPLS (www.smartpls.de), and the R package sem (https://cran.r-project.org/web/packages/sem/sem.pdf) provide graphical representation of the structural equations and the correlation values between the observed measured variables and the unobserved latent variables.

7.5.3   SERVQUAL

The SERVQUAL type models provide information that enable service managers to focus on specific action items. The characteristics of the InfoQ dimensions for such models are:

1. *Data resolution*: The data used in such models is based on questionnaires.

2. *Data structure*: The data typically used in such models is based on Likert scale anchored at 1–5.

3. *Data integration*: The questionnaire data is applied to the model without direct reference to service-related performance indicators such as response time and customer complaints.

4. *Temporal relevance*: SERVQUAL surveys are usually conducted once a year.

5. *Chronology of data and goal*: The approach is used on support of annual improvement plans.

6. *Generalizability*: The gap structure provides the basis for generalization of service characteristic, beyond the specific customer touchpoints investigated in a specific SERVQUAL questionnaire.

7. *Operationalization*: The gap structure provides the basis for specific service improvement actions.

8. *Communication*: The data analysis from such models is usually presented in tabular form.

7.5.4 Bayesian networks

The main characteristics of the BN in terms of InfoQ dimensions are:

1. *Data resolution*: The BNs can treat discretized continuous, nominal, and ordinal variables. Some BNs can also directly address continuous data.

2. *Data structure*: The data handled by BNs can include semantic data derived from text analysis ("bag of words") and operational data such as response time or number of customer complaints. In its simplest implementation, a BN is based on discretized data and CPDs between linked variables.

3. *Data integration*: BNs are particularly effective in integrating qualitative and quantitative variables.

4. *Temporal relevance*: BNs can be updated routinely by loading updated data and deriving updated posterior estimates. This ability, derived from the Bayesian context of BNs, provides for unique capabilities to ensure ongoing temporal relevance.

5. *Chronology of data and goal*: In a BN temporal variables like year or month can be used in the network and thereby allow for flexible conditioning providing enhanced chronology of data and goal.

6. *Generalizability*: The diagnostic and predictive capabilities of BNs provide generalizability to population subsets. The causality relationship provides further generalizability to other contexts such as organizational processes or specific job functions.

7. *Operationalization*: The use of a model with conditioning capabilities provides an effective tool to set up improvement goals and diagnose pockets of dissatisfaction.

8. *Communication*: The visual display of a BN makes it particularly appealing to decision makers who feel uncomfortable with mathematical models.

7.5.5   The Rasch model

The RM provides information at the individual and the item level. The characteristics of the InfoQ dimensions for this model are:

1. *Data resolution*: Rash models rely on questionnaires with specific items matching the customer touchpoints representing interactions determining the customer experience.

2. *Data structure*: The data used is based on the responses to questions, not to comments or any semantic information.

3. *Data integration*: The RM integrates item and individual specific characteristics. These two components can also be explained by using appropriate covariates.

4. *Temporal relevance*: Application of Rasch-based surveys is usually conducted periodically.

5. *Chronology of data and goal*: The information from the model cannot be updated unless a new survey is conducted.

6. *Generalizability*: The model is highly generalizable, as originally conceived by Georg Rasch under the concept of specific objectivity.

7. *Operationalization*: The model provides a clear distinction between individual tendencies and item specific satisfaction levels.

8. *Communication*: The model estimates can be presented visually with bar plots or otherwise. Its various diagnostic plots provide effective data presentation tools.

7.5.6   CUB models

The CUB models account for measurement uncertainty in evaluating customer levels of satisfaction. The characteristics of the InfoQ dimensions for such models are:

1. *Data resolution*: Data for CUB model analysis is derived from a questionnaire.

2. *Data structure*: CUB models do not explicitly handle textual comments or covariates.

3. *Data integration*: CUB models integrate the intensity of feeling toward a certain item with the response uncertainty. These two components can also be explained by using appropriate covariates.

4. *Temporal relevance*: Analysis using CUB is relevant to periodic or special purpose surveys.

5. *Chronology of data and goal*: The model does not provide for partial updating.

6. *Generalizability*: The model is not generalizable per se. However, its components offer interesting cognitive and psychological interpretations.

7. *Operationalization*: The model is mostly focused on explaining the outcomes of a survey. Insights on uncertainty and feelings can lead to interesting diverse initiatives.

8. *Communication*: The model estimates can be visually presented with bar plots or otherwise.

7.5.7   Control charts

The information provided by a control chart analysis of customer surveys is varied. The characteristics of the InfoQ dimensions for such an analysis are:

1. *Data resolution*: Control charts can handle continuous and categorical data.

2. *Data structure*: The data used in control chats can be univariate or multivariate.

3. *Data integration*: Control charts can be split by covariate values. Basic univariate control charts do not provide an effective data integration approach.

4. *Temporal relevance*: Event-driven surveys, analyzed with control charts, provide updated information on an ongoing basis.

5. *Chronology of data and goal*: Control charts provide for effective indication of change over time or difference between survey topics.

6. *Generalizability*: The analysis provides insights relevant to the data at hand without generalizable theory.

7. *Operationalization*: The findings clearly distinguish significant from random effects, thereby helping decision makers to effectively focus their improvement efforts.

8. *Communication*: The visual display of a control chart makes it very appealing for communication and visualization of the analysis.

7.6   Summary

Table 7.4 presents the ratings for each of the models described in Section 7.4, based on the discussion in Section 7.5, using the eight InfoQ dimensions. The ratings were obtained using a scale from 1 ("very poor") to 5 ("very good"). The overall InfoQ scores, by model, are computed using geometric means of desirability functions, in percentages. These range from 39% to 87% with the BN model producing the highest information quality. This assessment is subjective and is based on discussions with various experts. The models which scored the highest InfoQ scores are the BNs, the regression models, and the control charts.

The chapter presents seven types of models used in the analysis of customer surveys. Each model has unique characteristics that have been assessed using InfoQ dimensions. In analyzing customer surveys, an ensemble of models can enhance the InfoQ generated by individual models. Such an approach has been proposed by Kenett and Salini (2011a) with an application to a specific case study. The ability to integrate various models, with complementary strengths, presents an additional capability of InfoQ, the ability of integrating analysis from various models to increase overall InfoQ.

Appendix: A posteriori InfoQ improvement for survey nonresponse selection bias

Nonresponse is a critical issue in survey analysis. As discussed in Section 5.4, selection bias due to nonresponse is an a posteriori cause that can cause the set of completed surveys to not be representative of the population of interest, in the sense that some groups are over- or underrepresented in the sample. In this appendix we illustrate another use of InfoQ in customer surveys by examining a study that corrects for nonresponse in a posteriori data analysis.

In communicating customer survey results, the goal is to represent a frame that we see through the sample of questionnaire returns. In many examples, a link to a website questionnaire is sent to all listed customers and the survey is actually an attempted census. Ideally, we would like to get responses from all customers and collect the dataset X^*. In reality we are capturing a dataset X consisting only of responding customers so that $X \neq X^*$. Should we always weight the responses, as discussed in Section 5.4? This has negative implications in terms of estimator variance and the eighth InfoQ dimension of *communication*. Weighted results are more difficult to communicate to nontechnical managers who perceive this as a sort of data "fudging." This motivates us to first determine the need for weighting responses, hoping it can be avoided.

Consider a goal of estimating the level of customer satisfaction. To identify significant nonresponse patterns in customer surveys, Kenett (1991) proposed an approach based on comparing observed responses to expected responses by various customer classifications such as geographical location. The expected responses are derived from the distribution of the full list of customers by the relevant classification scheme.

Table 7.4 InfoQ score of various models used in the analysis of customer surveys.

InfoQ dimension	7.5.1 Regression	7.5.2 SEM	7.5.3 SERVQ	7.5.4 BN	7.5.5 Rasch	7.5.6 CUB	7.5.7 CC
Data resolution	5	3	3	5	3	3	5
Data structure	3	4	2	4	2	4	4
Data integration	3	3	2	4	2	2	4
Temporal relevance	5	3	3	4	3	3	3
Chronology of data and goal	5	2	2	5	2	3	4
Generalizability	4	5	5	5	5	4	3
Operationalization	3	3	5	4	3	3	3
Communication	3	4	3	5	2	3	5
InfoQ score	**68%**	**55%**	**46%**	**87%**	**39%**	**51%**	**69%**

Bold face represents aggregated scores.

Responses of customers, in a specific group, are analyzed using adjusted residuals and, to correct for multiple testing, critical values are derived from a Bonferroni-based test to determine the adjusted residual significance. If a significant nonresponse bias is determined, the model estimates may need to be evaluated by weighting of the responses using weights determined by the full list of customers (the target group). For more on weighting responses, see Section 5.4.

As an example, consider Table A that presents responses from a business to business (B2B) customer satisfaction survey aimed at a target group of 586 customers in six countries (Kenett and Salini, 2012). The survey was completed by 266 customers, and the adjusted residuals, by country, are listed in column Z. If $n=$ total number of returned surveys (here $n=266$), $K=$ number of categories (here $K=6$), $n_i=$ number of returned surveys in category i, and $p_i=$ proportion of category i in the sampling frame or target population, $i=1, ..., K$, then

$$Z_i = \frac{n_i - E_i}{S_i}, \quad i = 1,...,K$$

where $E_i=Np_i$ is the expected returns in group i and $S_i=(Np_i(1-p_i))^{1/2}$ is the standard deviation of the returns in group i, $i=1, ..., K$.

To determine the significance of Z_i, one applies the M-test that is based on a Bonferroni upper bound. If all adjusted residuals, Z_i, are smaller, in absolute value, than a critical value C, no significant bias is declared. Cells with values of Z_i, above C or below $-C$, are declared significantly different, and a follow-up effort, such as weighting of responses, is initiated. For $K=6$, $C=2.39$ for a p-value of 5%. For details, see Kenett and Zacks (2014).

We see that customers from France are significantly underrepresented. Given the overall response rate of 266/586, we expect 28 customer responses from France, but in practice only 15 responded. The adjusted residual of -2.61 being less than $C=-2.39$ indicates a significant underrepresentation.

This result requires a follow-up analysis to see if the overall satisfaction of respondents in France differs from respondents in other countries. If it does, weighting the responses on overall satisfaction is required so as to present an unbiased estimate

Table A   Postdata collection correction for nonresponse bias in a customer satisfaction survey using adjusted residuals.

Region	Population	P_i	Expected	Returns	Z	Significance
Benelux	64	0.11	29	26	-0.59	Ok
France	61	0.10	**28**	**15**	-2.61	**5% significance**
Germany	215	0.37	98	112	1.78	Ok
Israel	73	0.12	33	23	-1.86	Ok
Italy	78	0.13	35	39	0.72	Ok
United Kingdom	95	0.16	43	51	1.33	Ok
Total	586	1		266		

for the whole group. In other words, we would use $h(X^*)$, where h is the weighted sample for France. If there is no difference between the overall satisfaction of respondents in France and in other countries, such weighting is unnecessary, and the published estimates can be directly computed from the responses X.

References

Anderson, E.W. and Fornell, C. (2000) Foundations of the American customer satisfaction index. *Total Quality Management*, 11, pp. 869–882.

Andrich, D. (2004) Controversy and the Rasch model: a characteristic of incompatible paradigms? *Medical Care*, 42, pp. 1–16.

Baumgartner, H. and Homburg, C. (1996) Applications of structural equation modelling in marketing and consumer research: a review. *International Journal of Research in Marketing*, 13, pp. 139–161.

Ben Gal, I. (2007) Bayesian Networks, in *Encyclopaedia of Statistics in Quality and Reliability*, Ruggeri, F., Kenett, R.S. and Faltin, F. (editors in chief), John Wiley & Sons, Ltd, Chichester, UK.

Boari, G. and Cantaluppi, G. (2011) PLS Models, in *Modern Analysis of Customer Satisfaction Surveys*, Kenett, R.S. and Salini, S. (editors), John Wiley & Sons, Ltd, Chichester, UK.

Bollen, K.A. (1989) *Structural Equations with Latent Variables*. John Wiley & Sons, Inc., New York.

Chambers, R. (2015) Probability Sampling, in *Model Quality Report in Business Statistics*, Davies, P. and Smith, P. (editors), https://www.researchgate.net/publication/2389063_Model_Quality_Report_in_Business_Statistics (accessed 20 October 2015).

Cronin, J.J., Jr. and Taylor, S.A. (1992) Measuring service quality: a re-examination and extension. *Journal of Marketing*, 56, pp. 55–68.

Cugnata, F., Kenett, R.S. and Salini, S. (2014) Bayesian network applications to customer surveys and InfoQ. *Procedia Economics and Finance*, 17, pp. 3–9.

Cugnata, F., Kenett, R.S. and Salini, S. (2016) Bayesian networks in survey data: robustness and sensitivity issues. *Journal of Quality Technology*, 48, p. 3.

De Battisti, F., Nicolini, G. and Salini, S. (2012) The Rasch Model, in *Modern Analysis of Customer Satisfaction Surveys*, Kenett, R.S. and Salini, S. (editors), John Wiley & Sons, Ltd, Chichester, UK.

Godfrey, A.B. and Kenett, R.S. (2007) Joseph M. Juran, a perspective on past contributions and future impact. *Quality and Reliability Engineering International*, 23, pp. 653–663.

Iannario, M. and Piccolo, D. (2012) CUB Models: Statistical Methods and Empirical Evidence, in *Modern Analysis of Customer Satisfaction Surveys*, Kenett, R.S. and Salini, S. (editors), John Wiley & Sons, Ltd, Chichester, UK.

Jensen, F.V. (2001) *Bayesian Networks and Decision Graphs*. Springer, New York.

Kenett, R.S. (1991) Two methods for comparing Pareto charts. *Journal of Quality Technology*, 23, pp. 27–31.

Kenett, R.S. (2004) The Integrated Model, Customer Satisfaction Surveys and Six Sigma. *Proceedings of the First International Six Sigma Conference*, Center for Advanced Manufacturing Technologies, Wroclaw University of Technology, Wroclaw, Poland.

Kenett, R.S. and Baker, E. (2010) *Process Improvement and CMMI for Systems and Software*. Taylor & Francis, Auerbach CRC Publications, Boca Raton, FL.

Kenett, R.S. and Salini, S. (2009) New Frontiers: Bayesian networks give insight into survey-data analysis. *Quality Progress*, 42, pp. 31–36.

Kenett, R.S. and Salini, S. (2011a) Modern analysis of customer surveys: comparison of models and integrated analysis (with discussion). *Applied Stochastic Models in Business and Industry*, 27, pp. 465–475.

Kenett, R.S. and Salini, S. (2011b) *Modern Analysis of Customer Satisfaction Surveys: With Applications Using R*. John Wiley & Sons, Ltd, Chichester, UK.

Kenett, R.S. and Zacks, S. (2014) *Modern Industrial Statistics: With Applications Using R, MINITAB and JMP*, 2nd edition. John Wiley & Sons, Ltd, Chichester, UK.

Kenett, R.S., Deldossi, L. and Zappa, D. (2011a) Quality Standards and Control Charts Applied to Customer Surveys, in *Modern Analysis of Customer Satisfaction Surveys*, Kenett, R.S. and Salini, S. (editors), John Wiley & Sons, Ltd, Chichester, UK.

Kenett, R.S., Perruca, G. and Salini, S. (2011b) Bayesian Networks, in *Modern Analysis of Customer Satisfaction Surveys*, Kenett, R.S. and Salini, S. (editors), John Wiley & Sons, Ltd, Chichester, UK.

Kruskal, J.B. (1965) Analysis of factorial experiments by estimating monotone transformations of data. *Journal of the Royal Statistical Society, Series B*, 27, pp. 251–263.

MacDonald, M., Mors, T. and Phillips, A. (2003) Management system integration: can it be done? *Quality Progress*, 36, pp. 67–74.

Parasuraman, A., Zeithaml, V. and Berry, L. (1985) A conceptual model of service quality and its implications for future research. *Journal of Marketing*, 49, pp. 41–50.

Parasuraman, A., Zeithaml, V.A. and Berry, L.L. (1988) SERVQUAL: a multiple-item scale for measuring customer perceptions of service quality. *Journal of Retailing*, 64, pp. 11–40.

Parasuraman, A., Berry, L.L. and Zeithaml, V.A. (1991) Refinement and reassessment of the SERVQUAL scale. *Journal of Retailing*, 67, pp. 420–450.

Pearl, J. (1985) Bayesian Networks: A Model of Self-Activated Memory for Evidential Reasoning. *Proceedings of the Seventh Conference of the Cognitive Science Society*, University of California, Irvine, CA, pp. 329–334.

Pearl, J. (1988) *Probabilistic Reasoning in Intelligent Systems: Networks of Plausible Inference*. Morgan Kaufmann, San Francisco, CA.

Pearl, J. (2000) *Causality: Models, Reasoning, and Inference*. Cambridge University Press, Cambridge, UK.

Piccolo, D. (2003) On the moments of a mixture of uniform and shifted binomial random variables. *Quaderni di Statistica*, 5, pp. 85–104.

Rasch, G. (1960) *Probabilistic Models for Some Intelligence and Attainment Tests*. Danish Institute for Educational Research, Copenhagen, expanded edition (1980) with foreword and afterword by B.D. Wright. The University of Chicago Press, Chicago, IL.

Rucci, A., Kim, S. and Quinn, R. (1998) The employee-customer-profit chain at sears. *Harvard Business Review*, 76, pp. 83–97.

Salini, S. and Kenett, R.S. (2009) Bayesian networks of customer satisfaction survey data. *Journal of Applied Statistics*, 36(11), pp. 1177–1189.

Vives-Mestres, M., Daunis-i-Estadella, J. and Martín-Fernández, J.A. (2014) Individual T2 control chart for compositional data. *Journal of Quality Technology*, 2, pp. 127–139.

Vives-Mestres, M., Martín-Fernández, J.A. and Kenett, R. (2015) Exploring CoDa Contribution to a Survey Analysis: ABC Data. *Proceedings of CoDaWork-2015, The Sixth Compositional Data Analysis Workshop*, Thió-Henestrosa, S. and Martín-Fernández, J.A. (editors). 1–5 June 2015, University of Girona, Girona, Spain.

Vives-Mestres, M., Martín-Fernández, J.A. and Kenett, R. (2016) Compositional data methods in customer surveys. *Quality and Reliability Engineering International*, 10.1002/qre.2029.

Zanella, A. (1998) A statistical model for the analysis of customer satisfaction: some theoretical and simulation results. *Total Quality Management*, 9, pp. 599–609.

Zanella, A. (2001) Measures and Models of Customer Satisfaction: The Underlying Conceptual Construct and a Comparison of Different Approaches. *The Sixth World Congress for Total Quality Management, Business Excellence—What Is to Be Done, Proceedings*, Vol. 1, The Stockholm School of Economics, St Petersburg, pp. 427–441.

8

Healthcare

8.1 Introduction

Many physicians, patients, health journalists, and politicians do not understand health statistics. Unfortunately, the statistics community makes little effort to educate the public or healthcare providers in understanding health statistics. This collective issue in statistical illiteracy has resulted in serious consequences for health. We list in the following text a few examples described in more detail in Gigerenzer et al. (2007). See also Kenett (2012).

The British Committee on Safety of Medicines issued a warning that the third-generation oral contraceptive pill increased the risk of a thrombosis twofold—that is, by 100%. This caused great anxiety among women taking the pill, many of whom stopped using it. The studies on which the warning was based showed that among every 7000 women who took the previous generation pill, one had a thrombosis and that this number increased to two for women who took the third-generation pill. The relative risk increase was indeed 100%, but the absolute risk increase was one in 7000. The pill scare led to an estimated 13 000 abortions in the following year in England and Wales, resulting in a cost increase for the National Health Service estimated at about £4–6 million. The scare was due to the low level of information quality (InfoQ) contained in the official warning. A report of absolute increase would have had higher InfoQ, with a lower level of generated panic.

A second example comes from a study of 150 gynecologists, in which one third did not understand the meaning of a 25% risk reduction created by mammography screening. Most of them believed that if all women were screened, 25%, or 250 fewer

Information Quality: The Potential of Data and Analytics to Generate Knowledge,
First Edition. Ron S. Kenett and Galit Shmueli.
© 2017 John Wiley & Sons, Ltd. Published 2017 by John Wiley & Sons, Ltd.
Companion website: www.wiley.com/go/information_quality

women out of every 1000, would die of breast cancer, although the best evidence-based estimate is about 1 in 1000. This is another example of low InfoQ due to inadequate communication.

A third example comes from the former New York City mayor Rudi Giuliani who, while running for president of the United States, disclosed that he was treated and cured of prostate cancer. He then stated that the chance of surviving prostate cancer in the United States was 82% while in the United Kingdom it was 44%, implying that living in the United States doubled his chance of survival. In fact, prostate cancer mortality rates are the same in the United States and the United Kingdom, and survival rates are misleading statistics because of lead time bias. Consider two groups of men who die at age 70 of prostate cancer. The men in the first group do not participate in prostate-specific antigen (PSA) screening, and their cancer is detected from symptoms at age 67. The second group undergoes screening, and their cancers are detected at age 60. The five-year survival rate of the first group is 0% while for the second group it is 100%. The difference Giuliani referred to is largely due to the widespread use of PSA screening in the United States compared to the United Kingdom. This is again an example of low InfoQ due to misleading communication.

8.2   Institute of medicine reports

Healthcare represents 17.6% of the US economy and grows by 8–10% annually. The Institute of Medicine (IOM) published two high-impact reports that shed light on the phenomenon of medical errors in the United States: "To Err is Human: Building a Safer Health System" (1999) and "Crossing the Quality Chasm: A New Health System for the 21st Century" (2001). These reports show that at least two million hospital patients suffered dangerous infections and diseases during their hospital stays. Surgical errors, including improper surgical instruments or techniques, wrong-site surgery, improper anesthesia, and improper monitoring contribute to an estimated 98 000 deaths each year in US hospitals. Additionally, about 5% of all prescriptions are filled incorrectly. This alone results in about 7000 deaths a year in the United States.

The IOM committee recognized that simply calling on individuals to improve safety measures would be as misguided as blaming individuals for specific errors. Healthcare professionals have customarily viewed errors as a sign of an individual's incompetence or recklessness. As a result, rather than learning from such events and using information to improve safety and prevent new events, healthcare professionals have had difficulty admitting or even discussing adverse events or "near misses" for fear of professional censure, administrative blame, lawsuits, or personal feelings of shame (Leape, 2002). Acknowledging this, the IOM report put forth a four-part plan that applies to all who are, or will be, at the front lines of patient care: clinical administrators; regulating, accrediting, and licensing groups; boards of directors; industry and government agencies. It also suggested actions that patients and their families could take to improve safety. The committee emphasized the need to develop a new field of healthcare research, a new taxonomy of error, and new tools for addressing problems. It also understood that responsibility for taking action could not be borne by any single group or individual and had to be addressed by healthcare organizations and groups that influence regulation, payment, legal liability, education, and training, as well as patients and their families. It called on Congress to create a National Center for Patient Safety within the Agency for Healthcare Research and Quality to develop new tools and patient care systems that make it easier to do things correctly and more difficult to do them incorrectly.

Stelfox et al. (2006) assessed the impact of the IOM reports. They searched the medical literature database MEDLINE to identify English language articles on patient safety and medical errors published between November 1, 1994 and November 1, 2004 as well as US federal funding of patient safety research awards for the fiscal years 1995–2004. A total of 5514 articles on patient safety and medical errors were published during the ten-year study period. Table 8.1 presents the four InfoQ components for the IOM reports and for the Stelfox et al. (2006) study, as well as a potential study aimed at investigating the effect of the IOM reports on improving patient safety. The table highlights the differences between the three goals, datasets, analyses, and utilities.

Stelfox et al. (2006) found that the rate of patient safety publications increased from 59 to 164 articles per 100 000 MEDLINE publications following the release of the IOM report. Publications of original research increased from an average of 24–41 articles per 100 000 MEDLINE publications after the release of the report, while patient safety research awards increased from 5 to 141 awards per 100 000 federally funded biomedical research awards. The most frequent subject of patient safety publications before the IOM reports was malpractice (6% vs. 2%), while organizational culture was the most frequent subject (1% vs. 5%) after publication of the report. They concluded that publication of the report "To Err is Human" was associated with an increased number of patient safety publications and research awards.

Table 8.1   InfoQ components for IOM-related studies.

	IOM reports	Stelfox et al. (2006)	Potential study
Goal (*g*)	Quantify magnitude of patient safety rates	Did IOM reports raise awareness and stimulate research?	Did IOM reports improve patient safety?
Data (*X*)	Patient-level data at hospitals in the United States	Publications and research awards on patient safety and medical errors before/after IoM reports (1994–2004)	Patient-level hospital stay data before/after reports
Analysis (*f*)	Summary statistics	Compare before/after rates of publications	Compare before/after safety
Utility (*U*)	Accuracy	Size of improvement	Safety improvements, cost reduction in healthcare system

The IOM reports appear to have stimulated research and discussion about patient safety issues, but whether this will translate into safer patient care remains unknown. It appears that the assessment of InfoQ of the IOM reports is highly dependent on its original goals. In terms of stimulating research and increasing awareness, the InfoQ is high. Considering the goal of actually improving patient safety in US hospitals, the level of InfoQ, ten years later, seems less clear. The eight InfoQ dimensions provide a checklist of initiatives IOM could have taken or can take now to improve the quality of information generated by the IOM reports. These include temporal relevance, chronology of data and goal, data structure, and operationalization (see Table 8.2). Table 8.2 also provides subjective ratings for each of the InfoQ dimensions on a scale of 1 ("very poor") to 5 ("very good").

Achieving improvements in patient safety requires in-depth assessment with data resolution at the local hospital level, something not provided by IOM, and continuous monitoring capabilities for providing updated data. This gap, in the context of the IOM reports and follow-up activities, is affecting the level of InfoQ of these reports. Table 8.2 also presents ratings for the Stelfox et al. (2006) study on each of the InfoQ dimensions. The Stelfox analysis is descriptive in scope. No general models have been proposed. The size of the sample is, however, large enough to provide some statistical generalization.

8.3   Sant'Anna di Pisa report on the Tuscany healthcare system

The Laboratorio Management e Sanità of the Scuola Superiore Sant'Anna in Pisa has been monitoring 130 indicators for the regional health system of Tuscany since 2004 (Nuti et al., 2010). Tuscany is one of the 20 regions of Italy, with 12 local health authorities and five teaching hospitals. The reports on these indicators are one of the most comprehensive sources of data on healthcare systems available to the public.

Integrated with this data, a computer-assisted telephone interview-based survey was conducted to assess inpatient satisfaction, (see Murante et al., 2013). The questionnaire consisted of 28 questions covering the patient's relationship with doctors and nurses, communication process, information provided at discharge, and overall evaluation of care. Seven questions covered the patient's sociodemographic characteristics: age, gender, educational level, self-reported health status, employment, chronic diseases, and previous hospitalization. A stratified random sampling procedure was used to select participants for this study. The sampling frame was composed of inpatients discharged from Tuscan hospitals during the period September–December 2008. All 34 public general hospitals (excluding a pediatric hospital) were involved in the study. Patients hospitalized in medical, surgical, and obstetric–gynaecologic–pediatric wards were included, whereas newborn babies and patients treated in intensive care units or in a day hospital were excluded. When repeat admissions were recorded, only the last one was considered.

The goal of the survey was to explore determinants of public hospitals inpatients' satisfaction at the individual level and organizational level by applying risk adjusted

Table 8.2   InfoQ dimensions and ratings for Stelfox et al. (2006) data and for the IOM reports.

InfoQ dimension	Stelfox et al. (2006) study	InfoQ rating	IOM reports	InfoQ rating
Data resolution	Individual publications on patient safety and medical errors between 1994 and 2004 found by MEDLINE search. For each publication: subject and whether it won a research award	3	A breakthrough study that collected information on patient safety in a systematic way at high-level resolution	3
Data structure	Data based on reports derived from published articles. No cross validation with alternative data sources and no reference to data from social networks	3	Data organized on the basis of a range of reports	3
Data integration	No data integration, linking ongoing clinical data at the hospital or regional level with data published in journals on the basis of focused research initiatives	3	Data presented without integration on the basis of unique sources	2
Temporal relevance	Data available before and after IOM reports	5	At the time of publications, data was up to date	3
Generalizability	Analysis is descriptive in scope. No general models have been proposed. Sample size is large enough to provide some statistical generalization	4	Data presented descriptively without fitting models or running any data analysis that allows for generalization	3
Chronology of data and goal	Report not followed by ongoing analysis, so not updated	3	Data is now outdated for supporting current goals	3
Operationalization	Report operationalizes "safety" by listing specific corrective actions	5	Report conclusions have been the basis for many patient safety initiatives	3
Communication	Report has had general dissemination success	5	Report was disseminated in a print format publication	5

hierarchical models. The analysis focuses on the effect of hospital characteristics (such as self-discharges) on overall evaluations and across hospital variation in scores. Sociodemographic information, admission mode, place of residence, hospitalization ward, and continuity of care were found to be statistically significantly correlated with inpatient satisfaction. It was also observed that hospitals with a higher percentage of patients leaving against medical advice received lower scores. The latter result suggests that this measure provides a useful indicator of a hospital's inability to meet patient needs and a proxy indicator of patient dissatisfaction with hospital care. The four InfoQ components for this study are summarized in Table 8.3.

A total of 14 934 patients completed the phone interview, yielding a 61% response rate. In 2008, the 34 hospitals surveyed had on average about 360 beds (range 33–1645). With a score ranging from 0 to 100, patients rated hospital assistance and willingness to recommend a specific hospital on average, as 85 and 93, respectively. Satisfaction with communication had the highest score: patients received clear answers to their questions; adequate and concordant information was given; during consultation, their privacy was respected; and they were treated as individuals. The relationship with nurses was evaluated more positively than the relationship with doctors. The Tuscany study was used later as a benchmark for a study conducted in Israel (Kenett and Lavi, 2014).

In evaluating the InfoQ of the Tuscan patient satisfaction survey, we observe well-integrated data, combining qualitative and quantitative reports, in-depth analysis that permits operationalization of the findings, and high generalizability in terms of healthcare systems. In the related publication we found effective operationalization of the findings such as the actual introduction of an indicator tracking percentage of patients leaving against medical advice. Without that temporal relevance, chronology of data, and goal would be gradually affected by the growing elapsed time from the study. The eight InfoQ dimensions are summarized and rated in Table 8.4.

Table 8.3   InfoQ components for Sant'Anna di Pisa study.

Goal (g)	Explore determinants of public hospitals inpatients' satisfaction at the individual level and organizational level
Data (X)	130 indicators for Tuscany healthcare published by Scuola Superiore Sant'Anna and inpatient satisfaction survey data from 34 hospitals
Analysis (f)	Risk-adjusted hierarchical models
Utility (U)	Statistical significance of determinants

8.4   The haemodialysis case study

Physicians are interested in evaluating and forecasting adverse events that might provoke morbidity, mortality, or longer hospital stay for a patient; moreover, they want to quantify a patient's risk profile. The latter is the assessment of a patient's medical parameters by using probability distributions, given the patient's status and prior domain knowledge. Physicians typically summarize risk probability distributions by a percentile and decide on acceptability of risks by setting thresholds on these

Table 8.4   InfoQ dimensions and ratings on 5-point scale for Sant'Anna di Pisa study.

InfoQ dimension	Tuscany healthcare system	InfoQ rating
Data resolution	Inpatient-level data from survey, hospital-level data from healthcare indicators	5
Data structure	Combined data from reports and self-initiated surveys	4
Data integration	Data was not integrated in one study. Reports are separated for separate data sources	3
Temporal relevance	The system and the reports are updated yearly	5
Generalizability	Data was collected from five different hospitals. Some comparisons were made between the Tuscany region and other regions in Italy	4
Chronology of data and goal	Reports are annual and the goal is mostly explanatory and not predictive. From that perspective, the match between data analysis and goal is adequate	5
Operationalization	Patient satisfaction operationalized via combination of questions (patient's relationship with doctors and nurses, communication process, information provided at discharge, and overall evaluation of care)	4
Communication	Website provides adequate access with effective visualization, without considering statistical significance of outcomes	4

percentiles. Detection of unacceptable risks and the resulting risk mitigation analysis completes the risk management process.

At an upper level of a healthcare organization, economic losses and costs due to adverse events are evaluated, mainly to choose convenient forms of insurance schemes. Furthermore, for better governance it is useful to understand risk levels and how each risk contributes to economic losses.

When data is scarce, a good source of information is the experience of physicians. In such scenarios, Bayesian methodology can be used to estimate operational and clinical risk profiles. The approach described here involves healthcare of end-stage renal disease (ESRD). The following application is an example of a study that developed a decision support system designed to manage operational and clinical risks in healthcare environments. The main goal was to support nephrologists and risk managers who manage operational and clinical risk in healthcare. The four InfoQ components associated with this study are listed in Table 8.5.

Many statistical models applied to risk management estimate risks without consideration for decision making. Decision models must be considered in order to realize a fully integrated risk management process. The integration of risk estimation and decision making can be achieved with Bayesian networks (see also Chapter 7 on customer surveys).

Table 8.5   InfoQ components for the haemodialysis decision support system.

Goal (g)	Quantify patient-level risk profiles for improved medical treatment
Data (X)	1. Clinical and demographic data from $n = 10\,095$ monitored dialysis sessions collected from 47 patients over five years
	2. Expert opinions
Analysis (f)	Bayesian networks (BN)
Utility (U)	Mortality ratio, hospitalization ratio, and adherence to treatment. The BN was designed to minimize an acceptance measure Acc(P, M_l)

In this case study, a general definition of risk was used by analyzing the nonadherence of patients and healthcare providers to the treatment plan of nephrologists for ESRD.

The data for this study comes from patients without any residual renal function and with similar demographic characteristics (age, nationality, and race) from the Limited Assistance Centre of the hospital of Mede (PV), Italy. For more on this example and general risk management in healthcare, see Cornalba (2009) and Kenett (2012). The data was collected from a typical haemodialysis facility where nurses compose the permanent staff and three physicians supervise the dialysis sessions. Two data types were available: data collected from monitoring each patient's dialysis session and patient history data, such as age, gender, and number of days from the first dialysis session in the department. The sample size is $n = 10\,095$ monitored dialysis sessions, collected from 47 patients over a period of five years (from 2001 to 2006). Follow-up data was obtained at approximately three-month intervals. Among the large number of available medical parameters, only those affecting clinical and operational risks were selected. An appropriate set of risks was identified as follows:

- Clinical risks: hospitalization and mortality risks caused by bone disease, renal diet, and causes related to cardiovascular problems

- Operational risks: risks that involve both the process and the device

The study uses a Bayesian network, where the network structure is based on expert opinion information. In particular, results of experimental studies were used as known constraints for choosing prior probability distributions. The shape of prior probability distributions was subjectively introduced during model construction. The distribution with the greatest Shannon–Weaver entropy (Lee and Wright, 1994) was selected as a prior. Mixtures of conjugate priors were applied to determine some priors. These provide a sufficiently rich range of prior density "shapes" to enable a risk manager to approximate reasonably closely any particular prior belief function of interest. A mixture of beta densities was selected when expert information suggested that data was clustered around more than one mode. The model parameters were learned from the data. In the following subsections we provide more details about the network design, estimation, and deployment. For more on Bayesian networks see Kenett (2016). For software implementing Bayesian networks see bnlearn (2008).

8.4.1 Bayesian network design

The complex domain of ESRD is represented by the Bayesian network shown in Figure 8.1. The network includes $V = 34$ nodes denoted as $X_1, X_2, \ldots, X_{34}$. The number of vertexes results from a trade-off between a deeper description of dialysis and a

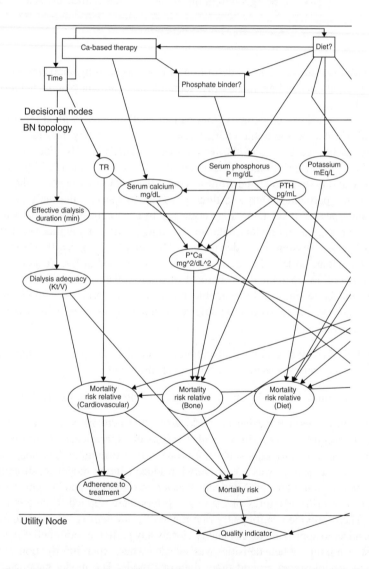

Figure 8.1   Bayesian network of patient haemodialysis treatment. Source: Cornalba (2009). Reproduced with permission from Springer.

super-exponential growth of possible networks. Each variable was classified into one group of causes, such as *Dialysis Quality Indexes* = {*Dialysis adequacy* (Kt/V), *PTH* pg/ml, *Serum albumin* g/dl} and *HD Department Performances* = {*Serum phosphorus PO4* mg/dl, *Potassium* mEq/l, *Serum calcium* mg/dl}.

Some dependence relationships among variables are defined on the basis of medical literature, such as Dialysis Outcome and Practice Pattern Study (DOPPS) (Kim et al., 2003), and guidelines of the DOQI Group (2003). DOPPS is one of the most important sources of international observational studies of haemodialysis practices

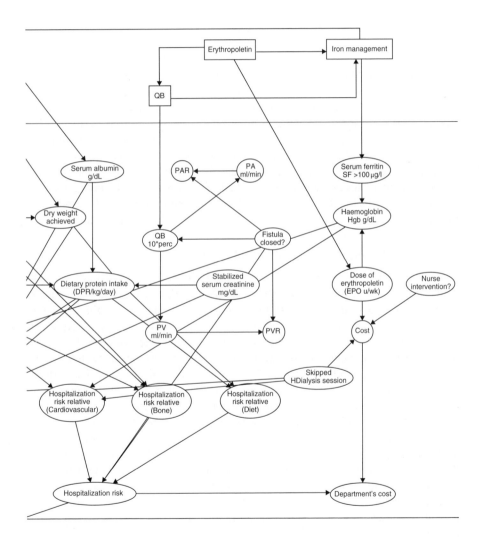

Figure 8.1   (Continued)

and outcomes based on countries with large populations of dialysis patients, such as France, Germany, Italy, Spain, Japan, the United Kingdom, and the United States.

Let M be a Bayesian network with vector of variables $\mathbf{X}$. For each variable X_k with parents $pa(X_k)$, we define $\mathrm{Sp}(X_k)$ to be the number of entries in $P(X_k \mid pa(X_k))$, and the size is $\mathrm{Size}(M) = \sum_{X_k \in X} \mathrm{Sp}(X_k)$. Let P denote the distribution over $\mathbf{X}$ taken from the sample of database of cases. Let M_i be the i-th candidate Bayesian network for P, and let P_i be the joint probability distribution determined by M_i.

A Euclidean distance was used to compare the two distributions:

$$\mathrm{Dist}_E\left(P, P_i\right) = \sum_{x \in X}\left(P(x) - P_i(x)\right)^2$$

The best Bayesian network minimizes the following acceptance measure: $\mathrm{Acc}(P, M_i) = \mathrm{Size}(M_i) + k\mathrm{Dist}(P, P_i)$, where k is a positive real number.

Given the network $V = \{X_1, \ldots, X_{34}\}$ and the set of directed edges E, four variables (serum calcium—Ca, parathyroid hormone—PTH, hemoglobin—Hgb, and serum ferritin—Fe) were left without fixing their causal links, since the available prior knowledge was compatible with different Bayesian network representations. Therefore, this part of the network was learned from the data on all available patients, using the model selection strategy described earlier. After the choice of the Bayesian network topology, batch learning was applied, thus determining the probabilities of the model and updating conditional probability distributions. Some of the probabilities can be also derived from the literature, as described in Cornalba (2009).

8.4.2   Deploying the Bayesian network

Given the set of dependences and data, it is possible to assess risk profiles for the j-th patient both by the conditional probability distribution $P(X \mid pa(X))$ and by the unconditional probability distribution $P(\mathbf{X})$. Moreover, working on marginal posterior distributions, it is possible to evaluate the single contribution of variable X_i for the j-th risk profile. Each prior distribution is assumed as a reference prior, that is, the probability distribution for an average patient. The latest is compared with the posterior probability distribution, which is learned from the patient's data. The distance between the distributions is measured using Euclidean distance, which is chosen for its symmetric property. Table 8.6 shows some marginal posterior distributions for the j-th patient's risk profile and related effects, given a decision and an action.

For example, during the first year, the patient has a *failure* risk profile better than the reference profile (99.41% vs. 95.68%), while during the first update (15 months), there is a worsening (94.13% vs. 95.68%) that has to be controlled by decision making. The action *add 30 minutes* restores the patient's risk profile to an acceptable level of risk (98.59%). Let X_k be the k-th variable which defines the risk profile for the j-th patient. The Euclidean distance between the j-th patient distribution and the reference distribution (R) is defined as

$$\mathrm{Dist}\left(P_J, P_R\right) = \sum_{x \in X_k}\left(P_J(x) - P_R(x)\right)^2$$

Table 8.6   Marginal posterior distributions for the j-th patient's risk profile (True = risk has materialized).

Variable state	Marginal posterior distribution in %			Decision
	Reference profile	First year	First update	
				Plus 30 minutes
False	95.68	99.41	94.13	98.59
True	4.32	0.59	5.87	1.41
		Hemoglobin, Hgb		Plus one dose of erythropoietin
0–9	22.55	8.56	21.66	20.11
9–10	31.22	50.61	24.09	31.71
10–11	19.08	31.17	51.82	20.03
11–12	13.01	4.58	1.15	13.56
12–13	8.67	3.2	0.82	8.77
13–14	3.3	1.13	0.28	3.49
...	...	...	...	...
		Erythropoietin dose		Plus one dose of erythropoietin
0–1 000	12.64	2.97	1.54	12.9
0–6 000	44.93	10.57	5.48	45.01
6 000–11 000	29.12	83.32	91.36	29.89
11 000–16 000	8.26	2.18	1.13	8.39
...	...	...	...	...

The score, defined by the distance, prioritizes variables on the basis of their importance to determining the risk profile. In this way nephrologists can cluster key medical variables for causes in a given future period. In this example, the most important adverse event is due to an incorrect dose of erythropoietin administered to the j-th patient so that resources must be better allocated in follow-up dialysis.

It is also possible to explore marginal posterior probability distributions of the target variables. For example, during the first update of both therapeutic protocol and data collection, the mortality risk of the j-th patient increases (the probability distribution shifts to the right). To restore the correct risk profile, the nephrologist can add a dose of erythropoietin.

8.4.3   Integrating the physician's expertise into the BN

To complete the risk management process, physicians must make a decision either on a patient's treatment or a device's substitution. This decision problem has been represented by an influence diagram (ID). The set of decisions D and the corresponding set of actions for each decision are the following:

d_1: "Time": keep or add 30 minutes to dialysis session.

d_2: "Ca-based therapy":

  *Treat hypercalcemia.

  *Continue current therapy.

  *Decrease vitamin D dose to achieve ideal Ca; decrease Ca-based phosphate binders.

  *Decrease or discontinue vitamin D dose/Ca-based phosphate binders; decrease Ca dialysate if still needed; assess trend in serum PTH as there may be low turnover.

d_3: "Phosphate binder":

  *Assess nutrition; discontinue phosphate binder if being used.

  *Begin dietary counseling and restrict dietary phosphate; start or increase phosphate binder therapy.

  *Begin short-term Al-based phosphate binder use, and then increase non-Al-based phosphate binder; begin dietary counseling and restrict dietary phosphate increase dialysis frequency.

d_4: "Diet": apply a "hypo" diet or keep his/her diet.

d_5: "QB": increase, keep, or decrease QB.

d_6: "Erythropoietin": keep, decrease, or increase (1 EPO) the current dose.

d_7: "Iron management": keep the treatment; iron prescription.

Nephrologists are provided an ordering of the decision nodes $(d_1 \ldots d_7)$. The shape of the loss function, $L(d, \theta)$, depends on individual views of risk: the risk attitude of nephrologists (or clinical governance) can be assumed as neutral, and a linear loss function is chosen. The quality indicator node summarizes $L(d, \theta)$ within the ID.

Mortality ratio, hospitalization ratio, and adherence to treatment are outcome measures that represent the weighted contribution to the loss function, defined on D and θ. For a distribution of these values in two patients, see Table 8.7. The risks for Patient 2 are higher than for Patient 5. Patient 5 requires a higher dose of erythropoietin. Analyzing the most important causes and the consequence of each action, it is possible to assess each scenario and prioritize actions that should be taken for the j-th patient. With such a decision support system, the risk manager can recommend the best treatment (Kenett, 2012).

Table 8.8 describes the eight InfoQ dimensions for this study, as well as ratings. The InfoQ of this healthcare example is relatively high, given its direct implementation as a decision support system in practice in ongoing patient treatment. This requires chronology of data and goal, operationalization, and communication. The application of Bayesian networks also has a positive effect in terms of data resolution, data structure, and data integration (see Kenett, 2016).

Table 8.7   Posterior distributions of outcome measures for two patients.

Pt. 2		Pt. 5	
Hospitalization Ratio		**Hospitalization Ratio**	
1.0E–3	0–1.5	1.0E–3	0–1.5
0.00	1.5–3	0.00	1.5–3
35.54	3–4.5	28.86	3–4.5
58.64	4.5–6	63.74	4.5–6
5.46	6–7.5	6.99	6–7.5
0.31	7.5–9	0.35	7.5–9
Mortality Ratio		**Mortality Ratio**	
4.82	0–3	9.71	0–3
43.74	3–6	30.65	3–6
16.19	6–9	19.54	6–9
21.02	9–12	14.02	9–12
5.43	12–15	11.59	12–15
3.95	15–18	8.91	15–18
2.38	18–21	4.10	18–21
1.23	21–24	0.87	21–24
Mortality risk relative (Ca.		**Mortality risk relative (Ca.**	
14.04	0–1	4.91	0–1
85.58	1–1.5	93.19	1–1.5
0.38	1.5–2	1.90	1.5–2
1.0E–3	2–2.5	1.0E–3	2–2.5
1.0E–3	2.5–3	1.0E–3	2.5–3
1.0E–3	3–3.5	1.0E–3	3–3.5
1.0E–3	3.5–8	1.0E–3	3.5–8
Mortality risk relative (B...		**Mortality risk relative (B...**	
18.00	0–1	76.83	0–1
73.68	1–1.5	19.20	1–1.5
0.59	1.5–2	0.14	1.5–2
Mortality risk relative (Diet		**Mortality risk relative (Diet**	
0.00	0–1	0.00	0–1
49.15	1–1.5	42.14	1–1.5
33.38	1.5–2	33.60	1.5–2
9.24	2–2.5	11.94	2–2.5
4.72	2.5–3	7.04	2.5–3
3.51	3–3.5	5.27	3–3.5
Dose of erythropoietin		**Dose of erythropoietin**	
12.63	0–1000	2.97	0–1000
44.93	1000–6000	10.57	1000–6000
29.12	6000–11000	83.32	6000–11000
9.25	11000–1600	2.18	11000–1600
3.95	16000–2100	0.93	16000–2100
0.09	21000–2600	0.02	21000–2600
0.02	26000–3100	0.00	26000–3100
0.01	31000–3600	0.00	31000–3600
0.00	36000–4100	4.89E–4	36000–4100

Source: Cornalba (2009). Reproduced with permission from Springer.

8.5   The Geriatric Medical Center case study

NataGer Medical Center, located near a midsize coastal town in central Israel, began in 1954 as a nursing home. The center was later converted into a recovery medical center for elderly patients. There are 365 beds in the center, with over 12 medical wards. Main services provided at NataGer are acute geriatrics, nursing and complex

Table 8.8   InfoQ dimensions and ratings on 5-point scale for haemodialysis study.

InfoQ dimension	The haemodialysis case study	InfoQ rating
Data resolution	Data is collected at a resolution needed for both tracking patient physiology and determining medical interventions	5
Data structure	Data is derived from a range of sources including dialysis sessions and patient demographics. There was no data on eating habits or irregular activity such as long rest or extra working hours	4
Data integration	The Bayesian network integrated expert knowledge (through priors and structure) with data-driven parameter estimation	5
Temporal relevance	Data collected in 2006 is still relevant since no major advances in ESRD have been implemented since	5
Generalizability	Analysis is generic and applies to any similar clinical service	5
Chronology of data and goal	The method provides the treating physician with an online tool for treating patients	5
Operationalization	Several tools provide the capability to operationalize the approach effectively and efficiently	5
Communication	The tool displays distribution values to physicians. Some physicians might prefer a binary dashboard triggering specific actions, when needed	4

nursing care, rehabilitation, long-term ventilation care, oncology–hospice care, and care for mentally frail. NataGer is comanaged by a chief medical officer (CMO) and a chief executive officer (CEO).

NataGer's patient safety manager was looking for opportunities to engage employees in patient safety issues through collaboration and data analysis. As a result, an intervention was designed in the context of a research project on application of integrated models in healthcare systems. After two meetings of the patient safety manager with hospital management (including the chief nursing officer (CNO)), two goals were chosen for the intervention process:

1. Patient fall reduction—Reducing patients' falls by 10%, without increasing medicinal treatment or physical constraints

2. Bedsores reduction—Reducing new occurrence of bedsores in addition to establishing a methodology for bedsores reporting and measurement

The patient safety manager was appointed to lead both projects with a professional mentoring assistance. Team members were chosen by management and included doctors, RNs, nurse aides, a pharmacist, a nutritionist, and a patient safety manager. Both teams were trained on tools and process improvement methodology. Both teams analyzed the available data, performed brainstorming, used analytic tools for prioritization of tasks and concept, performed a fishbone analysis, and were trained on improvement tools (Kenett and Lavi, 2014).

After completing the definition of a "problem statement," the teams presented their work plan to a steering committee consisting of the CMO, CEO, and CNO. A biweekly two hours meeting was scheduled for each team to ensure ongoing progress. Teams experienced some difficulties, such as uncooperative team members and personal relationship issues, as well as heavy workloads, which made leaving their work for the meetings problematic. The CMO and CNO were updated frequently on the progress of the projects and provided assistance. The teams met the steering committee again to present their findings and suggest improvement solutions. As a result of this meeting, the falls reduction team finalized a visual display documenting characteristics of the patient, to enable the hospital team to become more efficient and effective in handling patients (see Figure 8.2).

The second team began improvements in patient handling activity in order to reduce wasted time and provide time for actions designed to reduce bedsores occurrence. After six months, both projects reached the "Improve" phase in the Define–Measure–Analyze–Improve–Control (DMAIC) Six Sigma roadmap with significant improvements.

The bedsores team began by developing a solution to lost time of RNs during their shifts, due to walking distances, lost equipment, supplies, standardized work processes, and so on. Their goal was to help staff adopt position changes as a standard procedure, thereby preventing new bedsores in their patients. The team started the process with complaints of shortage of suitable equipment for bedsores prevention. Initially, exploring alternative directions was not considered. However, thanks to a strong team leader and professional mentoring, the team discussed possible causes for bedsores occurrence and rated the causes using an analytical tool presented in Figure 8.3. Each team member listed the causes and rated them according to ability to change, required investment, and potential outcome. Figure 8.3 presents the results of the rating process.

As a result of this analysis, the team learned that several causes must be addressed in their work, beyond special prevention equipment which requires a substantial investment. This allowed the team to make progress on issues under their influence such as placing patients in the right position, scheduled position changes for patients, replacement of wet clothing, and so on. The team performed observations and time measurements which were analyzed and used as a baseline for determining standard work specifications and full-time equivalent (FTE) allocation of employees. Bedsores data is now being collected regularly by a dedicated RN, and ongoing training of bedsores reporting and prevention is being held. In this healthcare system, management was fully engaged during the process, and more projects are scheduled to be launched. The four InfoQ components for each of the two projects are summarized in Table 8.9. The InfoQ of the data analysis performed in both the falls reduction and bedsores teams was considerable. Table 8.10 summarizes the eight InfoQ dimensions and scores.

Both project teams were able to collect relevant data, in a timely manner, and turn it into actionable interventions which proved effective. However, they were not able

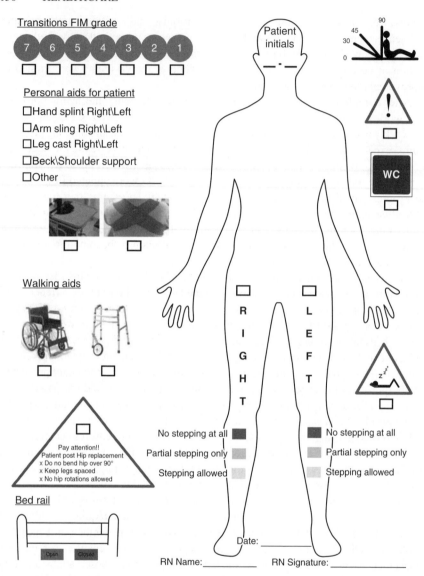

Figure 8.2   Visual board display designed to help reduce patients' falls. Source: Kenett and Lavi (2014). Courtesy of Yifat Lavi.

to correlate the reported data with other sources such as the use of pharmaceutical supplies or information from healthcare monitoring systems. While the data analysis in these projects was simple from a statistical perspective, the main sources of high InfoQ were the focused efforts to identify causes (goal definition), the clear definition of a utility, and a commitment to collect the right data and solve practical challenges in a timely manner.

H = High M = Medium L = Low	Scoring Table									
Topic	Ability to Change			Required Investment			Potential for Value Achieved			Score
	L	M	H	L	M	H	L	M	H	
Patient placed in bed in a wrong position	0	0	7	5	2	0	0	0	7	95
Patient is sitted in a wrong position	0	1	6	5	2	0	0	0	7	93
Patient is wearing wet clothes	0	0	7	5	1	1	0	0	7	93
Wrong treatment protocol for patient	0	0	7	2	5	0	0	0	7	88
Wet bed sheets	0	0	7	3	3	1	0	0	7	88
Bed sheets are not placed properly on bed	0	1	6	3	4	0	0	0	7	88
Loose clothing items	1	1	5	4	2	1	0	0	7	83
Protocol is not aligned with working process	0	1	6	0	6	1	0	0	7	79
Late instructions written in patients' protocol	0	1	6	2	4	1	0	2	5	79
Transfusion branoulli presures skin	1	1	4	6	0	1	0	1	6	79
Tight dressing	0	2	5	2	3	2	0	0	7	79
Tight cloths	1	1	5	4	2	1	1	1	5	76
Staff lacks awareness of risk factors	0	1	6	0	5	2	0	0	7	76
Old and unupdated patients' protocols	0	1	6	0	4	3	0	0	7	74
Reevaluation of patients' condition not performed	0	1	6	2	3	2	0	3	4	74

Figure 8.3   Prioritization tool for potential causes for bedsores occurrence.

Table 8.9   InfoQ components for the two NataGer projects data.

	Patient falls study	Bedsores reduction study
Goal (*g*)	Reduce number of patient falls by 10% without increasing chemical or physical constraints	Reduce new occurrence of bedsores; establish a methodology for bedsores reporting and measuring
Data (*X*)	Data on patient falls including location and time. Data on number of patients by wards	Data on patients with bedsores. Data on number of patients by wards
Analysis (*f*)	Control charts of falls per patient days	Control charts of bedsores per patient days
Utility (*U*)	Chronic level of patient falls (less is better)	Chronic level of bedsores (less is better)

Table 8.10   InfoQ dimensions and ratings on 5-point scale for the two NataGer projects.

InfoQ dimension	Patient falls study	InfoQ rating	Bedsores reduction study	InfoQ rating
Data resolution	Monthly data adequate for low-scale monitoring	3	Monthly data adequate for low-scale monitoring	3
Data structure	Self reports and accident reports	4	Routine reports by nurses	4
Data integration	Location and time of event is integrated in database	4	Data is available by patient, and additional work is needed to enter demographic and treatment data in database	3
Temporal relevance	Data is relevant on a monthly basis	3	Data is relevant on a monthly basis	3
Generalizability	Lessons learned can be generalized	4	Because of the poor data, integration generalization is difficult	2
Chronology of data and goal	For the goal of overall improvement, the analysis is adequate	5	For the goal of overall improvement, the analysis is adequate	5
Operationalization	Improve proactive activities designed to reduce the risks of falls	4	Operationalized RNs lost time in shifts: walking distances, lost equipment, supplies, and standardized work processes	5
Communication	Visual board display	5	Prioritization tool	3

8.6   Report of cancer incidence cluster

This case study is about a report by Rottenberg et al. (2013) on cancer incidence clusters in a specific industrial subdistrict geographical area. In presenting this case, we refer to a comprehensive discussion by Walter (2015) on analyzing and reporting such incidences in an unrelated geographical area. The paper by Walter is structured to address three conceptual dimensions:

1. Statistical aspects—Are the identified clusters statistically unusual, based on an underlying model of "no clustering?"

2. Epidemiological aspects—Do the observed patterns suggest the presence of unknown or suspected sources of increased risk, and, if so, do these patterns offer insight on potential causes of disease?

3. Public policy aspects (given the observed data and analytic results, is policy action called for?).

After introducing the case we will review it from an InfoQ perspective by applying some of the elements in the discussion in Walter (2015).

A cancer cluster is the occurrence of a greater than expected number of cancer cases among a group of people in a defined geographic area over a specific time period. A cancer cluster may be suspected when people report that several family members, friends, neighbors, or coworkers have been diagnosed with the same or related types of cancer. Cancer clusters can help scientists identify cancer-causing substances in the environment. However, most suspected cancer clusters turn out, on detailed investigation, not to be true cancer clusters. That is, no cause can be identified, and the clustering of cases appears to be a random occurrence. To assess a suspected cancer cluster accurately, investigators must determine whether the type of cancer involved is a primary cancer (a cancer that is located in the original organ or tissue where the cancer began) or a cancer that has metastasized (spread) to another site in the body from the original tissue or organ (a secondary cancer). Investigators consider only primary cancer when they investigate a suspected cancer cluster. A confirmed cancer cluster is more likely if it involves one type of cancer than if it involves multiple cancer types. This is because most carcinogens in the environment cause only a specific cancer type rather than causing cancer in general.

Many reported clusters include too few cancer cases for investigators to determine whether the number of cancer cases is statistically significantly greater than the expected number under chance occurrence and to confirm the existence of a cluster. Investigators must show that the number of cancer cases in the cluster is statistically significantly greater than the number of cancer cases expected given the age, sex, and racial distribution of the group of people who developed the disease. If the difference between the actual and expected number of cancer cases is statistically significant, the finding is unlikely to be the result of chance alone. However, it is important to keep in mind that even a statistically significant difference between actual and expected numbers of cases can arise by chance (a false positive). An important challenge in confirming a cancer cluster is accurately defining the group of people considered potentially at risk of developing the specific cancer (typically the total number of people who live in a specific geographic area). When defining a cancer cluster, there is a tendency to expand the geographic borders, as additional cases of the suspected disease are discovered. However, if investigators define the borders of a cluster based on where they find cancer cases, they may alarm people about cancers not related to the suspected cluster. Instead, investigators first define the "at-risk" population and geographic area and then identify cancer cases within those parameters. A confirmed cancer cluster may not be the result of any single external cause or hazard (also called an exposure). A cancer cluster could be the result of chance, an error in the calculation of the expected number of cancer cases, differences in how cancer cases were classified or a known cause of cancer such as smoking. Even if a cluster is confirmed, it can be very difficult to identify the cause. People move in and out of a geographic area over time, which can make it difficult for investigators to identify hazards or potential carcinogens to which they may have been exposed and to obtain medical records to confirm the diagnosis of cancer. Also, it typically takes a long time for cancer to develop, and any relevant exposure may have occurred in the past or in a different geographic area from where the cancer was diagnosed. For more on identifying and reporting cancer clusters in the United States, see www. cancer.gov/cancertopics/factsheet/Risk/clusters.

The Rottenberg et al. (2013) study is based on a retrospective cohort study using baseline measurements from the Israel Central Bureau of Statistics 1995 census on residents of the Haifa subdistrict, which houses major industrial facilities in Israel. The census database was linked with the Israel Cancer Registry (ICR) for cancer data. Smoking prevalence data was obtained from the Central Bureau of Statistics 1996/1997 and 1999/2000 health surveys. The objective of the study was to assess the association between organ-specific cancer incidence and living in the specific industrial subdistrict compared to other areas in Israel, after controlling for sociodemographic variables. The data used in the study consisted of a total of 175 704 persons reported as living in the industrial subdistrict, with a total of 8034 reported cancer cases. This cohort was compared to a random sample of the same size from the general population. The census database was linked (using the personal identification number, names, and other demographical data) with the ICR for cancer data, including date of cancer diagnoses up to December 2007 and International Classification of Diseases for Oncology (ICD-O)—Version 3 topography and morphology. The ICR receives compulsory notification from numerous sources of data, including pathologic reports, discharge summaries, and death certificates, and has been archiving cancer cases since 1982. Completeness of the registry was found to be about 95% for solid tumors. The entire cohort's data was merged with the Israeli Population Registry in order to obtain dates and causes of death. Smoking prevalence data was obtained from the CBS 1996/1997 and 1999/2000 health surveys. This survey was conducted by telephone interview and included nearly 29 000 people of ages 15 years and above. The survey addressed a wide range of topics including demographic, socioeconomic, and health variables.

Specific questions addressing smoking included the following: Do you smoke? At what age did you start smoking? How many cigarettes do you smoke a day? A sample size of 80 000 people was calculated according to the life expectancy of the study population and the expected cancer incidence rates in men and women designed in order to detect a hazard ratio (HR) of 1.15 with 90% power and a significance level of 0.05 when comparing the incidence of cancer among people living in the industrial subdistrict to that of people living in the rest of the country. The analysis included fitting a multivariate Cox proportional hazards model controlling for age, gender, ethnicity, continent of birth, education years, and income per person. Analysis of the 1996/1997 and 1999/2000 CBS health surveys demonstrates that in the industrial subdistrict, 21.5% and 21.8% people smoked versus 23.3% and 23.2% in the rest of Israel ($p = 0.068$ and $p = 0.192$, respectively, using a chi-square test). The study concluded that the industrial subdistrict, after various adjustments, experienced a significantly increased cancer risk (HR = 1.16, 95% CI: 1.11–1.21, $p < 0.001$) with specific increases in lung, head and neck, colorectal, gastric and esophagus, bladder, and cervical cancers. The report of a significant difference in incidence of cancer in the 0–14 age group, with 0.4% in the industrial subdistrict versus 0.2% in the rest of the country ($p = 0.009$), was particularly alarming and led to strong public responses. The InfoQ components of this use case are summarized in Table 8.11.

In evaluating the InfoQ of the cancer incidence study, we apply some of the principles described in Walter (2015). Walter warns the reader that a high number of observed cases in an area with a high number of individuals at risk may not indicate a local increase in individual risk, rather simply a larger local number at risk (e.g., many cancer cases at a retirement community need not suggest pollution, since the elderly are more prone to cancer). As mentioned in the beginning of this section, most statistical methods of cluster detection define the number of local cases expected in the absence of clustering based on the size and composition of the local at-risk population with respect to known risk factors (e.g., age, race, and sex of the at-risk population) and seek to detect local aggregations in space and time that are above and beyond those anticipated by local aggregations of such risk factors. In contrast to statistical analysis, the general motivation in epidemiology is

Table 8.11   InfoQ components of cancer incidence report.

Goal (g)	Assess the association between organ-specific cancer incidence and living in industrial subdistrict
Data (X)	175 704 persons reported as living in the industrial subdistrict, with a total of 8034 cancer cases
Analysis (f)	Hazard ratio to develop cancer comparing industrial subdistrict compared to the rest of the country after adjusting for age, gender, ethnic group, and continent of birth
Utility (U)	Statistical significance of organ-specific cancer incidence in industrial subdistrict

the accurate and reliable identification and quantification of risk factors associated with the onset, progression, treatment, and prevention of disease. With respect to cluster studies, the motivating epidemiologic question expands on the statistical question earlier: Does the observed pattern of disease reveal new insights regarding potential causes of local increases in disease risk? While both the statistical and epidemiologic questions are based on the observed pattern of cases, the epidemiologic question not only assesses whether the observed pattern is unusual (with respect to known risk factors) but also whether any observed anomalies reveal specific epidemiologic insight into local factors driving risk. That is, does a statistical excess of cases necessarily suggest a true underlying local increase in risk due to some local factor? This question is often specified even further to focus on potential local environmental exposures, that is, does the observed pattern of disease reveal new insights regarding potential local environmental exposures

Table 8.12   InfoQ dimensions and ratings of cancer incidence study by Rottenberg et al. (2013).

InfoQ dimension	Cancer incidence study	InfoQ rating
Data resolution	Data was collected at the individual level including date of cancer, topography, and morphology. Smoking statistics were available only at the subdistrict level	4
Data structure	Combined data from the Israel Cancer Registry (ICR), the Israeli census, and the national health survey	4
Data integration	Data was not integrated in one study by matching personal identification number and other demographical data. Smoking data was considered separately	3
Temporal relevance	Some data used precedes the publication date by over ten years	2
Generalizability	Some comparisons were made to similar studies in the United States	2
Chronology of data and goal	For policy makers, chronology of data and goal is a critical dimension. The lag in publication has a negative effect on information quality of the study	3
Operationalization	Study raised significant concern, especially with regard to cancer incidence in the 0–14 age group. Some concerns voiced over the statistical analysis of the study caused policy makers to take a cautionary view of the study	3
Communication	Study was amply quoted by the local press with many misinterpretations of the results	2

Table 8.13 Scoring of InfoQ dimensions for each of the four healthcare cases studies.

InfoQ dimension/use case	8.2 IOM	8.2 Stelfox	8.3 hospitals	8.4 ESRD	8.5 falls	8.5 bedsores	8.6 cancer
Data resolution	3	3	5	5	3	3	4
Data structure	3	3	4	4	4	4	4
Data integration	2	3	3	5	4	3	3
Temporal relevance	3	5	5	5	3	3	2
Chronology of data and goal	3	4	4	5	4	2	2
Generalizability	3	3	5	5	5	5	3
Operationalization	3	5	4	5	4	5	3
Communication	5	5	4	4	5	3	2
Use case score	**50**	**68**	**79**	**93**	**73**	**57**	**43**

Bold figures represent aggregated values.

causing local increases in disease risk? Epidemiology works best when one has a clearly defined population at risk with well-measured exposures, demographics, and disease outcomes. The use of statistical significance (p-values) is often a controversial topic in the epidemiologic literature, and accurately assessing significance of a suspected cluster is a challenging statistical issue. Some epidemiologic critics also question the logic of searching for potential causes of disease after identifying an area of higher than expected incidence, pointing out that some areas will have the highest local risk and it can be difficult to determine whether the highest rate is unusually high due to some local cause of increased risk. These critics point out that focusing only on the highest local rates for epidemiologic follow-up studies can lead to bias. Taking things to the next level, the motivating policy question is, does the observed pattern of disease suggest necessary or recommended policy responses? Potential policy responses may involve more detailed follow-up, for example, additional data collection, exposure surveys, or in-depth investigation of case histories within the cluster. For more on these considerations, see Walter (2015) and references therein.

In Table 8.12 we review the study by Rottenberg et al. (2013) from an InfoQ perspective, accounting for the comments and questions posed by Walter (2015) in similar epidemiologic studies. The eight InfoQ dimensions are summarized and rated in Table 8.12.

8.7   Summary

The chapter reviewed seven case studies of applied research in the context of healthcare systems. The first three examples relate to national or regional initiatives. The next three examples are more local in that they treat a specific type of patient and a particular medical center. The last example is about the important topic of cancer incidence reports. Table 8.13 presents the InfoQ assessment of the seven healthcare case studies on each of the eight InfoQ dimensions using ratings on a scale of 1 ("very poor") to 5 ("very good"). This assessment is subjective, based on consultations with people involved in the projects and studies.

From Table 8.13 we see that InfoQ scores ranged from 43% to 93% with the haemodialysis decision support system scoring highest (93%) and the cancer study scoring lowest (43%). Addressing the challenges of healthcare systems is uniquely complex and requires addressing individual, professional, technological, ethical, and organizational issues. Setting goals is particularly critical in healthcare applications. For an example about the treatment of low back pain, see Gardner et al. (2015). Providing high InfoQ in such studies should be considered a key objective.

References

bnlearn Package (2008) http://www.bnlearn.com/ (accessed May 5, 2016).

Cornalba, C. (2009) Clinical and operational risk: a Bayesian approach. *Methodology and Computing in Applied Probability*, 11, pp. 47–63.

DOQI Group (2003) K/DOQI clinical practice guidelines for bone metabolism and disease in chronic kidney disease. *American Journal of Kidney Disease*, 42(3), pp. 1–201.

Gardner, T., Refshauge, K., McAuley, J., Goodall, S., Hubscher, M. and Smith, L. (2015) Patient led goal setting in chronic low back pain—what goals are important to the patient and are they aligned to what we measure? *Patient Education and Counseling*, 98(8), pp. 1035–1038.

Gigerenzer, G., Gaissmaier, W., Kurz-Milcke, E., Schwartz, L.M. and Woloshin, S. (2007) Helping doctors and patients to make sense of health statistics. *Psychology Science in the Public Interest*, 8, pp. 53–96.

Institute of Medicine (IOM). (1999) *To Err Is Human. Building a Safer Health System*, in Kohn, L.T., Corrigan, J.M. and Donaldson, M.S. (editors), National Academy Press, Washington, DC.

Institute of Medicine (IOM). (2001) *Crossing the Quality Chasm: A New Health System for the 21st Century*, in Corrigan, J.M., Donaldson, M.S. and Kohn, L.T. (editors), National Academy Press, Washington, DC.

Kenett, R.S. (2012) Risk Management in Drug Manufacturing and Healthcare, in *Statistical Methods in Healthcare*, Faltin, F., Kenett, R.S. and Ruggeri, F. (editors), John Wiley & Sons, Ltd, Chichester, UK.

Kenett, R.S. (2016) On generating high InfoQ with Bayesian. *Quality Technology and Quantitative Management*, http://www.tandfonline.com/doi/abs/10.1080/16843703.2016.1 189182?journalCode=ttqm20 (accessed May 5, 2016).

Kenett, R.S. and Lavi, Y. (2014) Integrated management principles and their application to healthcare systems. *Sinergie*, 93(1), pp. 213–239.

Kim, J., Pisoni, R.L., Danese, M.D., Satayathum, S., Klassen, P. and Young, E.W. (2003) Achievement of proposed NKF-K/DOQI bone metabolism and disease guidelines: results from the dialysis outcomes and practice patterns study (DOPPS). *Journal of the American Society of Nephrology*, 14(25), pp. 269–270.

Leape, L.L. (2002) Reporting of adverse events. *New England Journal of Medicine*, 347(20), 1633–1638.

Lee, R.C. and Wright, W.E. (1994) Development of human exposure-factor distributions using maximum entropy inference. *Journal of Exposure Analysis and Environmental Epidemiology*, 4, pp. 329–341.

Murante, A., Seghieri, C., Brown, A. and Nuti, S. (2013) How do hospitalization experience and institutional characteristics influence inpatient satisfaction? A multilevel approach. *International Journal of Health Planning and Management*, 9(3), pp. 247–260.

Nuti, S., Vainieri, M. and Bonini, A. (2010) Disinvestment for reallocation: a process to identify priorities in healthcare. *Health Policy*, 95(2–3), pp. 137–143.

Rottenberg, Y., Zick, A., Barchana, M. and Peretz, T. (2013) Organ specific cancer incidence in an industrial sub-district: a population-based study with 12 years follow-up. *American Journal of Cancer Epidemiology and Prevention*, 1, pp. 13–22.

Stelfox, H., Palmisani, S., Scurlock, C., Orav, E. and Bates, D.W. (2006) The "To Err is Human" report and the patient safety literature. *Quality & Safety in Health Care*, 15(3), pp. 174–178.

Walter, L. (2015) Discussion: statistical cluster detection, epidemiologic interpretation, and public health policy. *Statistics and Public Policy*, 2(1), pp. 1–8.

9

Risk management

9.1 Introduction

Risk management has long been pursued by services, business, and industry; indeed, insurance companies and banks' activities are based on its successful application. Risk management is currently enjoying enhanced attention and renewed prominence due to the belief that nowadays we can do a better job of it. This perception is based on phenomenal developments in the area of data processing and data analysis. The challenge is to turn "data" into information, knowledge, and insights (Kenett, 2008; Kenett and Raanan, 2010). This chapter is about meeting this challenge by evaluating how risk management analytics increase information quality (InfoQ). The growing interest in risk management is driven partly by regulations on corporate governance in the private and public sectors. However, much of risk management is still based on informal scoring using subjective judgment. Modern risk management systems incorporate several activities including:

System mapping. Developing and revisiting organizational charts and process maps are standard practice in risk assessments. A more comprehensive approach is the application of business process modeling to map actors, roles, activities, and goals (Kenett, 2013).

Risk identification. Generating a list of main risk events classified by area and subarea. Risk identification does not stop after the first analysis but rather is a continuous process.

Risk measurement and risk scoring. Assessing the loss or impact of a risk event and its probability.

Information Quality: The Potential of Data and Analytics to Generate Knowledge,
First Edition. Ron S. Kenett and Galit Shmueli.
© 2017 John Wiley & Sons, Ltd. Published 2017 by John Wiley & Sons, Ltd.
Companion website: www.wiley.com/go/information_quality

Risk prioritization. Prioritization is done on the basis of the risk scores. A one-dimensional score might be too general; consequently, one often considers several dimensions and multiple objectives. Technically, this process allows to identify the main *risk drivers.*

Decision. Making decisions about mitigating actions to risk events. Any risk management system is basically a *decision support system.* Usually, decisions are not made automatically (exceptions include automatic trading in finance). In addressing risks, one needs to take decisions and try to keep a record of the decisions taken.

Action. Carrying out an action to address or prevent risks. An action is not the same as a decision. Actions take place following a decision. It is critical to record the actual action.

Risk control. The ongoing process of monitoring the effect of the actions, including rescoring the risks, introducing new risks (more prioritization), removing old risks from the list and making new or repeat decisions and actions. This process of feedback and iteration should be a continuous process.

Additional general risk management components include:

Risk champions. Facilitators who have ownership of the process or individual areas of risk.

Stakeholders. Employees, customers, experts, partners, shareholders, patients, students, the general public, special subgroups of the public (e.g., old, young, and handicapped), the government, regulatory agencies, etc. It is important to draw all the stakeholders into the risk management process.

Experts. Key personnel in risk management. However, one should not solely rely on the experts.

Senior management commitment. As in the case of quality, implementing risk management soundly requires champions at the board of director's level, not just a back office risk function. Ideally, risk is a regular item on board meeting agendas.

Risk communication. Communicating occurrence or prediction of risk events to all stakeholders.

Practical control. Limitations on actions, both legal and physical, for example, one cannot prevent a hurricane.

The potential of data-driven knowledge generation is endless when we consider both the increase in computational power and the decrease in computing costs. When combined with essentially inexhaustible and fast electronic storage capacity, it seems that our ability to solve the intricate problems of risk management has increased by several orders of magnitude. As a result, the position of chief risk officer (CRO) is gaining popularity in today's business world. Especially following the 2008 collapse in the financial markets, the idea that risk must be better managed than it had been in the past is now widely accepted. We suggest that providing insights from data analysis with high InfoQ is a critical competency of modern risk managers. Section 9.2 provides several examples. For more examples see NASA (2002) and Ruggeri et al. (2007).

9.2   Financial engineering, risk management, and Taleb's quadrant

Mathematical economics and financial engineering provide complementary aspects to risk management. They consist of discussions on complete/incomplete markets and rational expectation models and include *equilibrium theory*, *optimization theory*, *game theory*, and several other disciplines. The concept of a complete market has various technical definitions. In general, it means that every player in the market has the same information and that new information reaches every player at the same time and is instantaneously incorporated into prices, so that no player can get an advantage from an information point of view (as in insider trading). If the market is not in equilibrium, then there is opportunity for *arbitrage*, a something-for-nothing profit just by taking a buy or sell action. Rational expectations assume that an efficient market is in equilibrium (because if not efficient, some players would have an advantage and will be able to drive the market in a particular direction). Under rational expectation, one behaves according to the best prediction of the future market equilibrium point. Note that this is quite close to Bayes' rule which states that the conditional expectation $\theta_{est} = E(\theta|X)$ is the best rule for estimation of θ under quadratic loss:

$$L = \left(\theta_{est} - \theta\right)^2.$$

Rational expectation implies that the conditional expectation under all available (past) information should be the basis for (rational) decisions. These economic principles represent the goal of minimizing some risk metrics. Very roughly, if we like a particular risk metric and we fail to minimize it, there is a better rule than the one we are using. Even worse, someone else might be able to take advantage of us. Some modern economic theories are concerned with operating with *partial information*. Mathematical economics is an extension of the decision theory discussed earlier, with specific decisions taken at certain times. Typical actions are *buy*, *sell*, *swap*, and *price* (the act of setting a price). For more on these topics, see Tapiero (2004). For more on loss functions and mathematical models see Raiffa (1997) and Kenett and Zacks (2014).

The freedom to make a decision is called an *option*. Note the connection with risk management, where we may or may not have this freedom (we do not have the freedom to stop a hurricane, but we may have some prevention or mitigation options for the effect of the hurricane). The key point is that the option will have a time element: a freedom, for example, to buy or sell at or before a certain date and perhaps at a certain price. A *real option* is where the option is investment in a "real" object like a factory or technology. When time elapses, certain things happen: the value of assets may change or we may have to pay interest. The things that happen are usually random (stochastic) and we typically have no control over their occurrence. We do have control over the choice of what to buy or sell and the time at which to do it. Once we buy or receive an option, we are in the grip of the market until we sell. A whole body of work in economics and portfolio theory is connected to mean-risk or mean-variance theory. When dealing with operational issues, aspects of operational risks are evaluated using similar techniques, including value at risk (VaR) and related scoring methods. These techniques are beyond the scope of this introductory section. For more on operational risks, see Kenett and Raanan (2010).

In 2007, Nassim Taleb published a book that predicted the 2008 economic meltdown (Taleb, 2007). Taleb described the meltdown and similar major impact events, such as the 9/11 terrorist attack, the Katrina hurricane, and the rise of Google as "black swans." A black swan is a highly improbable event with three principal characteristics: it is unpredictable, it carries a massive impact and after the fact, we concoct an explanation that makes it appear less random, and more predictable, than it was. The Western world did not know that black swans existed before 1697 when they were first spotted in Australia. Up until then, swans were thought to be white. Why do we not acknowledge the phenomenon of black swans until they occur? Part of the answer, according to Taleb, is that humans are hardwired to learn specifics when they should be focused on generalities. We concentrate on things we already know and fail to take into consideration what we don't know. We are, therefore, unable to truly estimate opportunities, too vulnerable to the impulse to simplify and categorize and not open enough to reward those who can imagine the "impossible." Taleb has studied how we fool ourselves into thinking that we know more than we actually do. He claims that we focus our thinking on the irrelevant and inconsequential while large events continue to surprise us and shape our world. A black swan is a consequence of poor InfoQ (low InfoQ). In explaining the occurrence of black swans, Taleb refers indirectly to the generalization dimension of InfoQ.

Taleb proposed a mapping of randomness and decision making into a quadrant with two classes of randomness and two types of decisions (Taleb, 2009). Decisions referred to as "simple" or "binary" lead to data-driven answers such as "very true" or "very false" or "a product is fit for use or defective." In these cases, statements of the type "true" or "false" can be stated with confidence intervals and P-values. A second type of decisions is more complex, emphasizing both its likelihood of occurrence and its consequences. The dimensions of the Taleb quadrant characterize randomness. A first classification category is based on "forecastable events," implied by uncertainty described with finite variance (and thus from thin-tail probability distributions). A second category relates to "unforecastable events," defined by probability distributions with fat tails. In the first category, exceptions occur without significant consequences since they are predictable. The traditional random walk, converging to a Gaussian–Poisson process, provides such an example. In the second category, large consequential events are experienced but are also more difficult to predict. Fractals and infinite variance models provide such examples, (see Tapiero, 2009; Kenett and Tapiero, 2010). These are conditions with large vulnerability to black swans in the context of robust, fragile, and antifragile systems (Taleb, 2008, 2009, 2012). The Taleb analysis points out the need to consider operationalization of the data analysis outcomes. Taleb's work is partially related to InfoQ.

We consider the four InfoQ components of the Taleb quadrant approach in a generic sense and not in the context of a specific implementation:

Goal (g): Assess the combination of decision making with the type of data available. In this sense, the Taleb quadrant provides an example of decision theory discussed in Chapter 2.

Data (X): The data considered in the Taleb quadrant consists of a distribution with "fat tails" with finite variance.

Analysis (f): The analysis implied by the Taleb quadrant maps the type of decision to one of the four possible combinations along two dimensions: (i) binary, true/false decisions versus complex decisions and (ii) assessing the impact of uncertainty as related to either predictable (forecastable) events or unpredictable (unforecastable) events.

Utility (U): The utility considered by Taleb involves risk functions in terms of types of decisions and classification of uncertainty as described earlier.

9.3   Risk management of OSS

We present another methodological approach to risk management with a particular generic application to the adoption of open-source software (OSS). OSS is increasingly playing a leading role in information technology (IT) practices. Its pervasive adoption is not without risks for an industry that has experienced significant failures in product quality, timelines, and delivery costs. Inadequate risk management has been identified among the top deficiencies when implementing OSS-based solutions (Franch et al., 2013). A crucial aspect in managing and mitigating OSS adoption risks is that of understanding the behavior and dynamics of the OSS communities that provide the software components. This can be achieved via the analysis of big amounts of data related, for example, to community activeness, component reliability, or capacity of managing the OSS maintenance and evolution. The objective of this section is to present a risk management method and platform for managing OSS risks based on an approach presented in Kenett et al. (2014). The approach integrates different data sources that represent different aspects of OSS communities in the context of a business-oriented decision-making framework. This puts technical OSS adoption and deployment decisions in the context of organizational and business strategy. The objective is to reduce the gap between the concerns of companies and institutions using OSS and the possible risk of business and technical failures in adopting it. The ultimate goal of implementing a data-driven risk management approach to OSS adoption is to increase the global confidence of companies and institutions with respect to OSS. For a general introduction of risk management in system and software development see Kenett and Baker (2010).

In the following, we describe an approach to risk management of OSS by Kenett et al. (2014), and examine this approach in terms of InfoQ.

Proactive risk management in OSS is about the monitoring and flagging of changes in measurable properties of open-source artifacts that are indicative of the occurrence of software and business risks. Key risk indicators (KRIs) are associated with one or more measurements taken over one or more values of artifacts. Deriving insights from conceptual links between risk events, indicators, and measurements is nontrivial. Key methodological as well as theoretical questions need to be answered to derive risk-related insights from measurable data, such as what are the proper indicators for defining and for measuring risk events (the KRIs), how to operationalize an indicator into one or more specific metrics for measurement (e.g., using the goal–question–metric (GQM) paradigm or Bayesian networks (BNs); see Kenett et al. (2014)). Moreover, the predictive ability of measurements related to risk events needs to be validated. In analyzing data used to generate KRIs, we combine a variety of analytic techniques such as social networks, goal-oriented techniques, and the aforementioned BNs. Social network analysis provides the ability to track activeness, timeliness, forking probability, and other risk indicators. The network structure can reflect communication links between committers and contributors or between developers involved in testing and software releases. Metrics reflecting the dynamics of network structures include *closeness* and *betweenness centrality*. A general approach presented in Kenett et al. (2014) is based on a three-layered view to risk management in OSS projects. The layers include the gathering and aggregation of data from communities and projects (layer I), a set of risk indicator variables (layer II) and an analysis of the impact of the risk on business goals (layer III).

9.3.1   Layer I: Raw data

In this first layer, data is collected from OSS communities and projects that determine the risk drivers. We distinguish between raw data collected over time and risk drivers that represent summarized data over specific time windows. The data refers to the characteristics of the OSS components developed by the communities such as measures concerning the code: number of open bugs, number of files changed/commits (changes to the code), forum posts per day, mails per day, and amount of documentation. Other measures highlight the structure of the community in terms of its evolution, for example, changes in its roles or members and in the quality and quantity of relationships between them. These measures establish the "dynamic shape" of the community, recognizing, for example, the presence of a core of active members and how this core changes over time. The data sources for these measures include community repositories, versioning systems, mail lists, bug trackers, and forums. The corresponding measurement instruments are designed to implement a continuous monitoring process to report data to the statistical and reasoning engines that are used in the other layers. However, human intervention is needed because: data sources might be unavailable for a particular component or community, values to be calculated require a dedicated activity such as the evaluation of security or performance, and some values are not directly accessible or are very costly to compute, such as level of expertise of the organization's OSS development team.

9.3.2   Layer II: Risk indicators and risk models

This layer defines the set of indicators of potential risks and models that allow linking these risks to the possible objectives of the adopting organization. The indicators are variables extracted via the OSS community data analysis derived from OSS project measurements and OSS community measurements as described earlier or via expert assessment. Several categories of indicators can be observed. Here we refer to three of them:

1. *Risk indicators* related to the specific OSS projects. These can be grouped using criteria such as reliability and maintainability of the code.

2. *Community indicators* related to the OSS ecosystems that may be extracted from community measures. These allow us to build indicators such as *community activeness* or *community cohesion*.

3. *Contextual indicators* that can be extracted from objectives of the adopting organization such as its OSS business strategy or the type of project in which the OSS component is to be introduced.

Statistical analysis, BNs, and social network analysis can be used to determine risk indicators and business risks. In particular:

• Statistical analysis of data from OSS communities allows determining the trends and distributions of data such as the number of bugs opened or fixed in a given number of days for a particular OSS component.

• BNs are used to link community data gathered from community data sources and community risk metrics to component risk indicators and community risk indicators. Such links can reflect expert opinions based on experience in OSS adoption and community context. In particular, experts can be presented with simulated or real scenarios, described via the values of the OSS measures, and asked to rate the risk indicators with respect to the specific scenarios. The data collected in these rating exercises is then used to construct a BN linking risk drivers in Layer I with risk indicators in Layer II and business risks in Layer III. As an example, Figure 9.1 shows how the risk indicator "activeness" is linked to several risk drivers such as number of mails per day, number of testers, number of companies adopting the OSS component, etc. In the BN one can see that, with the specific data used to construct the network, the probability of very high activeness is 15% and very low activeness is 13%. The analysis was performed with the freeware GeNie (dslpitt.org/genie).

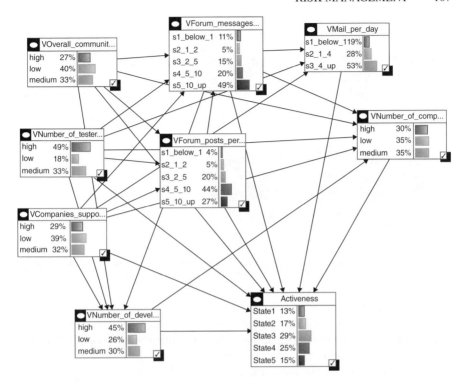

Figure 9.1   Bayesian network linking risk drivers with the activeness of risk indicators.

- The community measures can be analyzed via social network analysis (SNA) techniques and tools that can be used to understand the structure and evolution of the OSS community such as NodeXL (http://nodexl.codeplex.com). Figure 9.2 is an example of tracking email exchanges between contributors and committers of the XWiki open-source platform. The various network configurations are described by metrics derived from the network structure.

All the risk indicators contribute to the definition of a risk model. This model allows the representation of possible causes of risks, the risk indicators, and their connection to possible business risks for the adopter's organization as represented in Layer III.

9.3.3   Layer III: Business risks

Organizations that adopt OSS are exposed to several types of business risks, summarized in three categories:

1. *Strategic risks* related to the company's strategy and plan, such as pricing pressure, failures to comply with regulation, industry or sector downturn, or partner issues.

2. *Operational risks* such as poor capacity management or cost overrun.

3. *Financial risks* such as assets lost, debts, or accounting problems.

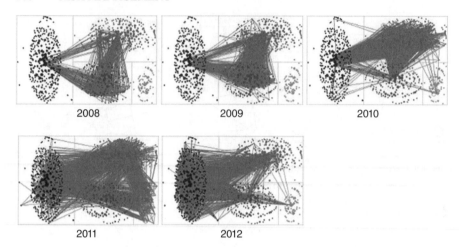

2008   2009   2010

2011   2012

Figure 9.2   Social network based on email communication between OSS contributors and committers.

These generic risks need to be tailored to specific application instances in order to provide added value to decision makers. Business goals are included in models that represent the ecosystems that blend together communities, OSS adopter organizations, and other key actors. The key relationships between these actors can be represented through dependencies in goal-oriented business process management models. For more on this model with examples, see Kenett et al. (2014).

This three-layered model provides a comprehensive analysis, providing decision makers involved in risk management monitoring and mitigation activities with dashboards linking raw data to organizational risks (Figure 9.1). The approach provides optimal coverage of the eight InfoQ dimensions. It requires the right data resolution and structure, provides for effective data integration and results in data with high temporal relevance and chronology of data and goal. Moreover, it offers high generalizability and operationalization and enables effective communication of findings.

InfoQ components in the Kenett et al. (2014) **three-layered approach to OSS risk management:**

Goal (g): Provide decision makers with data-driven indicators of current levels of risk exposure.

Data (X): Risk drivers that have been identified by application experts.

Analysis (f): The analysis is based on a three-layered model driven by BN (Figure 9.1).

Utility (U): The approach allows risk managers and decision makers to be responsive to trends and changes in the risk evaluation map and trigger risk mitigation strategies.

9.4   Risk management of a telecommunication system supplier

This section is based on experience gained in the management of a network of private branch exchange (PBX)'s telecommunication systems by a virtual network operator (VNO) of telecommunication services (Kenett and Raanan, 2010). The typical customers of the VNO are small and medium enterprises (SMEs) and large enterprises requiring both voice and data lines at their premises at different contractually agreed qualities of service. The customers outsource the maintenance of the PBXs and the actual management of the communication services to the VNO.

When a malfunction occurs, customers refer to the VNO's call center, which can act remotely on the PBX, for example, by rebooting the system. If the problem cannot be solved remotely, as in the case of hardware failure, a technician is sent on-site. The latter situation is clearly more expensive than remote intervention by the call center operator. Call center contacts and technician reports are logged in the VNO customer relationship management database. A PBX is doubly redundant, consisting of two independent communication systems and self-monitoring software. Automatic alarms produced by the equipment are recorded in the PBX system log. Call center operators can access the system log to control the status of the PBX. In addition, centralized monitoring software collects system logs from all the installed PBXs on a regular basis in round robin fashion. Among the operational risk events, PBX malfunctioning may have different impacts on the VNO. At one extreme, the call center operator can immediately solve the problem. At the other end, an on-site technical intervention is required, with the customer's offices inoperative meanwhile. In this case, there is a risk of lack of compliance with the contractual SLA. To record the impact of a malfunction, the severity level of the problem occurred is evaluated and documented by the technician. The customers of the VNO are business enterprises, such as offices, banks, industries, governments, insurance agencies, etc. The credit score of these customers is used to monitor financial risks. The aim of operational and financial data merging is to provide a single data source integrating financial information, such as information gathered from company balance sheets, with operational and service data. In this context, the extract, transform and load (ETL) processing creates a merged database of the available sources of data, including customer type, financial information, call center logs, technician reports, and PBX system logs.

Some of the variables typically part of such a database include (see Table 9.1 for a summary):

- Attribute Date consists of the problem's opening date and time, defined as the time the call center receives a customer call reporting a malfunctioning.

- PBX-ID represents a unique identifier of the PBX involved. If more than one PBX has been installed by a customer, this is determined by the call center operator based on the customer's description of the problem and the available configuration of PBXs installed at the customer's premise.

Table 9.1   Log of technicians' on-site interventions (techdb).

Attribute	Description
Date	Opening date of the intervention
PBX-ID	Unique ID of the private branch exchange
CType	Customer line of business
Tech-ID	Unique ID of technician's interventions
Severity	Problem severity recorded after problem solution:
	1 = low-level problem
	2 = medium, intermittent service interruptions
	3 = critical, service unavailable
Prob-ID	Problem type recorded after problem solution

- CType is the customer's line of business, including banks, health care, industry, insurance, and telecommunication businesses.

- Tech-ID is a unique identifier of the technician's interventions: during the same intervention, one or more problems may be tackled (see Table 9.1).

- Severity is a measure of the problem's impact. It is defined on a scale from 1 to 3.

- Prob-ID is a coding of the malfunction solved by the technician. Two hundred problem descriptions are codified.

- It is worth noting that the closing date of the intervention is missing.

Since 200 problem types might be too fine grained for operational risk analysis, problem types are organized as problems recorded in the technician's database and a classification of Basel II IT operational risk event categories: software, hardware, interface, security, and network communications.

For a given telecommunication system (PBX), the technician's log includes an identifier PBX-ID, a date–time TestDate when log is downloaded and the set of alarms (Alarms) since the previous log download. Sixteen distinct alarms are available in the data. The last data source includes financial indicators derived from balance sheets of customers interested in new services or in renewing contracts. Financial balance sheets as inputs for the integration are supposed to be in the machine-processable eXtensible Business Reporting Language (XBRL) or standard (XBRL). Moreover, since only a subset of the information encoded in the XBRL document is relevant as an output of the merging, we have to restrict our integration to a limited number of financial indicators. In other words, the selection of attributes is an outcome of the development of the statistical model and/or it is a parameter of the data merging service. Table 9.2 summarizes an example of the financial indicators available for a given customer of the service, determined by the PBX-ID attribute.

Table 9.2   Balance sheet indicators for a given costumer of the VNO (balance).

Attribute	Description
PBX-ID	Unique ID of the private branch exchange
Return on equity	Measures the rate of return on the ownership interest
Noncurrent assess	An asset which is not easily convertible to cash or not expected to become cash within the next year
Net cash from regular operations	Cash flow available for debt service—the payment of interest and principal on loans
Net profit	The gross profit minus overheads minus interest payable plus/minus one off items for a given time period
Current ratio	Measures whether or not a firm has enough resources to pay its debts over the next 12 months
Current assets	An asset on the balance sheet that is expected to be sold or otherwise used up in the near future
Current liabilities	Liabilities of the business that are to be settled in cash within the fiscal year or the operating cycle
Pre-Tax_Profit	The amount of profit a company makes before taxes are deducted

Assessing the various types of risks requires integrating data from heterogeneous sources. In our context, data includes the customer type, the amount of service calls, the related malfunctions and, as a final dimension, the financial strength of such a customer. The integration was performed with the open-source Pentaho software (www.pentaho.com).

The risk scores derived from the operational and financial data were integrated with a Bayesian weight. For more on the statistical analysis techniques applied to this case study, see Figini et al. (2010).

An additional analysis of this data was performed using association rules implementing the relative linkage disequilibrium (RLD) measure of association (see Kenett and Salini, 2008a, 2008b). The analysis of textual data with association rules is sometimes called "market basket analysis" since many of such techniques were developed to analyze customer's purchasing habits in supermarket stores. An association rule considers a set of purchased product items in order to identify patterns with high prevalence and rare occurrences.

Figure 9.3 shows the top ten association rules, in over 500 association pairs in the integrated VNO database, sorted by RLD, on a two-dimensional representation of the simplex not listing the fourth value in the 2×2 table (which is the complement of the listed three values). On the bottom left part of the simplex, there are rules with high support, point: $(1,0,0)$; on the bottom right there are rules with low support, point: $(0,0,1)$; and at the top are the rules with only the left hand side (LHS) being observed, without the right hand side (RHS), point: $(0,1,0)$. The table corresponding to the center point is $(0.25,0.25,0.25)$, that is, all combinations of A and B occur in equal proportions. In Figure 9.3, we see a medium size support for the association

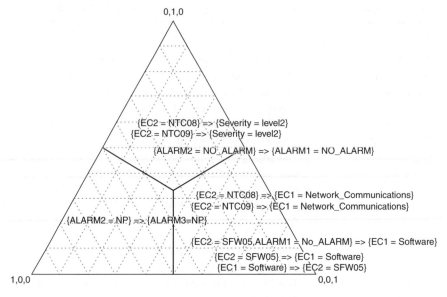

Figure 9.3   Simplex representation of association rules of event categories in telecom case study.

of the two risk events (NTC08 and NTC09) with events of severity two. The association with maximal support is between network protocol (NP) events with Alarm 2 and NP events with Alarm 3. The implication is that this association occurs frequently.

This case study provides an example of data integration, operationalization, and communication.

InfoQ components for the Kenett and Raanan (2010) telecom VNO risk management approach:

Goal (g): Improve risk management procedures by obtaining early warning signs and supporting the evaluation of risk mitigation strategies.

Data (X): Data from three main sources: technician problem report tickets, call center data, and financial reports of customers.

Analysis (f): Combine data sources using BN and identify patterns using association rules.

Utility (U): Cost function associated with risk events.

9.5   Risk management in enterprise system implementation

Enterprise systems (ES) are software packages that offer integrated solutions to companies' information needs. ES like enterprise resource planning (ERP), customer requirement management (CRM), and product data management (PDM) have gained great significance for most companies on operational as well as strategic levels. An ES implementation (ESI) process, as other system development processes, is a complex and dynamic process. It is characterized by occurrences of unplanned problems and events, which may lead to major restructuring of the process with severe implications to the entire company (Kenett and Lombardo, 2007; Kenett and Raphaeli, 2008; Kenett and Raanan, 2010).

Given the growing significance and high risk of ESI projects, much research has been undertaken to develop better understanding of such processes in various disciplines. However, the literature on ESI, IT, and organizational change management does not offer substantial and reliable generalizations about the process dynamics and the relationships between IT and organizational change.

IT projects carry important elements of risk, and it is probable that there will be deviation at some point in the project's life cycle. Risk in a project environment cannot be completely eliminated. The objective of a risk management process is to minimize the impact of unplanned incidents in the project by identifying and addressing potential risks before significant negative consequences occur.

IT risk management should be performed in such a way that the security of IT components, such as data, hardware, software, and the involved personnel, can be ensured. This can be achieved by minimizing external risks (acts of God and political restrictions) or internal risks (technical problems and access safety). Risk management as a strategic issue for the implementation of ERP systems is discussed in Tatsiopoulos et al. (2003).

Change management can be considered a special case of risk management. Both approaches rely, among other things, on survey questionnaires that generate data with similar structural properties. We will exploit this similarity and discuss data analysis techniques that are commonly applicable to both change management and risk management.

Assessing organizational readiness and risk profiles in an ESI is essentially a multivariate task. The approach taken in the literature has typically been to compile composite indices and generate graphical displays such as radar charts or risk maps. In this section we present data analysis techniques to sort and rank risk profiles and readiness assessments in a truly multivariate sense (see Kenett and Raphaeli, 2008).

9.5.1   Cause–event–action–outcome (CEAO) for mapping causality links

Events that occur within an enterprise system implementation process can be described in terms of chains called cause–event–action–outcome (CEAO). These chains have been classified according to a reference framework that identifies technical as well as organizational and human aspects. These processes are called dimensions and include the business process, the ES system, and the project management process:

- The *business process* for which the system is implemented. The focal process consists of all activities that will be supported or affected by the new ES. The business processes are permanent processes, possibly subject to continuous change. The word permanent is used to distinguish daily tasks from temporary tasks of an implementation project.

- The *design and tuning of the new ES*. The focal process consists of all activities needed to adapt or tune the system and align it with the business. Design and tuning of the ES is a temporary process but may extend beyond the implementation project.

- *Project management* of the ESI process. The focal process consists of all activities needed to plan and monitor the implementation process, select and perform the implementation strategy, select the system and implement it in the organization, compose a project team, manage project documents, etc. Project management is a temporary process.

In addition to the distinction between business processes, the ES system itself and the project management process dimensions, the CEAO chains are classified according to six organizational aspects: strategy and goals, management, structure, process, knowledge and skills, and social dynamics. The 18 cells created by the intersection of dimension and aspect are called focus cells (Wognum et al., 2004; Kenett and Lombardo, 2007). The six aspects are further explained as follows:

1. *Strategy and goals*. Strategy and goals are the medium- and long-term goals to be achieved and the plans for realizing these goals. The strategy and goals for the ES and the implementation project should match the business goals and strategy.

2. *Management*. The management aspect deals with setting priorities, assigning resources, and planning and monitoring processes.

3. *Structure*. Structure involves the relationships between elements of the organizational system, such as processes, people, and means. Structure includes tasks, authorities and responsibilities, team structures, process structure, and structure of the ES.

4. *Process*. Process involves the steps needed to perform the focus process of each dimension: the primary business process and relevant support and management processes, the project process, and the enterprise system design and adaptation process.

5. *Knowledge and skills*. This aspect refers to the knowledge and skills that are needed to perform the focus processes in each dimension.

6. *Social dynamics*. The aspect social dynamics refers to the behaviors of people, their norms, and rituals. Social dynamics often become visible in informal procedures and (lack of) communication.

A CEAO chain is a mapping of a problem and solution containing the following items: An *event* is defined as a problem created by decisions, actions, or by events outside the control of the organization. A *cause* is an underlying reason or action, leading to the event. For each event, it is possible to specify one or several causes, which are linked to the event through a parent–child relationship. An *action* is the solution taken to resolve the event; it includes the method of performing or means used. Each action is connected to outcomes. The mapping of causes of the CEAO chains into a reference framework leads to different clusters of CEAO chains. Each cluster belongs to a focal cell in the framework.

Context data provide a view of the company and ES, such as company size, type of ES, cultural region, etc. It is expected that ESI process execution is influenced by these characteristics. Context sensitivity analysis was done by experienced change management experts in an attempt to distinguish between local patterns (that occur only in specific situations due to the context characteristics), and generic patterns that can be generalized across ESI processes. For example, if we compare two different sized companies, SME (less than 250 employees) and large (more than 250 employees), it is expected that size-dependent patterns, such as greater project resources and higher complexity adoption processes in a large company, cause major differences in the ESI processes.

An assessment of a company's situation at the start of an ESI project enables the anticipation of problems that may occur and the reuse of knowledge gathered in CEAO chains from other ESIs.

SAP is one of the most successful ERP systems (www.sap.com). The SAP implementation methodology includes five phases, with a total typical duration of 12–24 months. These phases are:

1. *Project Preparation.* This provides initial planning and preparation of the SAP project including the resources allocated, detailed activities plans, and the technical environment setup.

2. *Business Blueprint.* A detailed documentation of the business process requirements and an executive summary is produced (To-Be), following a short description of the current situation (As-Is).

3. *Realization.* The purpose is to implement the requirements based on the business blueprint. The main activities are the customizing of the system and building the conversion, interface, and add-on programs.

4. *Final Preparation.* This includes the completion of the final preparations (including testing, end user training, system management, and cutover activities) and the final improvement of the company's readiness to go live.

5. *Go Live Support.* A support organization for end users is set up to optimize the overall system performance.

At each phase of the implementation, and separately for each organizational unit, a CEAO workshop is conducted. During these workshops, past, present, or future events are recorded and classified using CEAO chains. A sample CEAO chain is presented in Figure 9.4. In the example of Figure 9.4, the event "users unable to use the system at go live" is attributed to four causes classified to four different aspect categories. For example, "Training performed on dummy system" is classified as "Knowledge and Skills." In the example analyzed in the following text, 264 causes

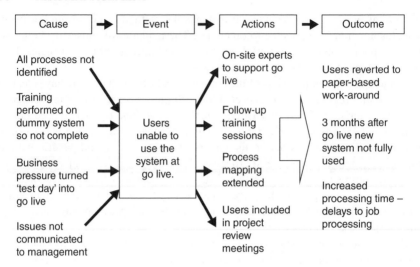

Figure 9.4   A sample CEAO chain.

were classified within the reference framework, following CEAO workshops in five divisions. The detailed statistics are presented in Table 9.3.

Kenett and Raphaeli (2008) analyze data gathered by such implementations using correspondence analysis. We briefly describe here the details of this case study and in the summary section, present an InfoQ assessment of this analysis.

Correspondence analysis is a statistical visualization method for representing associations between the levels of a two-way contingency table. The name is a translation of the French "Analyses des Correspondances," where the term "correspondance" denotes a "system of associations" between the elements of two sets. In a two-way contingency table, the observed association of two traits is summarized by the cell frequencies. A typical inferential aspect is the study of whether certain levels of one characteristic are associated with some levels of another. Correspondence analysis displays the rows and columns of a two-way contingency table as points in a low-dimensional space, so that the positions of the row and column points are consistent with their associations in the table. The goal is to obtain a global view of the data, useful for interpretation.

Simple correspondence analysis performs a weighted principal components' analysis of a contingency table. If the contingency table has r rows and c columns, the number of underlying dimensions is the smaller of $(r-1)$ or $(c-1)$. As with principal components, variability is partitioned, but rather than partitioning the total variance, simple correspondence analysis partitions the Pearson χ^2 statistic. Traditionally, for n observations in the contingency table, correspondence analysis uses χ^2/n, which is termed *inertia* or *total inertia*, rather than χ^2. The inertias associated with all of the principal components add up to the total inertia. Ideally, the first one, two, or three components account for most of the total inertia.

Table 9.3   Classification of 264 CEAO chains by aspect and division (output from MINITAB version 12.1).

Rows: aspect; Columns: division

	DivA	DivB	DivC	DivD	DivE	All
Knowledge and skills	7	1	16	6	9	39
	13.73	7.14	34.04	8.33	11.25	14.77
Management	16	4	11	13	12	56
	31.37	28.57	23.40	18.06	15.00	21.21
Process	0	2	1	15	23	41
	0.00	14.29	2.13	20.83	28.75	15.53
Social dynamics	16	4	12	15	9	56
	31.37	28.57	25.53	20.83	11.25	21.21
Strategy and goals	5	0	0	5	15	25
	9.80	0.00	0.00	6.94	18.75	9.47
Structure	7	3	7	18	12	47
	13.73	21.43	14.89	25.00	15.00	17.80
All	51	14	47	72	80	264
	100.00	100.00	100.00	100.00	100.00	100.00

Values indicate the number of CEAO chains in each cell, and their % of column.

Row profiles are vectors of length c and therefore lie in a c-dimensional space (similarly, column profiles lie in an r-dimensional space). Since this dimension is usually too high to allow easy interpretation, one tries to find a subspace of lower dimension (preferably not more than two or three) that lies close to all the row profile points (or column profile points). The profile points are then projected onto this subspace. If the projections are close to the profiles, we do not lose much information. Working in two or three dimensions allows one to study the data more easily and, in particular, allows for easy examination of plots. This process is analogous to choosing a small number of principal components to summarize the variability of continuous data. If $d =$ the smaller of $(r-1)$ and $(c-1)$, then the row profiles (or equivalently the column profiles) lie in a d-dimensional subspace of the full c-dimensional space (or equivalently the full r-dimensional space). Thus, there are at most "d" principal components. For more details on correspondence analysis, see Greenacre (1993).

An implementation of correspondence analysis to the SAP study (using MINITAB version 17.2, www.minitab.com) is presented in Figure 9.5. From that analysis, we can characterize division C as contributing mostly knowledge and skills CEAOs, division E is mostly affected by strategy and goals and processes, division D by structure, and divisions A and B by management and social dynamics. Such an analysis provides invaluable feedback to the organization's management in determining whether these signals are due to the personal management style of the division managers or specificities in the activities of the divisions. The insights gained through

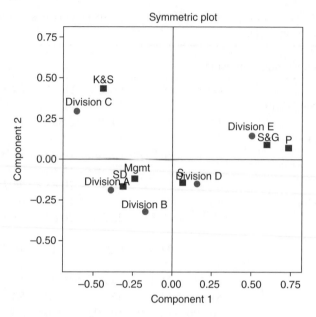

Figure 9.5   Correspondence analysis of CEAO chains in five divisions by aspect. K&S = knowledge and skills; Mgmt = management; P = process; S = structure; S&G = strategy and goals; SD = social dynamics.

such analysis should lead to proactive actions for improving the organization's preparedness to an ERP implementation.

A similar analysis can be conducted when risks are attributed to different risk categories, and risk elicitation is done in different parts of the organization or for different modules of the SAP system.

InfoQ components for the SAP change management study

Goal (g): Smooth deployment of an enterprise resource management (ERP) system affecting most processes in an organization.

Data (X): Data from CEAO chains recording events related to the ERP implementation process.

Analysis (f): Collect and classify CEAO chains. Apply correspondence analysis to the data.

Utility (U): Cost function of ERP implementation negative events.

9.6   Summary

We reviewed four case studies of applied research in the context of risk management. The first two examples present two general approaches to risk management: Taleb quadrants and adoption of OSS. The last two examples are focused on risk management infrastructure of a telecommunication company and a specific deployment of an enterprise resource management platform (ERP). Table 9.4 presents an InfoQ assessment of the four risk management case studies presented earlier by qualifying on a scale from 1 ("very poor") to 5 ("very good") the eight InfoQ dimensions. This assessment is subjective based on consultation with people involved in these projects.

From Table 9.4, we see that InfoQ scores ranged from 51% to 96% with the OSS risk management model scoring highest (96%) and the Telecom case study scoring lowest (63%). Addressing the challenges of risk management in decision support systems is uniquely complex and requires data at the right resolution, with continuous updates and integrated analysis (Kenett and Raanan, 2010).

Table 9.4   Scoring of InfoQ dimensions of the five risk management cases studies.

InfoQ dimension/use case	9.2 Taleb's quadrant	9.3 OSS	9.4 telecom	9.5 ESI
Data resolution	2	5	3	2
Data structure	2	5	4	4
Data integration	3	5	4	3
Temporal relevance	5	4	3	3
Chronology of data and goal	4	5	2	3
Generalizability	5	5	2	5
Operationalization	5	5	4	4
Communication	5	5	4	5
Use case score	**63**	**96**	**51**	**60**

InfoQ: Values in bold face represent aggregated values on a 0–100 scale.

References

Figini, S., Kenett, R.S. and Salini, S. (2010) Integrating operational and financial risk assessments. *Quality and Reliability Engineering International*, 26(8), pp. 887–897.

Franch, X., Susi, A., Annosi, M.C., Ayala, C.P., Glott, R., Gross, D., Kenett, R., Mancinelli, F., Ramsay, P., Thomas, C., Ameller, D., Bannier, S., Nili Bergida, N., Blumenfeld, Y., Bouzereau, O., Costal, D., Dominguez, M., Haaland, K., López, L., Morandini, M., Siena, A. (2013) Managing Risk in Open Source Software Adoption. *ICSOFT*, Reykjavík, pp. 258–264.

Greenacre, M.J. (1993) *Correspondence Analysis in Practice*. Harcourt, Brace and Company/ Academic Press, London.

Kenett, R.S. (2008) From data to information to knowledge. *Six Sigma Forum Magazine*, 11, pp. 32–33.

Kenett, R.S. (2013) Implementing SCRUM using business process management and pattern analysis methodologies. *Dynamic Relationships Management Journal*, 2(2), pp. 29–48.

Kenett, R.S. and Baker, E. (2010) *Process Improvement and CMMI for Systems and Software: Planning, Implementation, and Management*. Taylor & Francis/Auerbach Publications, New York/Boca Raton, FL.

Kenett, R.S. and Lombardo, S. (2007) The Role of Change Management in IT Systems Implementation, in *Handbook of Enterprise Systems Architecture in Practice*, Saha, P. (editor), National University of Singapore, Idea Group Inc., Singapore.

Kenett, R.S. and Raanan, Y. (2010) *Operational Risk Management: A Practical Approach to Intelligent Data Analysis*. John Wiley & Sons, Ltd, Chichester, UK.

Kenett, R.S. and Raphaeli, O. (2008) Multivariate methods in enterprise system implementation, risk management and change management. *International Journal of Risk Assessment and Management*, 9(3), pp. 258–276.

Kenett, R.S. and Salini, S. (2008a) Relative Linkage Disequilibrium: A New Measure for Association Rules, in *Advances in Data Mining: Medical Applications, E-Commerce, Marketing, and Theoretical Aspects*, Perner, P. (editor), Lecture Notes in Computer Science, Springer-Verlag, Vol. 5077. ICDM 2008, Leipzig, Germany, July 2008.

Kenett, R.S. and Salini, S. (2008b) Relative linkage disequilibrium applications to aircraft accidents and operational risks. *Transactions on Machine Learning and Data Mining*, 1(2), pp. 83–96.

Kenett, R.S. and Tapiero, C. (2010) Quality, risk and the Taleb quadrants. *Risk and Decision Analysis*, 1(4), pp. 231–246.

Kenett, R.S. and Zacks, S. (2014) *Modern Industrial Statistics: With Applications Using R, MINITAB and JMP*, 2nd edition. John Wiley & Sons, Ltd, Chichester, UK.

Kenett, R.S., Franch, X., Susi, A. and Galanis, N. (2014) Adoption of Free Libre Open Source Software (FLOSS): A Risk Management Perspective. *Proceedings of the 38th Annual IEEE International Computer Software and Applications Conference (COMPSAC)*, Västerås, Sweden.

NASA – National Aeronautics and Space Administration (2002) Probabilistic Risk Assessment Procedures Guide for NASA Managers and Practitioners. Version 1.1. Office of Safety and Mission Assurance. Washington, DC.

Raiffa, H. (1997) *Decision Analysis: Introductory Readings on Choices Under Uncertainty*. McGraw Hill, New York.

Ruggeri, F., Kenett, R.S. and Faltin, F. (2007) *Encyclopaedia of Statistics in Quality and Reliability*, 4 Volume Set, 1st edition. John Wiley & Sons, Ltd, Chichester, UK.

Taleb, N.N. (2007) *The Black Swan: The Impact of the Highly Improbable*. Random House, New York.

Taleb, N.N. (2008) The Fourth Quadrant: A Map of the Limits of Statistics. *Edge*, http://www.edge.org/3rd_culture/taleb08/taleb08_index.html (accessed May 6, 2016).

Taleb, N.N. (2009) Errors, robustness and the fourth quadrant. *International Journal of Forecasting*, 25(4), pp. 744–759.

Taleb, N.N. (2012) *Antifragile*. Random House and Penguin, New York.

Tapiero, C. (2004) *Risk and Financial Management: Mathematical and Computational Methods*. John Wiley & Sons, Inc., New York.

Tapiero, C. (2009) The Price of Safety and Economic Reliability, NYU-Poly Technical Report. www.ssrn.com/abstract=1433477 (accessed May 6, 2016).

Tatsiopoulos, I.P., Panayiotou, N.A., Kirytopoulos, K. and Tsitsiriggos, K. (2003) Risk management as a strategic issue for implementing ERP systems: a case study from the oil industry. *International Journal of Risk Assessment and Management*, 4(1), pp. 20–35.

Wognum, P., Krabbendam, J., Buhl, H., Ma, X. and Kenett, R.S. (2004) Improving enterprise system support – a case-based approach. *Journal of Advanced Engineering Informatics*, 18(4), pp. 241–253.

10

Official statistics

10.1 Introduction

Official statistics are produced by a variety of organizations including central bureaus of statistics, regulatory healthcare agencies, educational systems, and national banks. Official statistics are designed to be used, and increasing their utility is one of the overarching concepts in official statistics. An issue that can lead to misconception is that many of the terms used in official statistics have specific meanings not identical to their everyday usage. Forbes and Brown (2012) state: "All staff producing statistics must understand that the conceptual frameworks underlying their work translate the real world into models that interpret reality and make it measurable for statistical purposes.... The first step in conceptual framework development is to define the issue or question(s) that statistical information is needed to inform. That is, to define the objectives for the framework, and then work through those to create its structure and definitions. An important element of conceptual thinking is understanding the relationship between the issues and questions to be informed and the definitions themselves."

In an interview-based study of 58 educators and policymakers, Hambleton (2002) found that the majority misinterpreted the official statistics reports on reading proficiency that compare results across school grades and across years. This finding was particularly distressing since policymakers rely on such reports for funding appropriations and for making other key decisions. In terms of information quality (InfoQ), the quality of the information provided by the reports was low. The translation from statistics to the domain of education policy was faulty.

Information Quality: The Potential of Data and Analytics to Generate Knowledge,
First Edition. Ron S. Kenett and Galit Shmueli.
© 2017 John Wiley & Sons, Ltd. Published 2017 by John Wiley & Sons, Ltd.
Companion website: www.wiley.com/go/information_quality

The US Environmental Protection Agency, together with the Department of Defense and Department of Energy, launched the Quality Assurance Project Plan (see EPA, 2005), which presents "steps … to ensure that environmental data collected are of the correct type and quality required for a specific decision or use." They used the term data quality objectives to describe "statements that express the project objectives (or decisions) that the data will be expected to inform or support." These statements relate to descriptive goals, such as "Determine with greater than 95% confidence that contaminated surface soil will not pose a human exposure hazard." These statements are used to guide the data collection process. They are also used for assessing the resulting data quality.

Central bureaus of statistics are now combining survey data with administrative data in dynamically updated studies that have replaced the traditional census approach, so that proper integration of data sources is becoming a critical requirement. We suggest in this chapter and related papers that evaluating InfoQ can significantly contribute to the range of examples described earlier (Kenett and Shmueli, 2016).

The chapter proceeds as follows: Section 10.2 reviews the InfoQ dimensions in the context of official statistics research studies. Section 10.3 presents quality standards applicable to official statistics and their relationship with InfoQ dimensions, and Section 10.4 describes standards used in customer surveys and their relationship to InfoQ. We conclude with a chapter summary in Section 10.5.

10.2   Information quality and official statistics

We revisit here the eight InfoQ dimensions with guiding questions that can be used in planning, designing, and evaluating official statistics reports. We accompany these with examples from official statistics studies.

10.2.1   Data resolution

Data resolution refers to the measurement scale and aggregation level of the data. The measurement scale of the data should be carefully evaluated in terms of its suitability to the goal *(g)*, the analysis methods used *(f)*, and the required resolution of the utility *(U)*. Questions one should ask to determine the strength of this dimension include the following:

- Is the data scale used aligned with the stated goal of the study?

- How reliable and precise are the data sources and data collection instruments used in the study?

- Is the data analysis suitable for the data aggregation level?

A low rating on data resolution can be indicative of low trust in the usefulness of the study's findings. An example of data resolution, in the context of official statistics, is the Google Flu Trends (www.google.org/flutrends) case. The goal of the original application by Google was to forecast the prevalence of influenza on the basis of the type and extent of Internet search queries. These forecasts were shown to strongly correlate with the official figures published by the Centers for Disease Control and Prevention (CDC) (Ginsberg et al., 2009). The advantage of Google's tracking system over the CDC system is that data is available immediately and forecasts have only a day's delay, compared to the week or more that it takes for the CDC to assemble a picture based on reports from confirmed laboratory cases. Google is faster because it tracks the outbreak by finding a correlation between what users search for online and whether they have flu symptoms. In other words, it uses immediately available data on searches that correlate with symptoms. Although Google initially claimed very high forecast accuracy, it turned out that there were extreme mispredictions and other challenges (David et al., 2014; Lazer et al., 2014). Following the criticism, Google has reconfigured Flu Trends to include data from the CDC to better forecast the flu season (Marbury, 2014). Integrating the two required special attention to the different resolutions of the two data sources.

10.2.2   Data structure

Data structure refers to the type(s) of data and data characteristics such as corrupted and missing values due to the study design or data collection mechanism. As discussed in Chapter 3, data types include structured, numerical data in different forms (e.g., cross-sectional, time series, and network data) as well as unstructured, nonnumerical data (e.g., text, text with hyperlinks, audio, video, and semantic data). Another type of data generated by official statistics surveys, called *paradata*, is related to the process by which the survey data was collected. Examples of paradata include the times of day on which the survey interviews were conducted, how long the interviews took, how many times there were contacts with each interviewee or attempts to contact the interviewee, the reluctance of the interviewee and the mode of communication (phone, Web, email, or in person). These attributes affect the costs and management of a survey, the findings of a survey, evaluations of interviewers, and analysis of nonresponders.

Questions one should ask to determine the data structure include the following:

- What types of variables and measurements does the data have?

- Are there missing or corrupted values? Is the reason for such values apparent?

- What paradata or metadata are available?

10.2.3   Data integration

With the variety of data sources and data types available today, studies sometimes integrate data from multiple sources and/or types to create new knowledge regarding the goal at hand. Such integration can increase InfoQ, but in other cases, it can reduce InfoQ, for example, by creating privacy breaches.

Questions that help assess data integration levels include the following:

- Is the dataset from a single source or multiple sources? What are these sources?

- How was the integration performed? What was the common key?

- Does the integrated data pose any privacy or confidentiality concerns?

One example of data integration and integrated analysis is the calibration of data from a company survey with official statistics data (Dalla Valle and Kenett, 2015) where Bayesian networks (BNs) were used to match the relation between variables in the official and the administrative data sets by conditioning. Another example of data integration between official data and company data is the Google Flu Trends case described in Section 10.2.1, which now integrates data from the CDC. For further examples of data integration in official statistics, see also Penny and Reale (2004), Figini et al. (2010), and Vicard and Scanu (2012).

10.2.4   Temporal relevance

The process of deriving knowledge from data can be put on a timeline that includes data collection, data analysis, and results' usage periods as well as the temporal gaps between these three stages. The different durations and gaps can each affect InfoQ. The data collection duration can increase or decrease InfoQ, depending on the study goal, for example, studying longitudinal effects versus a cross-sectional goal. Similarly, if the collection period includes uncontrollable transitions, this can be useful or disruptive, depending on the study goal.

Questions that help assess temporal relevance include the following:

- What is the gap between data collection and use?

- How sensitive are the results to the data collection duration and to the lag-time between collection and use? Are these durations and gaps acceptable?

- Will the data collection be repeated, and if so, is the frequency of collection adequate for users' goals?

A low rating on temporal relevance can be indicative of an analysis with low relevance to decision makers due to data collected in a different contextual condition. This can happen in economic studies, with policy implications based on old data. The Google Flu Trends application, which now integrates CDC data, is a case where temporal relevance is key. The original motivation for using Google search of flu-related keywords in place of the official CDC data of confirmed cases was the delay in obtaining laboratory data. The time gap between data collection and its availability is not an issue with Google search, as opposed to CDC data. Yet, in the new Google Flu Trends application, a way was found to integrate CDC data while avoiding long delays in forecasts. In this case, data collection, data analysis, and deployment (generating forecasts) are extremely time sensitive, and this time sensitivity was the motivation for multiple studies exploring alternative data sources and algorithms for detecting influenza and other disease outbreaks (see Goldenberg et al. (2002) and Shmueli and Burkom (2010)).

10.2.5   Chronology of data and goal

The choice of variables to collect, the temporal relationship between them, and their meaning in the context of the goal at hand affect InfoQ. This is especially critical in predictive goals and in inferring causality (e.g., impact studies).

Questions that help assess this dimension include the following:

- For a predictive goal, will the needed input variables be available at the time of prediction?

- For a causal explanatory goal, do the variables measuring the causes temporally precede the measured outcomes? Is there a risk of reverse causality?

A low rating on chronology of data and goal can be indicative of low relevance of a specific data analysis due to misaligned timing. For example, consider the reporting of the consumer price index (CPI). This index measures changes in the price level of a market basket of consumer goods and services purchased by households and is used as a measure of inflation with impact on wages, salaries, and pensions. A delayed reporting of CPI will have a huge economic impact which affects temporal relevance. In terms of chronology of data and goal, we must make sure that the household data on consumer goods and services, from which the CPI index is computed, is available sufficiently early for producing the CPI index.

10.2.6   Generalizability

The utility of $f(X \mid g)$ is dependent on the ability to generalize to the appropriate population. Official statistics are mostly concerned with statistical generalizability. Statistical generalizability refers to inferring f from a sample to a target population. Scientific generalizability refers to generalizing an estimated population pattern/ model f to other populations or applying f estimated from one population to predict individual observations in other populations.

While census data are, by design, general to the entire population of a country, they can be used to estimate a model which can be compared with models in other countries or used to predict outcomes in another country, thereby invoking scientific generalizability. In addition, using census data to forecast future values is also a generalization issue, where forecasts are made to a yet unknown context.

Questions that help assess the generalization type and aspects include the following:

- What population does the sample represent?

- Are estimates accompanied with standard errors or statistical significance? This would imply interest in statistical generalizability of f.

- Is out-of-sample predictive power reported? This would imply an interest in generalizing the model by predicting new individual observations.

10.2.7   Operationalization

We consider two types of operationalization: construct operationalization and action operationalization. Constructs are abstractions that describe a phenomenon of theoretical interest. Official statistics are often collected by organizations and governments to study constructs such as poverty, well-being, and unemployment. These constructs are carefully defined and economic or other measures as well as survey questions are crafted to operationalize the constructs for different purposes. Construct operationalization questions include the following:

- Is there an unmeasurable construct or abstraction that we are trying to capture?

- How is each construct operationalized? What theories and methods were used?

- Have the measures of the construct changed over the years? How?

Action operationalizing refers to the following three questions posed by W. Edwards Deming (1982):

1. What do you want to accomplish?

2. By what method will you accomplish it?

3. How will you know when you have accomplished it?

A low rating on operationalization indicates that the study might have academic value but, in fact, has no practical impact. In fact, many statistical agencies see their role only as data providers, leaving the dimension of action operationalization to others. In contrast, Forbes and Brown (2012) clearly state that official statistics "need to be used to be useful" and utility is one of the overarching concepts in official statistics. With this approach, operationalization is a key dimension in the InfoQ of official statistics.

10.2.8   Communication

Effective communication of the analysis and its utility directly impacts InfoQ. Communicating official statistics is especially sensitive, since they are usually intended for a broad nontechnical audience. Moreover, such statistics can have important implications, so communicating the methodology used to reach such results can be critical. An example is reporting of agricultural yield forecasts generated by government agencies such as the National Agricultural Statistics Service (NASS) at the US Department of Agriculture (USDA), where such forecasts can be seen as political, leading to funding and other important policies and decisions:

> NASS has provided detailed descriptions of their crop estimating and forecasting procedures. Still, market participants continue to demonstrate a lack of understanding of NASS methodology for making acreage, yield, and production forecasts and/or a lack of trust in the objectives

of the forecasts… Beyond misunderstanding, some market participants continue to express the belief that the USDA has a hidden agenda associated with producing the estimates and forecasts. This "agenda" centers on price manipulation for a variety of purposes, including such things as managing farm program costs and influencing food prices

(Good and Irwin, 2011).

In education, a study of how decision makers understand National Assessment of Educational Progress (NAEP) reports was conducted by Hambleton (2002) and Goodman and Hambleton (2004). Among other things, the study shows that a table presenting the level of advanced proficiency of grade 4 students was misunderstood by 53% of the respondents who read the report. These readers assumed that the number represented the percentage of students in that category when, in fact, it represented the percentage in all categories, up to advanced proficiency, that is, basic, proficient, and advanced proficiency. The implication is that the report showed a much gloomier situation than the one understood by more than half of the readers.

Questions that help assess the level of communication include the following:

- Who is the intended audience?

- What presentation methods are used?

- What modes of dissemination are being used?

- Do datasets for dissemination include proper descriptions?

- How easily can users find and obtain (e.g., download) the data and results?

10.3 Quality standards for official statistics

A concept of *Quality of Statistical Data* was developed and used in European official statistics and international organizations such as the International Monetary Fund (IMF) and the Organization for Economic Cooperation and Development (OECD). This concept refers to the usefulness of summary statistics produced by national statistics agencies and other producers of official statistics. Quality is evaluated, in this context, in terms of the usefulness of the statistics for a particular goal. The OECD uses seven dimensions for quality assessment: *relevance, accuracy, timeliness and punctuality, accessibility, interpretability, coherence,* and *credibility* (see Chapter 5 in Giovanini, 2008). Eurostat's quality dimensions are *relevance of statistical concepts, accuracy of estimates, timeliness and punctuality in disseminating results, accessibility and clarity of the information, comparability, coherence, and completeness.* See also Biemer and Lyberg (2003), Biemer et al. (2012), and Eurostat (2003, 2009).

In the United States, the National Center for Science and Engineering Statistics (NCSES), formerly the Division of Science Resources Statistics, was established within the National Science Foundation with a general responsibility for statistical data. Part of its mandate is to provide information that is useful to practitioners, researchers, policymakers, and the public. NCSES prepares about 30 reports per year based on surveys.

The purpose of survey standards is to set a framework for assuring data and reporting quality. Guidance documents are meant to help (i) increase the reliability and validity of data, (ii) promote common understanding of desired methodology and processes, (iii) avoid duplication and promote the efficient transfer of ideas, and (iv) remove ambiguities and inconsistencies. The goal is to provide the clearest possible presentations of data and its analysis. Guidelines typically focus on technical issues involved in the work rather than issues of contract management or publication formats.

Specifically, NCSES aims to adhere to the ideals set forth in "Principles and Practices for a Federal Statistical Agency." As a US federal statistical agency, NCSES surveys must follow guidelines and policies as set forth in the Paperwork Reduction Act and other legislations related to surveys. For example, NCSES surveys must follow the implementation guidance, survey clearance policies, response rate requirements, and related orders prepared by the Office of Management and Budget (OMB). The following standards are based on US government standards for statistical surveys (see www.nsf.gov/statistics/). We list them with an annotation mapping to InfoQ dimensions, when relevant (Table 10.1 summarizes these relationships) See also Office for National Statistics (2007).

10.3.1   Development of concepts, methods, and designs

10.3.1.1   Survey planning

Standard 1.1: Agencies initiating a new survey or major revision of an existing survey must develop a written plan that sets forth a justification, including goals and objectives, potential users, the decisions the survey is designed to inform, key survey estimates, the precision required of the estimates (e.g., the size of differences that need to be detected), the tabulations and analytic results that will inform decisions and other uses, related and previous surveys, steps taken to prevent unnecessary duplication with other sources of information, when and how frequently users need the data and the level of detail needed in tabulations, confidential microdata, and public-use data files.

This standard requires explicit declaration of goals and methods for communicating results. It also raises the issue of data resolution in terms of dissemination and generalization (estimate precision).

10.3.1.2   Survey design

Standard 1.2: Agencies must develop a survey design, including defining the target population, designing the sampling plan, specifying the data collection instruments and methods, developing a realistic timetable and cost estimate, and selecting samples using generally accepted statistical methods (e.g., probabilistic methods that

Table 10.1 Relationship between NCSES standards and InfoQ dimensions. Shaded cells indicate an existing relationship.

NCSES standard		Data resolution	Data structure	Data integration	Temporal relevance	Data–goal chronology	Generalizability	Operationalization	Communication
I. Development of concepts, methods, and design	Survey planning	▨					▨		▨
	Survey design		▨	▨			▨	▨	
	Survey response rate							▨	
	Survey pretesting	▨							
II. Collection of data	Developing sampling frames					▨		▨	
	Required notifications to potential survey respondents							▨	
	Data collection methodology							▨	
III. Processing and editing of data	Data editing								
	Nonresponse analysis and response rate calculation						▨		
	Coding								
	Data protection								
	Evaluation								▨

(*Continued*)

Table 10.1 (*Continued*)

NCSES standard		Data resolution	Data structure	Data integration	Temporal relevance	Data–goal chronology	Generalizability	Operationalization	Communication
IV. Production of estimates and projections	Developing estimates and projections						▓		
V. Data analysis	Analysis and report planning						▓		
	Inference and comparisons								
VI. Review procedures	Review of information products								
VII. Dissemination of information products	Releasing information					▓			▓
	Data protection and disclosure avoidance for dissemination								▓
	Survey documentation								
	Documentation and release of public-use microdata								

can provide estimates of sampling error). Any use of nonprobability sampling methods (e.g., cutoff or model-based samples) must be justified statistically and be able to measure estimation error. The size and design of the sample must reflect the level of detail needed in tabulations and other data products and the precision required of key estimates. Documentation of each of these activities and resulting decisions must be maintained in the project files for use in documentation (see Standards 7.3 and 7.4).

This standard advises on data resolution, data structure, and data integration. The questionnaire design addresses the issue of construct operationalization, and estimation error relates to generalizability.

10.3.1.3   Survey response rates

Standard 1.3: Agencies must design the survey to achieve the highest practical rates of response, commensurate with the importance of survey uses, respondent burden, and data collection costs, to ensure that survey results are representative of the target population so that they can be used with confidence to inform decisions. Nonresponse bias analyses must be conducted when unit or item response rates or other factors suggest the potential for bias to occur.

The main focus here is on statistical generalization, but this standard also deals with action operationalization. The survey must be designed and conducted in a way that encourages respondents to take action and respond.

10.3.1.4   Pretesting survey systems

Standard 1.4: Agencies must ensure that all components of a survey function as intended when implemented in the full-scale survey and that measurement error is controlled by conducting a pretest of the survey components or by having successfully fielded the survey components on a previous occasion.

Pretesting is related to data resolution and to the question of whether the collection instrument is sufficiently reliable and precise.

10.3.2   Collection of data

10.3.2.1   Developing sampling frames

Standard 2.1: Agencies must ensure that the frames for the planned sample survey or census are appropriate for the study design and are evaluated against the target population for quality.

Sampling frame development is crucial for statistical generalization. Here we also ensure chronology of data and goal in terms of the survey deployment.

10.3.2.2   Required notifications to potential survey respondents

Standard 2.2: Agencies must ensure that each collection of information instrument clearly states the reasons why the information is planned to be collected, the way such information is planned to be used to further the proper performance of the functions of the agency, whether responses to the collection of information are voluntary or mandatory (citing authority), the nature and extent of confidentiality to be provided, if any (citing authority), an estimate of the average respondent burden together with a request that the public direct to the agency any comments concerning the accuracy of this burden estimate and any suggestions for reducing this burden, the OMB control number, and a statement that an agency may not conduct and a person is not required to respond to an information collection request unless it displays a currently valid OMB control number.

This is another aspect of action operationalization.

10.3.2.3   Data collection methodology

Standard 2.3: Agencies must design and administer their data collection instruments and methods in a manner that achieves the best balance between maximizing data quality and controlling measurement error while minimizing respondent burden and cost.

10.3.3   Processing and editing of data

The standards in Section 10.3.3 are focused on the data component and, in particular, assuring data quality and confidentiality.

10.3.3.1   Data editing

Standard 3.1: Agencies must edit data appropriately, based on available information, to mitigate or correct detectable errors.

10.3.3.2   Nonresponse analysis and response rate calculation

Standard 3.2: Agencies must appropriately measure, adjust for, report, and analyze unit and item nonresponse to assess their effects on data quality and to inform users. Response rates must be computed using standard formulas to measure the proportion of the eligible sample that is represented by the responding units in each study, as an indicator of potential nonresponse bias.

This relates to generalizability.

10.3.3.3   Coding

Standard 3.3: Agencies must add codes to the collected data to identify aspects of data quality from the collection (e.g., missing data) in order to allow users to appropriately analyze the data. Codes added to convert information collected as text into a form that permits immediate analysis must use standardized codes, when available, to enhance comparability.

10.3.3.4   Data protection

Standard 3.4: Agencies must implement safeguards throughout the production process to ensure that survey data are handled confidentially to avoid disclosure.

10.3.3.5   Evaluation

Standard 3.5: Agencies must evaluate the quality of the data and make the evaluation public (through technical notes and documentation included in reports of results or through a separate report) to allow users to interpret results of analyses and to help designers of recurring surveys focus improvement efforts.

This is related to communication.

10.3.4   Production of estimates and projections

10.3.4.1   Developing estimates and projections

Standard 4.1: Agencies must use accepted theory and methods when deriving direct survey-based estimates, as well as model-based estimates and projections that use survey data. Error estimates must be calculated and disseminated to support assessment of the appropriateness of the uses of the estimates or projections. Agencies must plan and implement evaluations to assess the quality of the estimates and projections.

This standard is aimed at statistical generalizability and focuses on the quality of the data analysis (deriving estimates can be considered part of the data analysis component).

10.3.5   Data analysis

10.3.5.1   Analysis and report planning

Standard 5.1: Agencies must develop a plan for the analysis of survey data prior to the start of a specific analysis to ensure that statistical tests are used appropriately and that adequate resources are available to complete the analysis.

This standard is again focused on analysis quality.

10.3.5.2   Inference and comparisons

Standard 5.2: Agencies must base statements of comparisons and other statistical conclusions derived from survey data on acceptable statistical practice.

10.3.6   Review procedures

10.3.6.1   Review of information products

Standard 6.1: Agencies are responsible for the quality of information that they disseminate and must institute appropriate content/subject matter, statistical, and methodological review procedures to comply with OMB and agency InfoQ guidelines.

10.3.7   Dissemination of information products

10.3.7.1   Releasing information

Standard 7.1: Agencies must release information intended for the general public according to a dissemination plan that provides for equivalent, timely access to all users and provides information to the public about the agencies' dissemination policies and procedures including those related to any planned or unanticipated data revisions.

This standard touches on chronology of data and goal and communication. It can also affect temporal relevance of studies that rely on the dissemination schedule.

10.3.7.2   Data protection and disclosure avoidance for dissemination

Standard 7.2: When releasing information products, agencies must ensure strict compliance with any confidentiality pledge to the respondents and all applicable federal legislation and regulations.

10.3.7.3   Survey documentation

Standard 7.3: Agencies must produce survey documentation that includes those materials necessary to understand how to properly analyze data from each survey, as well as the information necessary to replicate and evaluate each survey's results (see also Standard 1.2). Survey documentation must be readily accessible to users, unless it is necessary to restrict access to protect confidentiality. Proper documentation is essential for proper communication.

10.3.7.4   Documentation and release of public-use microdata

Standard 7.4: Agencies that release microdata to the public must include documentation clearly describing how the information is constructed and provide the metadata necessary for users to access and manipulate the data (see also Standard 1.2). Public-use microdata documentation and metadata must be readily accessible to users.This standard is aimed at adequate communication of the data (not the results).

These standards provide a comprehensive framework for the various activities involved in planning and implementing official statistics surveys. Section 10.4 is focused on customer satisfaction surveys such as the surveys on service of general interest (SGI) conducted within the European Union (EU).

10.4   Standards for customer surveys

Customer satisfaction, according to the ISO 10004:2010 standards of the International Organization for Standardization (ISO), is the "customer's perception of the degree to which the customer's requirements have been fulfilled." It is "determined by the gap between the customer's expectations and the customer's perception of the product [or service] as delivered by the organization".

ISO describes the importance of standards on their website: "ISO is a nongovernmental organization that forms a bridge between the public and private sectors. Standards ensure desirable characteristics of products and services such as quality, environmental friendliness, safety, reliability, efficiency, and interchangeability—and at an economical cost."

ISO's work program ranges from standards for traditional activities such as agriculture and construction, to mechanical engineering, manufacturing, and distribution, to transport, medical devices, information and communication technologies, and standards for good management practice and for services. Its primary aim is to share concepts, definitions, and tools to guarantee that products and services meet expectations. When standards are absent, products may turn out to be of poor quality, they may be incompatible with available equipment or they could be unreliable or even dangerous

The goals and objectives of customer satisfaction surveys are clearly described in ISO 10004. "The information obtained from monitoring and measuring customer satisfaction can help identify opportunities for improvement of the organization's strategies, products, processes, and characteristics that are valued by customers, and serve the organization's objectives. Such improvements can strengthen customer confidence and result in commercial and other benefits."

We now provide a brief description of the ISO 10004 standard which provides guidelines for monitoring and measuring customer satisfaction. The rationale of the ISO 10004 standard—as reported in Clause 1—is to provide "guidance in defining and implementing processes to monitor and measure customer satisfaction." It is intended for use "by organizations regardless of type, size, or product provided," but it is related only "to customers external to the organization."

The ISO approach outlines three phases in the processes of measuring and monitoring customer satisfaction: planning (Clause 6), operation (Clause 7), and maintenance and improvement (Clause 8). We examine each of these three and their relation to InfoQ. Table 10.2 summarizes these relationships.

10.4.1 Planning

The planning phase refers to "the definition of the purposes and objectives of measuring customer satisfaction and the determination of the frequency of data gathering (regularly, on an occasional basis, dictated by business needs or specific events)." For example, an organization might be interested in investigating reasons for customer complaints after the release of a new product or for the loss of market share. Alternatively it might want to regularly compare its position relative to other organizations. Moreover, "Information regarding customer satisfaction might be obtained indirectly from the organization's internal processes (e.g., customer complaints handling) or from external sources (e.g., reported in the media) or directly from customers." In determining the frequency of data collection, this clause is related to chronology of data and goal as well as to temporal relevance. The "definition of … customer satisfaction" concerns construct operationalization. The collection of data from different sources indirectly touches on data structure and resolution. Yet, the use of "or" for choice of data source indicates no intention of data integration.

10.4.2 Operation

The operation phase represents the core of the standard and introduces operational steps an organization should follow in order to meet the requirements of ISO 10004. These steps are as follows:

a. identify the customers (current or potential) and their expectations

b. gather customer satisfaction data directly from customers by a survey and/or indirectly examining existing sources of information, after having identified the main characteristics related to customer satisfaction (product, delivery, or organizational characteristic)

c. analyze customer satisfaction data after having chosen the appropriate method of analysis

d. communicate customer satisfaction information

e. monitor customer satisfaction at defined intervals to control that "the customer satisfaction information is consistent with, or validated by, other relevant business performance indicators" (Clause 7.6.5)

Statistical issues mentioned in ISO 10004 relate to the number of customers to be surveyed (sample size), the method of sampling (Clause 7.3.3.3 and Annex C.3.1, C3.2), and the choice of the scale of measurement (Clause 7.3.3.4 and Annex C.4). Identifying the population of interest and sample design is related to generalization. Communication is central to step (d). Step (e) refers to data integration and the choice of measurement scale is related to data resolution.

10.4.3   Maintenance and improvement

The maintenance and improvement phase includes periodic review, evaluation, and continual improvement of processes for monitoring and measuring customer satisfaction. This phase is aimed at maintaining generalizability and temporal relevance as well as the appropriateness of construct operationalization ("reviewing the indirect indicators of customer satisfaction"). Data integration is used to validate the information against other sources. Communication and actionable operationalization are also mentioned (See Table 10.2).

10.5   Integrating official statistics with administrative data for enhanced InfoQ

As mentioned in the introduction to this chapter, official statistics are produced by a variety of organizations including central bureaus of statistics, regulatory healthcare agencies, educational systems, and national banks. A common trend is the integration of official statistics and organizational data to derive insights at the local and global levels. An example is provided by the Intesa Sanpaolo Bank in Italy that maintains an integrated database for supporting analytic research requests by management and various decision makers (Forsti et al., 2012). The bank uses regression models applied to internal data integrated with data from a range of official statistics providers such as:

- Financial statements (CEBI)

- EPO patents (Thomson Scientific)

Table 10.2 Relationship between ISO 10004 guidelines and InfoQ dimensions. Shaded cells indicate an existing relationship.

ISO 10004 phase	Data resolution	Data structure	Data integration	Temporal relevance	Data-goal chronology	Generalizability	Operationalization	Communication
Planning		■			■		■	
Operation	■		■	■		■		■
Maintenance and improvement				■		■		■

- Foreign direct investment (Reprint)

- ISO certificates (ACCREDIA)

- Trademarks (UIBM, OIHM, USPTO, and WIPO)

- Credit ratings (CEBI and Intesa Sanpaolo)

- Corporate group charts (Intesa Sanpaolo)

In this example, the competitiveness of an enterprise was assessed using factors such as innovation and R&D; intangibles (e.g., human capital, brands, quality, and environmental awareness); and foreign direct investment. Some of the challenges encountered in order to obtain a coherent integrated database included incomplete matching using "tax ID No." as the key since patent, certification, and trademark archives contain only the business name and address of the enterprise. As a result, an algorithm was developed for matching a business name and address to other databases containing both the same information and the tax ID No. With this approach different business names and addresses may appear for the same enterprise (for instance, abbreviated names, acronyms with or without full stops, presence of the abbreviated legal form, etc.). The tax ID No. of an enterprise may also change over the years. Handling these issues properly is key to the quality of information generated by regression analysis. For more details, see Forsti et al. (2012).

This section is about information derived from an analysis of official statistics data integrated with administrative datasets such as the Intesa bank example. We describe the use of graphical models for performing the integration. The objective is to provide decision makers with high-quality information. We use the InfoQ concept framework for evaluating the quality of such information to decision makers or other stakeholders. We refer here to two case studies analyzed in Dalla Valle and Kenett (2015), one from the field of education and the other from accident reporting. The case studies demonstrate a calibration methodology developed by Dalla Valle and Kenett (2015) for calibrating official statistics with administrative data. For other examples, see Dalla Valle (2014).

10.5.1   Bayesian networks

BNs are directed acyclic graphs (DAGs) whose nodes represent variables and the edges represent causal relationships between the variables. These variables are associated with conditional probability functions that, together with the DAG, are able to provide a compact representation of high-dimensional distributions. For an introduction and for more details about the definitions and main properties of BNs, see Chapters 7 and 8. These models were applied in official statistics data analysis by Penny and Reale (2004) who used graphical models to identify relevant components in a saturated structural VAR model of the quarterly gross domestic product that aggregates a large number of economic time series. More recently, Vicard and Scanu (2012) also applied BNs to official statistics, showing that the use of poststratification allows integration and missing data imputation. For general applications of BNs, see Kenett and Salini (2009), Kenett and Salini (2012), and Kenett (2016). For an application of BNs in healthcare, see Section 8.4.

10.5.2   Calibration methodology

The methodology proposed by Dalla Valle and Kenett (2015) increases InfoQ via data integration of official and administrative information, thus enhancing temporal relevance and chronology of data and goal. The idea is in the same spirit of external benchmarking used in small area estimation (Pfeffermann, 2013). In small area estimation, benchmarking robustifies the inference by forcing the model-based predictors to agree with a design-based estimator. Similarly, the calibration methodology by Dalla Valle and Kenett (2015) is based on qualitative data calibration performed via conditionalizing graphical models, where official statistics estimates are updated to agree with more timely administrative data estimates. The calibration methodology is structured in three phases:

Phase 1: Data structure modeling. This phase consists of conducting a multivariate data analysis of the official statistics and administrative datasets, using graphical models such as vines and BNs. Vines are a flexible class of multivariate copulas based on the decomposition of a multivariate copula using bivariate (conditional) copulas as building blocks. Vines are employed to model the dependence structure among the variables, and the results are used to construct the causal relationships with a BN.

Phase 2: Identification of the calibration link. In the second phase, a calibration link, in the form of common correlated variables, is identified between the official statistics and the administrative data.

Phase 3: Performing calibration. In the last phase, the BNs of both datasets are conditioned on specific "target" variables in order to perform calibration, taking into account the causal relationship among the variables.

These phases are applied to each of the datasets to be integrated. In Sections 10.5.3 and 10.5.4, we demonstrate the application of these three phases using case studies from education and transportation.

10.5.3   The Stella education case study

10.5.3.1   Stella dataset

The dataset used here was collected by the Italian Stella association in 2009. Stella is an interuniversity initiative aiming at cooperation and coordination of the activities of supervision, statistical analysis, and evaluation of the graduate and postgraduate paths. The initiative includes universities from the north and the center of Italy. The Stella dataset contains information about postdoctoral placements after 12 months for graduates who obtained a PhD in 2005, 2006, and 2007. The dataset includes 665 observations and eight variables.

The variables are as follows:

1. yPhD: year of PhD completion

2. ybirth: year of birth

3. unistart: starting year of university degree

4. hweek: working hours per week

5. begsal: initial net monthly salary in euro

6. lastsal: last net monthly salary in euro

7. emp: number of employees

8. estgrow: estimate of net salary rise in 2011 by percent

10.5.3.2   Graduates dataset

The second dataset contains information collected through an internal small survey conducted locally in a few universities of Lombardy, in Northern Italy, and in Rome. The sample survey on university graduates' vocational integration is based on interviews with graduates who attained the university degree in 2004. The survey aims at detecting graduates' employment conditions about four years after graduation. From the initial sample, the researchers only considered those individuals who concluded their PhD and are currently employed. After removing the missing values, they obtained a total of 52 observations. The variables in this dataset are as follows:

- mdipl: diploma final mark

- nemp: number of employees

- msalary: monthly net salary in euros

- ystjob: starting year of employment

10.5.3.3   Combining the Stella and Graduates datasets

We now describe the approach taken by Dalla Valle and Kenett (2015) for integrating the Stella data with the Graduates data for the purpose of updating the 2004 survey data with 2009 data from the Stella data.

Operations on Stella dataset:

1. *Data structure modeling.* A vine copula was applied to the Stella dataset in order to explore the dependence structure of the marginal distributions. The strongest dependencies were between begsal and lastsal. Moreover, the last salary (lastsal) is associated with the estimate of salary growth (estgrow) and with the number of employees of the company (emp). We notice two groups

of variables: one group includes variables regarding the company (begsal, lastsal, estgrow, and emp) and the other group includes variables regarding the individual (yPhD, ybirth, and unistart). The variable hweek is only dependent on lastsal conditional on emp. The vine model will help to determine the conditional rank correlations, which are necessary to define the corresponding BN. The BN represented in Figure 10.1 is the best network obtained by model validation, where statistical tests support the validity of the BN.

2. *Identification of the calibration link.* The calibration link used here is the last-sal variable. This decision was reached by discussion with education experts and is therefore subjective and context related.

3. *Performing calibration.* For calibration purposes, the Stella dataset is conditioned on a lower value of lastsal, similar to the average salary value of the Graduates dataset. In order to reach this lower value of salary, begsal, estgrow, and emp had to be decreased, as shown in Figure 10.2.

Furthermore, the Stella dataset is conditioned on a low value of begsal and emp and for a high value of yPhD. The variable yPhD is considered as a proxy of the starting year of employment, used in the Graduates dataset. In this case,

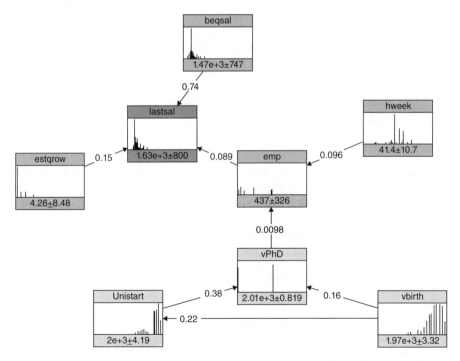

Figure 10.1   BN for the Stella dataset. Source: Dalla Valle and Kenett (2015). Reproduced with permission of John Wiley & Sons, Inc.

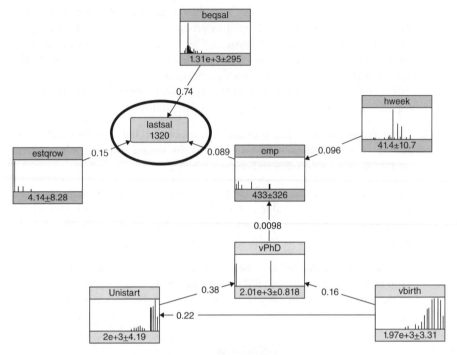

Figure 10.2   BN is conditioned on a value of lastsal which is similar to the salary value of the Graduates dataset. Source: Dalla Valle and Kenett (2015). Reproduced with permission of John Wiley & Sons, Inc.

the average salary decreases to a figure similar to the average value of the Graduates dataset, as in Figure 10.3.

Operations on Graduates dataset:

1. *Data structure modeling.* The researchers applied a vine copula to the Graduates dataset to explore the dependence structure of the marginals. Here the monthly salary is associated with the diploma final mark. The monthly salary is also associated with the number of employees only conditionally on the diploma final. They then applied the BN to the Graduates data, and the network in Figure 10.4 is the best one obtained by model validation.

2. *Identification of the calibration link.* The calibration link is the msalary variable, which is analogous to lastsal in the official dataset.

3. *Performing calibration.* For calibration purposes, the Graduates dataset is conditioned on a high value of msalary, similar to the average salary value of the Stella dataset. In order to reach this higher value of salary, both mdipl and nemp need to be increased, as in Figure 10.5. Finally, the Graduates dataset is conditioned on a high value of mdipl and nemp and for a low value of ystjob. In this case, the average salary increases, as in Figure 10.6.

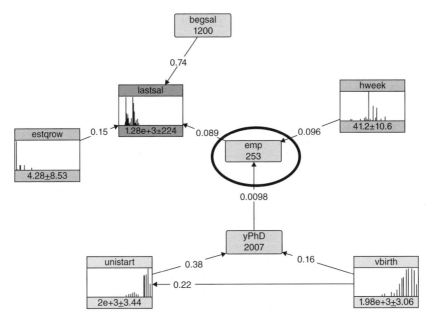

Figure 10.3   BN is conditioned on a low value of begsal and emp and for a high value of yPhD. Source: Dalla Valle and Kenett (2015). Reproduced with permission of John Wiley & Sons, Inc.

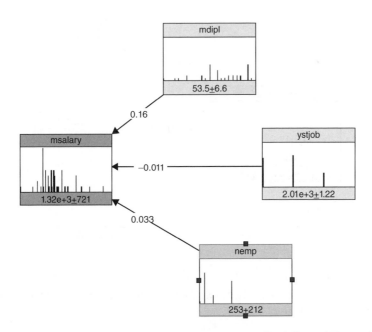

Figure 10.4   BN for the Graduates dataset. Source: Dalla Valle and Kenett (2015). Reproduced with permission of John Wiley & Sons, Inc.

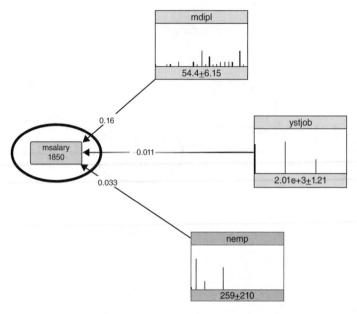

Figure 10.5   BN is conditioned on a high value of msalary. Source: Dalla Valle and Kenett (2015). Reproduced with permission of John Wiley & Sons, Inc.

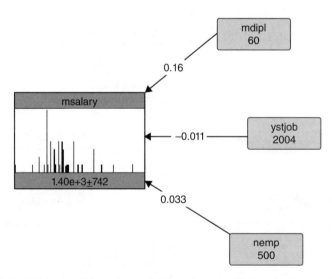

Figure 10.6   BN is conditioned on a high value of mdipl and nemp and for a low value of ystjob. Source: Dalla Valle and Kenett (2015). Reproduced with permission of John Wiley & Sons, Inc.

10.5.3.4   Stella education case study: InfoQ components

Goal (*g*): Evaluating the working performance of graduates and understanding the influence on the salary of company-related variables and individual-related variables.

Data *(X)*: Combined survey data with official statistics on education.

Analysis *(f)*: Use of vines and BNs to model the dependence structure of the variables in the data set and to calculate the conditional rank correlations.

Utility *(U)*: Assisting policymakers to monitor the relationship between education and the labor market, identifying trends and methods of improvement.

10.5.3.5   Stella education case study: InfoQ dimensions

1. *Data resolution.* Concerning the aggregation level, the data is collected at the individual level in the Graduates and Stella datasets, aligned with the study goal. The Graduates dataset contains information collected through an internal small survey conducted locally in a few Italian universities. Although the data collection should comply with a good standard of accuracy, detailed information about the data collection is not available for this dataset. Stella data is monitored by the authority of a consortium of a large number of universities, guaranteeing the reliability and precision of the data produced. The Graduates survey was produced for a specific study conducted internally by a few universities and data is not collected on a regular basis, while Stella produces periodic annual reports about education at a large number of Italian institutions. Considering the analysis goal of regularly monitoring the performance of the graduates in the labor market, there is still room for improvement for the level of InfoQ generated by this dimension, especially on the data resolution of the Graduates dataset.

2. *Data structure.* The type and structure of education data of both datasets are perfectly aligned with the goal of understanding the influence of different factors on the graduates' salary. Both datasets include continuous and categorical information based on self-reporting. Although the Stella data integrity is guaranteed by educational institution authorities, the dataset contained a small percentage of missing data, which was removed before implementing the methodology. The Graduates dataset contained a certain number of missing data, which was removed from the dataset. Information describing the corruptness of the data is unavailable. The level of InfoQ could be improved, especially for the Graduates dataset.

3. *Data integration.* This methodology allows the integration of multiple sources of information, that is, official statistics and survey data. The methodology performs integration through data calibration, incorporating the dependence structure of the variables using vines and BNs. Multiple dataset integration creates

new knowledge regarding the goal of understating the influence of company and individual variables on graduates' salary, thereby enhancing InfoQ.

4. *Temporal relevance.* To make the analysis effective for its goals, the time gaps between the collection, analysis, and deployment of this data should be of short duration, and the time horizon from the first to the last phase should not exceed one year. Stella data is updated annually and made available within a few months, while for the Graduates data, we do not have specific information about the time period between the data collection and deployment. Considering the goal of assisting policymakers to annually monitor the labor market and to allocate education resources, this dimension produces a reasonably good level of InfoQ, which could be increased by a more timely availability of the Graduates survey data. Moreover, the analysis of year 2009 could be made relevant to recent years (e.g., 2014) by calibrating the data with dynamic variables, which would enable updating the study information and in turn enhancing InfoQ.

5. *Chronology of data and goal.* Vines allow calculating associations among variables and identifying clusters of variables. Moreover, nonparametric BNs allow predictive and diagnostic reasoning through the conditioning of the output. Therefore, the methodology is highly effective for obtaining the goal of identifying and understanding the causal structure between variables.

6. *Generalizability.* The diagnostic and predictive capabilities of BNs provide generalizability to population subsets. The Graduates survey is generalized by calibration with the Stella dataset to a large population including several universities. However, we could still improve InfoQ by calibrating the data with variables referring to other institutions in order to make the study fully generalizable at a national level.

7. *Operationalization.* The methodology allows monitoring the performance on the labor market of graduates, describing the causal relationships between the salary and variables related to individuals and companies. Moreover, via conditioning, it allows us to calibrate the results on education obtained by small surveys with the results obtained by official sources. Therefore, the outputs provided from the model are highly useful to policymakers. The use of a model with conditioning capabilities provides an effective tool to set up improvement goals and to detect weaknesses in the education system and in its relationship with industries.

8. *Communication.* The graphical representations of vines and BN are particularly effective for communicating to technical and non-technical audiences. The visual display of a BN makes it particularly appealing to decision makers who feel uneasy with mathematical or other nontransparent models.

Based on this analysis, we summarize the InfoQ scores for each dimension as shown in Table 10.3. The overall InfoQ score for this study is 74%.

Table 10.3   Scores for InfoQ dimensions
for Stella education case study.

InfoQ dimension	Score
Data resolution	3
Data structure	3
Data integration	5
Temporal relevance	3
Chronology of data and goal	5
Generalizability	4
Operationalization	5
Communication	5

Scores are on a 5-point scale.

10.5.4   NHTSA transport safety case study

The second case study is based on the Vehicle Safety dataset. The National Highway Traffic Safety Administration (NHTSA), under the US Department of Transportation, was established by the Highway Safety Act of 1970, as the successor to the National Highway Safety Bureau, to carry out safety programs under the National Traffic and Motor Vehicle Safety Act of 1966 and the Highway Safety Act of 1966. NHTSA also carries out consumer programs established by the Motor Vehicle Information and Cost Savings Act of 1972 (www.nhtsa.gov). NHTSA is responsible for reducing deaths, injuries, and economic losses resulting from motor vehicle crashes. This is accomplished by setting and enforcing safety performance standards for motor vehicles and motor vehicle equipment and through grants to state and local governments to enable them to conduct effective local highway safety programs. NHTSA investigates safety defects in motor vehicles; sets and enforces fuel economy standards; helps states and local communities to reduce the threat of drunk drivers; promotes the use of safety belts, child safety seats, and air bags; investigates odometer fraud; establishes and enforces vehicle anti-theft regulations; and provides consumer information on motor vehicle safety topics. NHTSA also conducts research on driver behavior and traffic safety to develop the most efficient and effective means of bringing about safety improvements.

10.5.4.1   Vehicle Safety dataset

The Vehicle Safety data represents official statistics. After removing the missing data, we obtain a final dataset with 1241 observations, where each observation includes 14 variables on a car manufacturer, between the late 1980s and the early 1990s. The variables are as follows:

1. HIC: Head Injury, based on the resultant acceleration pulse for the head centre of gravity

2. T1: Lower Boundary of the time interval over which the HIC was computed

3. T2: Upper Boundary of the time interval over which the HIC was computed

4. CLIP3M: Thorax Region Peak Acceleration, the maximum three-millisecond "clip" value of the chest resultant acceleration

5. LFEM: Left Femur Peak Load Measurement, the maximum compression load for the left femur

6. RFEM: Right Femur Peak Load Measurement, the maximum compression load for the right femur

7. CSI: Chest Severity Index

8. LBELT: Lap Belt Peak Load Measurement, the maximum tension load on the lap belt

9. SBELT: Shoulder Belt Peak Load Measurement, the maximum tension load on the shoulder belt

10. TTI: Thoracic Trauma Index, computed on a dummy from the maximum rib and lower spine peak accelerations

11. PELVG: Pelvis Injury Criterion, the peak lateral acceleration on the pelvis

12. VC: Viscous Criterion

13. CMAX: Maximum Chest Compression

14. NIJ: Neck Injury Criterion

10.5.4.2   Crash Test dataset

The Crash Test dataset contains information about vehicle crash tests collected by a car manufacturer company for marketing purposes. The data contains variables measuring injuries of actual crash tests and is collected following good accuracy standards. We consider this data as administrative or organizational data.

The dataset is a small sample of 176 observations about vehicle crash tests. A range of US-made vehicles containing dummies in the driver and front passenger seats were crashed into a test wall at 35 miles/hour and information was collected, recording how each crash affected the dummies. The injury variables describe the extent of head injuries, chest deceleration, and left and right femur load. The data file also contains information on the type and safety features of each vehicle. A brief description of the variables within the data is provided as follows:

- Head_IC: Head injury criterion

- Chest_decel: Chest deceleration

- L_Leg: Left femur load

- R_Leg: Right femur load

- Doors: Number of car doors in the car

- Year: Year of manufacture

- Wt: Vehicle weight in pounds

10.5.4.3   Combining the Vehicle Safety and Crash Test datasets

We now describe the graphical approach for combining the official and administrative datasets.

Operations on Vehicle Safety dataset

1. *Data structure modeling.* Dalla Valle and Kenett (2015) applied a Gaussian regular vine copula to the Vehicle Safety dataset to explore the dependence structure of the marginal distributions. There are strong dependencies among most of the variables, for example, between CMAX and CSI. The variable CLIP3M is only dependent on CMAX conditional on LBELT. The vine model helps determine the conditional rank correlations, necessary for defining the corresponding BN. The researchers then applied a BN to the Vehicle Safety data. Figure 10.7 represents the BN for the Vehicle Safety data, where the nodes RFEM, HIC, LFEM, and CSI represent the variables also present in the Crash Test dataset. The BN in Figure 10.7 is the best network obtained by model validation.

2. *Identification of the calibration link.* The calibration links are the HIC, CSI, LFEM, and RFEM variables, which are analogous, respectively, to Head_IC, Chest_decel, L_Leg, and R_Leg in the Crash Test dataset.

3. *Performing calibration.* For calibration purposes, the Vehicle Safety dataset is conditioned on a low value of RFEM, similar to the average values of the Crash Test dataset, and a slightly higher value of CLIP3M. When changing the right femur and thorax region load values, HIC decreases, while CSI increases, becoming very similar to the corresponding values of the Crash Test dataset, as in Figure 10.7. Furthermore, the Vehicle Safety dataset is conditioned on a high value of CSI and for a low value of HIC, similar to the corresponding values of the Crash Test dataset (Figure 10.7). In this case, the left and right femur loads decrease, becoming closer to the corresponding average values of the Crash Test dataset.

Operations on Crash Test dataset

1. *Data structure modeling.* The researchers applied a vine copula to the Crash Test dataset to explore the dependence structure of the marginal distributions. Here the chest deceleration is associated with the head injury criterion and

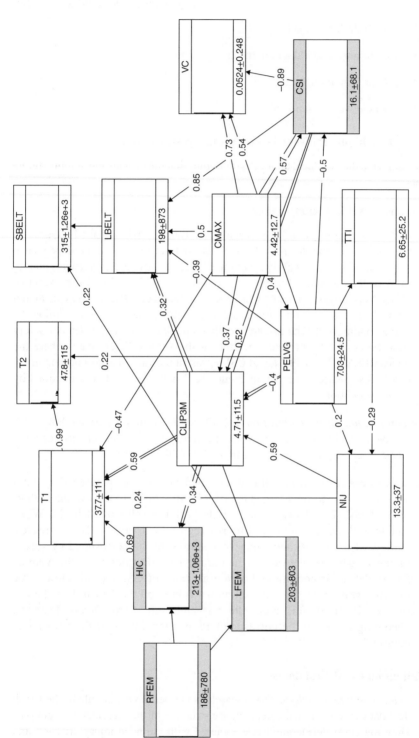

Figure 10.7 BN for the Vehicle Safety dataset. Source: Dalla Valle and Kenett (2015). Reproduced with permission of John Wiley & Sons, Inc.

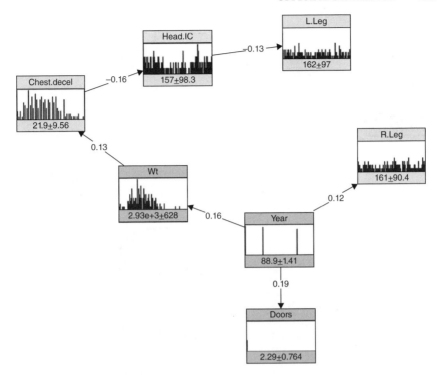

Figure 10.8   BN for the Crash Test dataset. Source: Dalla Valle and Kenett (2015). Reproduced with permission of John Wiley & Sons, Inc.

with the vehicle weight. The chest deceleration is associated with the left femur load only conditionally on the head injury criterion.

2. *Identification of the calibration link.* The researchers applied the BN to the Crash Test data. Figure 10.8 displays the best network obtained by model validation for the Crash Test data. The nodes Year, Doors, and Wt denote car type variables, and the other nodes denote injury variables.

3. *Performing calibration.* For calibration purposes, the Crash Test dataset is conditioned on a high value of Wt and Year (diagnostic reasoning). We notice from the changes in Chest.decel and Head.IC that a recent and lighter vehicle is safer and causes less severe injuries (Figure 10.9). Finally, the Crash Test dataset is conditioned on a low value of Wt and Year. An older and heavier vehicle is less safe and causes more severe injuries (Figure 10.10).

10.5.4.4   NHTSA safety case study: InfoQ components

Goal *(g)*: Assessing motor vehicle's safety and evaluating the severities of injuries resulting from motor vehicle crashes.

Data *(X)*: Combined small sample of Crash Test data with motor vehicle's safety official statistics.

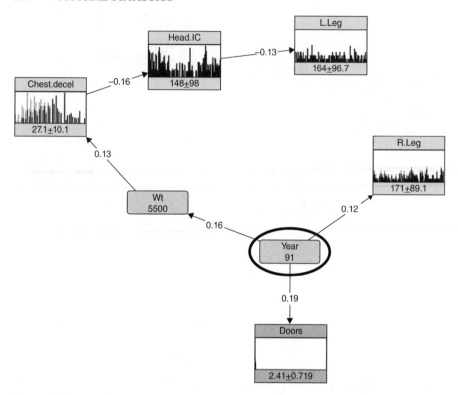

Figure 10.9   BN for the Crash Test dataset is conditioned on a high value of Wt and Year. Source: Dalla Valle and Kenett (2015). Reproduced with permission of John Wiley & Sons, Inc.

Analysis *(f)*: Use of vines and BNs to model the dependence structure of the variables in the data set and to calculate the conditional rank correlations.

Utility *(U)*: Assisting policymakers to set safety performance standards for motor vehicles and motor vehicle equipment, aiming at improving the overall safety of vehicles and therefore reducing deaths, injuries, and economic losses resulting from motor vehicle crashes.

10.5.4.5   NHTSA transport safety case study: InfoQ dimensions

1. *Data Resolution.* Concerning the aggregation level, the data is collected at crash test level in the Vehicle Safety as well as in the Crash Test dataset, aligned with the study goal. The Crash Test dataset contains information about vehicle crash tests collected by a car manufacturer company for marketing purposes. The data contains variables measuring injuries of actual crash tests and is collected following good accuracy standards. The Vehicle Safety data is monitored by the US NHTSA, guaranteeing the reliability and precision of the data. The Crash Test dataset was produced for a specific study

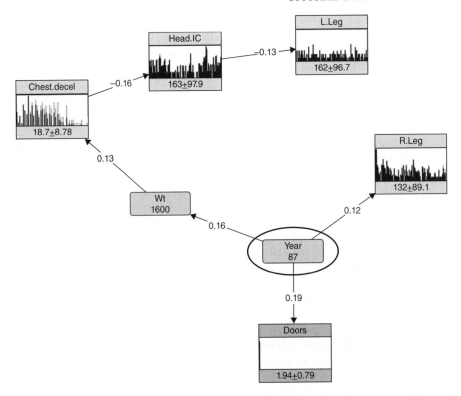

Figure 10.10   BN for the Crash Test dataset is conditioned on a low value of Wt and Year. Source: Dalla Valle and Kenett (2015). Reproduced with permission of John Wiley & Sons, Inc.

conducted internally by a car company and the data is not collected on a regular basis, while the NHTSA produces periodic reports about transport safety of a large range of motor vehicles. Considering the analysis goal of regularly evaluating motor vehicle safety, there is still room for improvement for the level of InfoQ generated by this dimension, especially on the data resolution of the Crash Test dataset.

2. *Data structure.* The type and structure of car safety data of both datasets are perfectly aligned with the goal of assessing motor vehicle safety and evaluating the severities of injuries resulting from motor vehicle crashes. The data includes continuous and categorical variables. Although the Vehicle Safety data integrity is guaranteed by the NHTSA, the dataset contained a small percentage of missing data, which was removed before implementing the methodology. The Crash Test dataset did not contain missing data and an exploratory data analysis revealed no corruptness in the data. The level of InfoQ for this dimension is therefore high.

3. *Data integration.* The calibration methodology allows the integration of multiple sources of information, that is, official statistics and marketing data. The methodology performs integration though data calibration, incorporating the dependence structure of the variables using vines and BNs. Multiple dataset integration creates new knowledge regarding the goal of assessing motor vehicle safety and evaluating the severities of injuries resulting from motor vehicle crashes, thereby enhancing InfoQ.

4. *Temporal relevance.* The time gaps between the collection, analysis, and deployment of this data should be of short duration, to make the analysis effective for its goals. Vehicle Safety data is updated quarterly and made available within a few months, while for the Crash Test data we do not have specific information about the time period between the data collection and deployment. Considering the goal of assisting policymakers to set safety performance standards for motor vehicles and motor vehicle equipment, this dimension produces a reasonably good level of InfoQ, which could be increased by a more timely availability of the Crash Test data. Moreover, the analysis of the 1987–1992 cars could be easily made relevant to recent years (e.g., 2014) by updating the data to include more recent vehicle models, thereby enhancing InfoQ.

5. *Chronology of data and goal.* Vines allow calculating associations among variables and identifying clusters of variables. Moreover, nonparametric BNs allow predictive and diagnostic reasoning though the conditioning of the output. Therefore, the methodology is highly effective for obtaining the goal of identifying and understanding the causal structure between variables.

6. *Generalizability.* The diagnostic and predictive capabilities of BNs provide generalizability to population subsets. The Crash Test survey is generalized by calibration with the Vehicle Safety dataset to a large range of motor vehicles population, enhancing InfoQ.

7. *Operationalization.* The construct *safety* is operationalized in both datasets using similar measurements. The methodology allows assessing motor vehicle safety and evaluating the severities of injuries resulting from motor vehicle crashes, describing the causal relationships between injuries in various parts of the body, in case of an accident, and the type of vehicle. Moreover, via conditioning, it allows us to calibrate the results on vehicle safety obtained by small datasets with the results obtained by official sources. Therefore, the outputs provided from the model are highly useful to policymakers, to set guidelines aiming at improving the overall safety of vehicles. The use of a model with conditioning capabilities provides an effective tool to set up improvement goals and to detect weaknesses in transport safety.

8. *Communication.* The graphical representations of vines and BNs are particularly effective for communicating with technical as well as non-technical audiences. The visual display of a BN makes it particularly appealing to decision makers who feel uneasy with mathematical or other nontransparent models.

Table 10.4   Scores for InfoQ dimensions for the NHTSA safety case study.

InfoQ dimension	Score
Data resolution	3
Data structure	4
Data integration	5
Temporal relevance	3
Chronology of data and goal	5
Generalizability	5
Operationalization	5
Communication	5

Scores are on a 5-point scale.

Based on this analysis, we summarize the InfoQ scores for each dimension as shown in Table 10.4. The overall InfoQ score for this study is 81%.

10.6   Summary

With the increased availability of data sources and ubiquity of analytic technologies, the challenge of transforming data to information and knowledge is growing in importance (Kenett, 2008). Official statistics play a critical role in this context and applied research, using official statistics, needs to ensure the generation of high InfoQ (Kenett and Shmueli, 2014, 2016). In this chapter, we discussed the various elements that determine the quality of such information and described several proposed approaches for achieving it. We compared the InfoQ concept with NCSES and ISO standards and also discussed examples of how official statistics data and data from internal sources are integrated to generate higher InfoQ.

The sections on quality standards in official statistics and ISO standards related to customer surveys discuss aspects related to five of the InfoQ dimensions: data resolution, data structure, data integration, temporal relevance, and chronology of data and goal. Considering each of these InfoQ dimensions with their associated questions can help in increasing InfoQ. The chapter ends with two examples where official statistics datasets are combined with organizational data in order to derive, through analysis, information of higher quality. An analysis using BNs permits the calibration of the data, thus strengthening the quality of the information derived from the official data. As before, the InfoQ dimensions involved in such calibration include data resolution, data structure, data integration, temporal relevance, and chronology of data and goal. These two case studies demonstrate how concern for the quality of the information derived from an analysis of a given data set requires attention to several dimensions, beyond the quality of the analysis method used. The eight InfoQ dimensions provide a general template for identifying and evaluating such challenges.

References

Biemer, P. and Lyberg, L. (2003) *Introduction to Survey Quality*. John Wiley & Sons, Inc., Hoboken.

Biemer, P.P., Trewin, D., Bergdahl, H., Japec, L. and Pettersson, Å. (2012) A Tool for Managing Product Quality. *European Conference on Quality in Official Statistics*, Athens.

Dalla Valle, L. (2014) Official statistics data integration using copulas. *Quality Technology and Quantitative Management*, 11(1), pp. 111–131.

Dalla Valle, L. and Kenett, R.S. (2015) Official statistics data integration to enhance information quality. *Quality and Reliability Engineering International*. 10.1002/qre.1859.

David, L., Kennedy, R., King, G. and Vespignani, A. (2014) The parable of Google Flu: traps in big data analysis. *Science*, 343(14), pp. 1203–1205.

Deming, W.E. (1982) *Out of the Crisis*. MIT Press, Cambridge, MA.

EPA (2005) Uniform Federal Policy for Quality Assurance Project Plans: Evaluating, Assessing, and Documenting Environmental Data Collection and Use Programs. https://www.epa.gov/sites/production/files/documents/ufp_qapp_v1_0305.pdf (accessed May 20, 2016).

Eurostat (2003) *Standard Quality Report*. Eurostat, Luxembourg.

Eurostat (2009) *Handbook for Quality Reports*. Eurostat, Luxembourg.

Figini, S., Kenett, R.S. and Salini, S. (2010) Integrating operational and financial risk assessments. *Quality and Reliability Engineering International*, 26, pp. 887–897.

Forbes, S. and Brown, D. (2012) Conceptual thinking in national statistics offices. *Statistical Journal of the IAOS*, 28, pp. 89–98.

Forsti, G., Guelpa, F. and Trenti, S. (2012) Enterprise in a Globalised Context and Public and Private Statistical Setups. *Proceedings of the 46th Scientific Meeting of the Italian Statistical Society (SIS)*, Rome.

Ginsberg, J., Mohebbi, M.H., Patel, R.S., Brammer, L., Smolinski, M.S. and Brilliant, L (2009) Detecting influenza epidemics using search engine query. *Nature*, 457, pp. 1012–1014.

Giovanini, E. (2008) *Understanding Economic Statistics*. OECD Publishing, Paris, France.

Goldenberg, A., Shmueli, G., Caruana, R.A. and Fienberg, S.E. (2002) Early statistical detection of anthrax outbreaks by tracking over-the-counter medication sales. *Proceedings of the National Academy of Sciences*, 99 (8), pp. 5237–5240.

Good, D. and Irwin, S. (2011) USDA Corn and Soybean Acreage Estimates and Yield Forecasts: Dispelling Myths and Misunderstandings, Marketing and Outlook Briefs, Farmdoc, http://farmdoc.illinois.edu/marketing/mobr/mobr_11-02/mobr_11-02.html (accessed May 20, 2016).

Goodman, D. and Hambleton, R.K. (2004) Student test score reports and interpretive guides: review of current practices and suggestions for future research. *Applied Measurement in Education*, 17(2), pp. 145–220.

Hambleton, R.K. (2002) How Can We Make NAEP and State Test Score Reporting Scales and Reports More Understandable?, in *Assessment in Educational Reform*, Lissitz, R.W. and Schafer, W.D. (editors), Allyn & Bacon, Boston, MA, pp. 192–205.

Kenett, R.S. (2008) From data to information to knowledge. *Six Sigma Forum Magazine*, November 2008, pp. 32–33.

Kenett, R.S. (2016) On generating high InfoQ with Bayesian networks. *Quality Technology and Quantitative Management*, http://www.tandfonline.com/doi/abs/10.1080/16843703. 2016.1189182?journalCode=ttqm20 (accessed May 5, 2016).

Kenett, R.S. and Salini, S. (2009) New Frontiers: Bayesian networks give insight into survey-data analysis. *Quality Progress*, August, pp. 31–36.

Kenett, R.S. and Salini, S. (2012) *Modern Analysis of Customer Satisfaction Surveys: With Applications Using R*. John Wiley & Sons, Ltd, Chichester, UK.

Kenett, R.S. and Shmueli, G. (2014) On information quality (with discussion). *Journal of the Royal Statistical Society, Series A*, 177(1), pp. 3–38.

Kenett, R.S. and Shmueli, G. (2016) From quality to information quality in official statistics, *Journal of Official Statistics*, in press.

Lazer, D., Kennedy, R., King, G. and Vespignan, A. (2014) The parable of Google flu: traps in big data analysis. *Science*, 343, pp. 1203–1205.

Marbury, D. (2014) Google Flu Trends collaborates with CDC for more accurate predictions. *Medical Economics*, November 5, http://medicaleconomics.modernmedicine.com/medical-economics/news/google-flu-trends-collaborates-cdc-more-accurate-predictions (accessed May 20, 2016).

Office for National Statistics (2007) Guidelines for Measuring Statistical Quality. Office for National Statistics, London.

Penny, R.N. and Reale, M. (2004) Using graphical modelling in official statistics. *Quaderni di Statistica*, 6, pp. 31–48.

Pfeffermann, D. (2013) New important developments in small area estimation. *Statistical Science*, 28 (1), pp. 40–68.

Shmueli, G. and Burkom, H (2010) Statistical challenges facing early outbreak detection in biosurveillance. *Technometrics*, 52(1), 39–51.

Vicard, P. and Scanu, M. (2012) Applications of Bayesian Networks in Official Statistics, in *Advanced Statistical Methods for the Analysis of Large Data-Sets*, Di Ciaccio, A., Coli, M. and Angulo Ibanez, J.M. (editors), Springer, Heidelberg, Germany, pp. 113–123.

Part III
IMPLEMENTING InfoQ

11

InfoQ and reproducible research

11.1 Introduction

The scientific press appears to be witnessing a confusion of terms, with *reproducibility*, *repeatability*, and *replicability* being given different and sometimes conflicting meanings (Kenett and Shmueli, 2015). Statistical work is focused on addressing the needs of the scientific, industrial, business, and service communities, and lack of clarity in terms is certainly an issue that needs to be addressed.

Reproducible research has been a topic of interest for many years, with famous controversies such as the water memory papers. In a 1988 paper published in *Nature*, Benveniste et al. reported that white blood cells called basophils, which control the body's reaction to allergens, can be activated to produce an immune response by solutions of antibodies that have been diluted so much that they contain none of these biomolecules at all. It was as though the water molecules somehow retained a memory of the antibodies that they had previously been in contact with, so that a biological effect remained when the antibodies were no longer present. This seemingly validated the claims made for highly diluted homeopathic medicines. The editor at the time, John Maddox, prefaced the paper with an editorial comment entitled "When to believe the unbelievable," which admitted: "There is no objective explanation of these observations." The editor questioned the quality of the information provided by the original research, but went ahead with the publication itself. For more on this controversial research see Nature (2004).

Information Quality: The Potential of Data and Analytics to Generate Knowledge,
First Edition. Ron S. Kenett and Galit Shmueli.
© 2017 John Wiley & Sons, Ltd. Published 2017 by John Wiley & Sons, Ltd.
Companion website: www.wiley.com/go/information_quality

Ioannidis et al. (2009) present a much discussed initiative to reproduce results from 20 papers on microarray gene expression that were published in *Nature Genetics* between January 2005 and December 2006. The work was done by four teams of three to six scientists each who worked together to evaluate each article. Results could be reproduced in two papers; results were reproduced with some discrepancies in six papers; in ten papers results could not be reproduced; and in two studies data could not be processed. For more on such assessments, see Ioannidis (2005, 2008).

From 2011, the "Reproducibility Project" (https://osf.io/ezcuj/wiki/home/) recruited many experimenters, with the aim of trying to reproduce results from 100 articles published in three well-established psychology journals in 2008. The results replicated with half the magnitude of original effects, representing a substantial decline. 97% of original studies reported significant results ($p < .05$). Only 36% of replications had significant results; 47% of original effect sizes were in the 95% confidence interval of the replication effect size and 39% of effects were rated by a survey to have replicated the original result; (Open Science Collaboration, 2015).

In another example from *Nature* (2014), a team of researchers led by Haruko Obokata of the Riken Center for Developmental Biology in Kobe, Japan, presented two papers claiming that a method using a simple acid-bath method is able to reprogramme mature mammalian cells back to an embryonic state. The paper was criticized for irregularities and apparently duplicated images. Numerous scientists also had difficulty reproducing the supposedly simple method. The team responded with the promise of corrections and a list of tips to help other scientists reproduce the results. However, the *Nature* paper was found to also contain two images apparently duplicated from Obokata's doctoral dissertation which reported experiments dealing with cells derived from a different process in an altogether different experiment. This led Teruhiko Wakayama, a cloning expert at the University of Yamanashi and a corresponding author on one of the papers, to say in a TV interview: "I have lost faith in the paper. Overall there are now just too many uncertainties about it. I think we have to wait for some confirmation... To check the legitimacy of the paper, we should retract it, prepare proper data and images, and then use those to demonstrate, with confidence, that the paper is correct." Wakayama expressed the need to reproduce the results.

These examples characterize the requirement for scientific research to advance science using results that various groups can derive independently. This requirement is fundamental to ensure the information quality (InfoQ) of published research. To achieve reproducible research requires proper documentation, with full visibility into assumptions, clear experimental design considerations, and access to collected data. The ability to perform an exact repetition is, however, not the main characteristic of reproducible research. This distinction is related to differences that arise when using the terms *reproducibility*, *repeatability*, and *replicability*. We use "recreating" for a generic verb when a general term is needed.

In the next four sections, we describe the definitions of repeatability, reproducibility, and replicability in different areas. Our goal is to show the diversity of definitions, which sometimes differ across fields and sometimes even within a field. The key point to note is that the definition of a term is typically based on a list of the conditions that remain constant versus the conditions that are varied in the study. We later revisit the terminology from an InfoQ lens, and specifically, we use the generalizability dimension to provide clarity on the three "R" terms.

11.2   Definitions of reproducibility, repeatability, and replicability

The terms reproducibility and repeatability are used with different goals and their generalizability leads to different directions. The literature provides some mixed and sometimes contradictory views on the meaning of these terms. We list below some examples from the literature.

Pearl (2013) states that "Science is about generalization, and generalization requires transportability. Conclusions that are obtained in a laboratory setting are transported and applied elsewhere, in an environment that differs in many aspects from that of the laboratory." Pearl's concept of transportability addresses recreating conclusions in multiple contexts or studies.

Drummond (2009) writes: "Reproducibility requires changes; replicability avoids them. A critical point of reproducing an experimental result is that irrelevant things are intentionally not replicated. One might say, one should replicate the result not the experiment." According to Drummond (2009), the key difference between reproducibility and replicability is thus the recreation of the scientific conclusions and insights, as opposed to recreating the exact same numerical results.

McNutt (2014) states: "Recently, the scientific community was shaken by reports that a troubling proportion of peer-reviewed preclinical studies are not reproducible." The clarification of what are considered reproducible studies is key to addressing this concern. McNutt stipulates: "There are a number of reasons why peer-reviewed pre-clinical studies may not be reproducible. The system under investigation may be more complex than previously thought, so that the experimenter is not actually controlling all independent variables. Authors may not have divulged all of the details of a complicated experiment, making it irreproducible by another lab. It is also expected that through random chance, a certain number of studies will produce false positives." This description appears to reflect what is considered in industry as repeatability (see Section 11.3).

Banks (2011) considers reproducibility from an editorial point of view, explaining why research in social sciences is difficult to reconstruct in view of hidden assumptions and analytic complexities. He emphasizes the need for proper documentation. It appears that Banks is more concerned with recreating the detailed results than recreating the overall discovery or effect.

A growing number of scientific journals requires that data used in published papers is uploaded to a repository with all required information about statistical analysis and the usage of software programmes. The focus of this requirement is on recreating the actual numerical results. Publishing a computer program is, however, not enough, even from a consideration of repeatability. Quoting Knuth (1984), "Let us change our traditional attitude to the construction of programs: Instead of imagining that our main task is to instruct a computer what to do, let us concentrate rather on explaining to humans what we want the computer to do. The practitioner of literate programming can be regarded as an essayist, whose main concern is with exposition and excellence of style. Such an author, with thesaurus in hand, chooses the names of variables carefully and explains what each variable means. He or she strives for a program that is comprehensible because its concepts have been introduced in an order that is best for human understanding, using a mixture of formal and informal methods that reinforce each other."

An attempt to implement the idea of Knuth in the classroom is provided by the STAT 157 course by Philip Stark and Aaron Culich at Berkeley (2014). Students in this class work in team projects, sharing all results using a GitHub repository. Knuth and Stark are more focused on properly communicating the work done so that others can continue it or redo it and derive similar results. This is part of the scientific aspect of reproducibility addressed by Pearl and Drummond, see also ICERM (2012). Finally, the "replicability research team" at Tel Aviv University has been addressing these issues and provides some clarifications as to the role of replicability in genome-wide association (GWA) studies (Heller and Yekutiely, 2014; TAU, 2014). The team provides particularly important contributions to the evaluation of repeated statistical testing and multiple comparisons in the context of false discovery rates (FDR).

In the following sections we describe the terminology used in different areas of science and technology with respect to reproducibility, replicability, and repeatability. In Section 11.6 we consider the role of generalizability and how it relates to the discussed types of recreation/repeats.

11.3   Reproducibility and repeatability in GR&R

Reproducibility and repeatability are terms used in industry in the context of gauge repeatability and reproducibility (GR&R). Reproducibility is a routine part of a GR&R study conducted by industry to assess the measurement error generated by various testers and test equipment. In these experiments, several testers are asked to retest a marked set of items. Differences between testers are used to evaluate the reproducibility of the testing system (e.g., see Kenett and Zacks, 2014). Another characteristic of the test equipment is variability induced by a repetition of the same tests, on the same test items, using the same equipment. This variability is called repeatability and is part of GR&R studies. In short, repeated evaluations under a variety of conditions attempt to estimate reproducibility, while repeated evaluations under identical conditions are aimed at estimating repeatability (see Table 11.1). These aspects of measurement uncertainty are of great concern in industry, as they contribute to false-positive and false-negative decisions with (sometimes very) harmful consequences. In industry, GR&R is used to evaluate the contributions of the measurement process and the measurement technology to the measurement uncertainty.

A typical GR&R study is conducted by focusing on a process measure or a quality characteristic of an object of interest. A certain type of gauge or measurement instrument is chosen as the measuring device to be analyzed. The GR&R study provides information

Table 11.1   Terminology in GR&R studies.

Term	What is held constant	What is varied
Reproducibility	Item, equipment	Operator
Repeatability	Item, equipment, operator	

on the device itself and how it is used. A set of J operators typically using the device are selected to participate in the study. In addition, I parts are chosen and prepared for the study. Each of the J operators is asked to measure the characteristic (size) of each of the I parts for r times (repeatedly measure the same part r times). The variation among the replications of the given parts measured by the same operator is the repeatability of the gauge. The variability among operators is the reproducibility. GR&R studies determine how much of your observed process variation is due to measurement system variation. The overall variation is broken down into three categories: part-to-part, repeatability and reproducibility. The reproducibility component can be further broken down into its operator and, operator by part, components. Stevens et al. (2013) discuss the use of *baseline data*—ongoing process data that can be used to estimate the total degree of variability. Using baseline data offers substantial gains in precision for estimating repeatability and reproducibility. The best plans to use along with baseline data often have very few parts.

Browne et al. (2009) ask whether the parts should be sampled at random. They focus on a system with just one operator and on estimating the ratio of measurement variance to total variance. The study has two phases. The first is a random sample of b parts, each measured once. The second phase uses a sample of n parts from the first phase, chosen to include parts with extreme values, but an average that is similar to the phase 1 average. Each part is measured K times. Why sample extreme parts? The idea is that the variance ratio of interest can be thought of as a regression coefficient, and it can be estimated most accurately when the regression problem focuses on parts with extreme values. Simulation studies show that in this setting leveraged plans are often much more efficient than standard plans.

These examples of different types and magnitudes of repeating an operation for different purposes show how a GR&R study is adapted to different conditions. Disclosing the intended *generalization* of the GR&R study permits us to evaluate the quality of the information derived from the study (see Section 11.6).

11.4   Reproducibility and repeatability in animal behavior studies

The concepts of reproducibility and repeatability in animal behavior studies have been mostly addressed by attempts at standardization in animal experiments. However, Richter et al. (2009) claim that "Because experimental treatments may interact with environmental conditions, experiments conducted under highly standardized

Table 11.2   Terminology in animal experiments.

Term	What is held constant	What is varied
Standardization	Laboratory, measurement conditions	–
Heterogenization	The underlying mechanism of action being measured	Laboratory

conditions may reveal local "truths" with little external validity." The authors show that introducing systematic variation of experimental conditions (which they call "heterogenization") may reduce spurious results and improve reproducibility (Richter et al., 2010). Table 11.2 maps the difference between standardization and heterogenization strategies.

11.5   Replicability in genome-wide association studies

Replicability is the term used in genome-wide association (GWA) studies, an area which assesses interactions between genetic variants and phenotypes. Published GWA studies often combine results of primary studies and of follow-up studies. Reporting the two separate studies gives a sense of the replicability of the results (Pei et al., 2014). Hence, in GWAs, replicability means repeating the study by the same lab or researchers, with a different technology or a different dataset (typically a sub-population at a later period). This differs from "independent replication," where the study is repeated by a different lab (Statblog, 2012). Rosenberg and Van Liere (2009), pointing to the genetic commonality behind such experiments, called this "pseudo-replicability". Table 11.3 maps the terminology used in GWA studies.

11.6   Reproducibility, repeatability, and replicability: the InfoQ lens

As we noted earlier, the definitions of the different concepts in a specific application area are typically based on a list of the conditions that are kept constant and a list of the conditions that are varied during the study. However, such lists differ dramatically between areas and are very application-specific. We therefore propose an alternative, simpler approach. The concepts of reproducibility, repeatability, and replicability in a specific application domain can be clarified by applying the *generalizability* dimension of the InfoQ framework. Specifically, the utility of $f(X \mid g)$ is dependent on the ability to generalize f to the appropriate population.

Two types of generalizability are statistical generalizability and scientific generalizability. Statistical generalizability refers to inferring from a sample to a target population. Scientific generalizability refers to applying a model based on a particular target population to other populations. This can mean either generalizing an estimated population pattern/model f to other populations or applying f, estimated

Table 11.3   Terminology in genome-wide association studies.

Term	What is held constant	What is varied
Replicability	Laboratory, researchers	Technology, data or period
Independent replication		Laboratory

from one population, to predict individual observations in other populations. Determining the level of generalizability requires careful characterization of g. Recall that a low level of generalizability reflects a study with relatively lower InfoQ.

The three concepts of reproducibility, repeatability, and replicability are aimed at assuring generalizability. However, a closer look reveals that they aim at generalizability of different types. To better distinguish between the three terms, rather than focusing solely on what is replicated, it is useful to consider *what we are trying to generalize to*. Is it generalizing an effect from a random sample to a larger population (statistical generalization)? Is it generalizing an effect from a particular context to other, different contexts (scientific generalization)? Is the replication needed for purposes of robustness checking and assuring the same effect occurs under different conditions, or is the replication needed for assessing noise or differences between different conditions?

By focusing on *what is generalized from a study and for what purpose*, it is easier to differentiate the three terms of reproducibility, repeatability, and replicability. The researcher (or consumer of a research study) should have in mind the goal of the study, the type of generalization(s) needed and the approach and replications performed to assure that the generalization is sound. As an example, consider these terms in the context of software development:

- Repeatability is technical in scope. It considers, in this context, how we test a new software version to ensure technical compatibility of older versions. This is sometimes called "regression testing." The generalization here is to consider performance on other software platforms.

- Reproducibility is functional in scope. In this context, the testing of a new software version is conducted to ensure that the functionality of the older version remains the same. The generalization here is to ensure that the "look and feel" is preserved.

- Replicability can be considered testing of software on new processors in order to ensure that outcomes and performance are not changed.

Explicit presentation of how the findings are generalized clarifies the terms. One approach providing such a presentation is to apply methods used in formative assessment of education programmes discussed in Chapter 6. Formative assessment can be performed with MERLO items designed to assess conceptual understanding using alternative representations that map a boundary of meaning (BOM); for details see Shafrir and Kenett (2010, 2015) and Shafrir and Etkind (2014). In the context of reproducible research, the suggestion is to require a statement of the qualitative findings of a specific research, with alternative representations representing versions with meaning equivalence (within the BOM) and surface similarity (outside the BOM). Such a presentation provides a clear generalization that helps communicate the findings to researchers interested in reproducing the findings and to nonspecialists who need to understand the essential elements to the findings. In the latter case *generalization* is a key element in effective *communication*, providing a link between two InfoQ dimensions.

Let us return to the three examples presented in Sections 11.3–11.5. In GR&R studies, the generalization of the findings implies improvement requirements for either the measurement technology or the measurement process as a whole, or both. The focus of GR&R is on statistical generalization that is used to determine the adequacy of a certain measurement capability in terms of the specification of product variability. A typical industry standard is to consider a cap of 10% on variability from overall GR&R in terms of the process range. When the GR&R exceeds 30% of the specification, the measurement uncertainty is considered too high as it will generate an excessive amount of false-positive and false-negative decisions. Our detailed description of the GR&R process and variations is aimed to show how statistical generalization is the reason for the different settings (what is kept constant and what is varied) and analyses. It is therefore useful to use the language of generalization to explain what is meant by reproducibility and repeatability in GR&R.

Now consider scientific generalization. One key goal is scientific discovery, sometimes described as uncovering a mechanism of action or a first principle. These insights provide significant scientific advancement and impact the path of future research. The water memory publication of Benveniste and the acid-bath stressing of stem cells at Riken were proposed to be such advances. In both cases the findings could not be reproduced and the claimed results were retracted. The problems incurred in these studies had to do with both fundamental aspects and the technicalities of the study.

In animal studies, considering the claim of Richter et al. from the standpoint of generalization clarifies the issue at hand. Standardized animal behavior experiments are differently generalizable from experiments with induced systematic variation of experimental conditions. In particular, standardization restricts generalization to a single lab (statistical generalization to a specific lab), while heterogenization allows for scientific generalization beyond a specific lab.

Considering generalization, replicability in GWA studies is aimed at generalizing results across technologies or populations, often in the sense of generalizing to the same population in the future.

Finally, another area where these considerations apply is big data analytics. In this domain, in which very large samples with many measurements are used in scientific studies, it is especially important to determine what type of generalization is needed and which of the Rs should be considered. With large and rich datasets, it is easier to find small and complex effects and to quantify noise. This not only offers new opportunities but also poses new threats. Statistical inference, a major tool for statistical generalization, encounters multiple challenges when it comes to large samples (Lin et al., 2013) and to testing hypotheses based on many measurements. With a large number of records, methods such as validation on random holdout sets can help assure generalization of an effect and its magnitude across multiple samples (statistical generalization).

11.7 Summary

This chapter provides a clarification of three concepts fundamental to scientific and industrial applications of analytic methods: reproducibility, repeatability, and replicability. We examine their definitions and uses in three different areas. We then invoke the generalizability dimension of InfoQ to characterize reproducibility, repeatability, and replicability in terms of the intended generalization. The generalizability dimension of InfoQ is key to clarify the meaning of these terms in specific situations.

Appendix: Gauge repeatability and reproducibility study design and analysis

A typical GR&R study is conducted by focusing on a process measure or a quality characteristic of an object of interest. A certain type of gauge or a certain instrument is chosen as the measuring device to be analyzed. The GR&R study provides information on the device itself and how it is used. A set of J operators typically using the device are selected to participate in the study. Usually $J = 2$ or 3. In addition, I parts are chosen and prepared for the study with usually $I = 10$. Each of the J operators is asked to measure the characteristic (size) of each of the I parts for r times (repeatedly measure the same part r times). The variation among the replications of the given parts measured by the same operation is the repeatability of the gauge. The variability among operators is the reproducibility. GR&R studies determine how much of your observed process variation is due to measurement system variation. The overall variation is broken down into three categories: part-to-part, repeatability, and reproducibility. The reproducibility component can be further broken down into its operator and, operator by part, components.

Let y_{ijk} be the k-th measurement by the j-th operator of the i-th part. A common statistical model is

$$y_{ijk} = \mu + \alpha_i + \beta_j + \alpha\beta_{ij} + \varepsilon_{ijk}, \quad i = 1,\ldots,I; \quad j = 1,\ldots,J; \quad k = 1,\ldots,K$$

where

$\mu =$ mean part size

$\alpha_i =$ part effect due to interpart variation (σ_P^2)

$\beta_j =$ operator effect due to interoperator variation (σ_O^2)

$\alpha\beta_{ij} =$ part by operator interaction (σ_{PO}^2)

$\varepsilon_{ijk} =$ error term due to repeatability variation (σ^2)

With this notation, we can write formulas for reproducibility and repeatability as defined in GR&R:

$$\text{Reproducibility} = \sigma_O^2 + \sigma_{PO}^2$$
$$\text{Repeatability} = \sigma^2$$
$$\text{GR\&R} = \sigma_O^2 + \sigma_{PO}^2 + \sigma^2$$

If the GR&R study includes a sample of the possible operators using the device, then we need to think how the sample represents the population of all possible operators. A common model, called a random effects model, is

$$\beta_j \sim N(0, \sigma_O^2)$$
$$\alpha\beta_{ij} \sim N(0, \sigma_{PO}^2)$$

With the random effects model, we can use ANOVA mean squares or sample ranges to estimate the variances of interest. Table A shows the expected values for each estimated variance.

The standard range estimators are based on the within-cell ranges R_{ij} and the range of the per operator averages, $R_O = \text{Max}\bar{y}_{.j} - \text{Min}\bar{y}_{.j}$.

The range estimator of reproducibility is $\dfrac{R_O}{d_2(J)}$, with expected value $\sqrt{\sigma_O^2 + (1/I)\sigma_{PO}^2 + (1/IK)\sigma^2}$.

The constant $d_2(J)$ is derived from an approximation of the distribution of the range (see Kenett and Zacks, 2014).

If the GR&R study includes all possible operators using the measurement device (the entire population of operators), then we need to characterize that specific team of operators, and the sampling assumption for operators is no longer reasonable. In this case the operator terms are called *fixed effects*. However, the part by operator interaction still involves sampling, so it remains a random effect. This leads to a *mixed effects* model.

In the mixed effects model, $E\{\text{MS operators}\} = (IK/J-1)\sum \beta_j^2 + K\sigma_{PO}^2 + \sigma^2$ and $\sigma_O^2 = 1/(J-1)\sum \beta_j^2$, assuming $\sum \beta_j = 0$.

Table A   ANOVA table of GR&R experiments.

Mean square (MS)	Expected value
MS error	σ^2
MS parts by operators	$K\sigma_{PO}^2 + \sigma^2$
MS operators	$IK\sigma_O^2 + K\sigma_{PO}^2 + \sigma^2$
MS parts	$JK\sigma_P^2 + K\sigma_{PO}^2 + \sigma^2$

In this case, the ability to estimate σ_0^2 depends on J, the number of operators sampled. For a fixed total sample size, using only two or three operators is not efficient. It is better to increase J and decrease I and K.

When the sole goals are repeatability and reproducibility, Vardeman and Van Valkenburg (1999) found that the best designs typically use a single part, measured two or three times by many operators. More repeats are desirable when repeatability is much larger than reproducibility.

The standard GR&R plan is balanced—each part by operator pair has the same number of measurements. Stevens et al. (2010) look at the benefits of unbalanced plans. They assume the goal is to estimate the ratio of measurement variance (repeatability and reproducibility) to total variance and that all operators are included. Their suggested plan consists of two steps:

1. Begin with a small standard plan.

2. Augment the plan with further data.

In step 2 there are two types of augmentation in conducting the GR&R study:

Type A augmentation: each operator measures a new set of parts once each.

Type B augmentation: each operator measures the same set of parts once each.

The benefits of both augmentation options are that measuring more parts leads to better estimates of the interpart variance.

- With just one operator (e.g., automated measurement systems), the best plans are standard plans with two measurements per part.

- With more than one operator, and no part by operator interaction, type A plans are the best, typically by 7–20% relative to the best standard plan.

- When there is part by operator interaction, the best plans depend on the number of operators.

- With two operators, plans of type B with a very small standard plan are most efficient.

- With three or four operators, plans of type A with a small standard plan are most efficient.

For more on a comparison of GR&R studied, see Steinberg (2014).

References

Banks, D. (2011) Reproducible research: a range of response. *Statistics, Politics, and Policy*, 2(1), Article 4.

Berkeley. (2014) http://berkeleysciencereview.com/reproducible-collaborative-data-science (Sarah Hillenbrand, June 11, 2014) (accessed May 23, 2016).

Browne, R., Steiner, S. and MacKay, J. (2009) Two-stage leveraged measurement system assessment. *Technometrics*, 51(3), pp. 239–249.

Drummond, C. (2009) Replicability Is Not Reproducibility: Nor Is It Good Science. *Proceeding of the Evaluation Methods for Machine Learning Workshop at the 26th ICML*, Montreal, Canada.

Heller, R. and Yekutiely, D. (2014) Replicability analysis for genome-wide association studies. *The Annals of Applied Statistics*, 8(1), pp. 481–498.

ICERM. (2012) icerm.brown.edu/tw12-5-rcem (accessed May 23, 2016).

Ioannidis, J.P.E. (2005) Why most published research findings are false. *PLoS Medicine*, 2(8), pp. 696–701.

Ioannidis, J.P.E. (2008) Why most discovered true associations are inflated. *Epidemiology*, 19(5), pp. 640–648.

Ioannidis, J.P.E., Allison, D., Ball, C., Coulbaly, I., Cui, X., Culhane, A., Falchi, M., Furlanello, C., Game, L., Jurman, G., Mangion, J., Mehta, T., Nitzberg, M., Page, G., Petretto, E., and van Noort, V. (2009) Repeatability of published microarray gene expression analyses. *Nature Genetics*, 41, pp. 149–155.

Kenett, R.S. and Shmueli, G. (2015) Clarifying the terminology that describes scientific reproducibility. *Nature Methods*, 12(8), p. 699.

Kenett, R.S. and Zacks, S. (2014) *Modern Industrial Statistics: With Applications in R, MINITAB and JMP*, 2nd edition. John Wiley & Sons, Ltd, Chichester, UK.

Knuth, D. (1984) Literate programming. *The Computer Journal (British Computer Society)*, 27(2), pp. 97–111.

Lin, M., Lucas, H. Jr. and Shmueli, G. (2013) Too big to fail: large samples and the P-value problem. *Information Systems Research*, 24(4), pp. 906–917.

McNutt, M. (2014) Reproducibility. *Science*, 343, p. 229.

Nature. (2004) www.nature.com/news/2004/041004/full/news041004-19.html (accessed May 23, 2016).

Nature. (2014) blogs.nature.com/news/2014/03/call-for-acid-bath-stem-cell-paper-to-be-retracted.html (accessed May 23, 2016).

Open Science Collaboration. (2015) Estimating the reproducibility of psychological science. *Science*, 349(6251), pp. 943–951.

Pearl, J. (2013) Transportability Across Studies: A Formal Approach, Working Paper R-372. UCLA Cognitive Science Laboratory.

Pei, Y.F., Zhang, L., Papasian, C.J., Wang, Y.P. and Deng, H.W. (2014) On individual genome-wide association studies and their meta-analysis. *Human Genetics*, 133(3), pp. 265–279.

Richter, S., Garner, J. and Wurbel, H. (2009) Environmental standardization: cure or cause of poor reproducibility in animal experiments? *Nature Methods*, 6, pp. 257–261.

Richter, S., Garner, J., Auer, C., Kunert, J. and Wurbel, H. (2010) Systematic variation improves reproducibility of animal experiments. *Nature Methods*, 7, pp. 167–168.

Rosenberg, N.A. and Van Liere, J.L. (2009) Replication of genetic associations as pseudoreplication due to shared genealogy. *Genetic Epidemiology*, 33, pp. 479–487.

Shafrir, U. and Etkind, M. (2014) Concept Science: Content and Structure of Labeled Patterns in Human Experience. Version 32.0.

Shafrir, U. and Kenett, R.S. (2010) Conceptual thinking and metrology concepts. *Accreditation and Quality Assurance*, 15(10), pp. 585–590.

Shafrir, U. and Kenett, R.S. (2015) Concept Science Evidence-Based MERLO Learning Analytics, in *Handbook of Applied Learning Theory and Design in Modern Education*, IGI Global, Hershey, PA.

Statblog. (2012) http://www.statsblogs.com/2012/07/03/replication-and-validation-in-omics-studies-just-as-important-as-reproducibility (accessed May 23, 2016).

Steinberg, D. (2014) Statistical Models for Gage R&R Studies. *The 4th Jerusalem Conference on Quality by Design (QbD) and Pharma Sciences*, May 20–22, Jerusalem, Israel. https://medicine.ekmd.huji.ac.il/schools/pharmacy/En/home/news/Pages/QbD2014.aspx (accessed May 23, 2016).

Stevens, N., Browne, R., Steiner, S. and MacKay, J. (2010) Augmented measurement system assessment. *Journal of Quality Technology*, 42(4), pp. 388–399.

Stevens, N., Browne, R., Steiner, S. and MacKay, J. (2013) Gauge R&R studies that incorporate baseline information. *IIE Transactions*, 45(11), pp. 1166–1175.

TAU. (2014) www6.tau.ac.il/benjamini/index.php/replicability-in-science/replicability-vs-reproducibility.html (accessed May 23, 2016).

Vardeman, S. and Van Valkenburg, E. (1999) Two-way random-effects analyses and gauge in R&R studies. *Technometrics*, 41(3), pp. 202–211.

12

InfoQ in review processes of scientific publications

12.1 Introduction

Publication of research in academic journals is an important component of scientific advancement as well as a contribution to personal professional development. In the publication process, reviewers play a critical role. They are the main advisors to the gatekeepers (the editor and associate editors), and they also provide feedback to the authors that can be valuable in improving the work. A good reviewer is able to see the contribution of the paper and judge its level and suitability relative to the standard of the particular journal. Yet, this process is usually carried out in an unstructured way with inherent variability between reviewers and even within the same reviewer. Quoting Gewin (2011): "many graduate and postdoctoral students were never taught how to review a manuscript; most peer reviewers learn journals' needs and the reviewer's role only through trial and error. Editors' expectations differ according to their fields, but most agree that simply writing thorough, respectful, and helpful reviews is the best way for early career scientists to find their footing and avoid mistakes."

In a recent controversial paper, eventually retracted from *Nature*,[1] one of the commenters wrote, "I feel that it would be great if *Nature* would find a way to publish the reviewers' comments on this manuscript as well as the editorial procedure. As long as the reviewers agree, it could be very beneficial. For instance, it might enable the community to see where things went awry in finding the concerns that

---

[1] blogs.nature.com/news/2014/03/call-for-acid-bath-stem-cell-paper-to-be-retracted.html

Information Quality: The Potential of Data and Analytics to Generate Knowledge,
First Edition. Ron S. Kenett and Galit Shmueli.
© 2017 John Wiley & Sons, Ltd. Published 2017 by John Wiley & Sons, Ltd.
Companion website: www.wiley.com/go/information_quality

have been discussed since the publication of these studies. It has been the common practice in EMBO since 2009 (http://bit.ly/1iAVP5i). As it might be problematic to apply this policy retrospectively, this case enhances the need to adopt this model going forward."

Francois (2015) analyzes data from an experiment designed to assess the effect of variability in the review process. In the experiment, 10% of the manuscripts (166 manuscripts) submitted for publication in a conference proceedings went through the review process twice. Arbitrariness was measured as the conditional probability for an accepted submission to get rejected if examined by the second committee. This number was equal to 60%, for a total acceptance rate of 22.5%. The author applied Bayesian analysis to integrate information on "quality" by introducing a hidden parameter which measures the probability that a submission meets basic quality criteria. The quality criteria considered in this study included novelty, clarity, reproducibility, correctness, and no ethical misconduct. These criteria were met by a large proportion of the submitted manuscripts. The Bayesian estimate for the hidden parameter was 56% with a 95% credibility interval of (0.34–0.83). As a result of this analysis, the author concluded that the arbitrariness probability is high and suggested that the acceptance rate for manuscripts at the conference should be increased in order to decrease arbitrariness in future review processes.

We are cognizant of the many statisticians in academia and in industry who regularly review papers for journals without a general framework to guide the review process. As in other disciplines, it is typically left to the reviewer's experience and good sense to determine acceptance of the paper. Together with the associate editor and editor's opinions, one assumes that the "wisdom of the review team" will reveal the value of a paper in a reliable and reproducible way. Moreover, in recent years, there has also been a concerted effort by many journals to expedite the reviewing cycle in order to make new knowledge available in a timely fashion. For example, *The Journal of Business & Economic Statistics* instituted the following policy:

> The Journal of Business & Economic Statistics has a policy that after the first round of revisions, papers must be either rejected or accepted subject to specific minor revisions.

The requirement to reduce the number of review cycles of a paper creates an even more urgent need to improve the first-round assessment of submitted manuscripts.

The InfoQ framework is recommended for reviewing papers to partially solve the aforementioned problems. The InfoQ framework ties the data (X), the data analysis (f), the analysis goal (g), and the utility measured using specific metric(s) (U). By examining each of the components and their relationships, we can learn about the contribution of a given project, study or paper. For example, the contribution can be a new research question, and/or a novel dataset, and/or a new analysis method or approach. Journals typically focus on contributions along one of these three directions. Generally speaking, methodological or theoretical statistics and data mining journals are interested in contributions to f and U, while applied journals (in statistics or other scientific areas) are more interested in novelties in g and X. For illustration, following are the guidelines for authors for *The Journal of the American Statistical Association (JASA) Applications and Case Studies*[2]: The *Applications and Case Studies* section publishes original articles that do one or more of the following:

1. For real datasets, present analyses that are statistically innovative as well as scientifically and practically relevant.

2. Contribute substantially to a scientific field through the use of sound statistical methods.

3. Present new and useful data, such as a new life table for a segment of the population or a new social or economic indicator.

4. Using empirical tests, examine or illustrate for an important application the utility of a valuable statistical technique.

5. Evaluate the quality of important data sources.

And for *Science* magazine[3]:

Research Articles should report a major breakthrough in a particular field. They should be in the top 20% of the papers that Science publishes and be of strong interdisciplinary interest or unusual interest to the specialist.

In comparison, the guidelines for *JASA Theory and Methods* state:

The Theory and Methods section publishes articles that make original contributions to the foundations, theoretical development, and methodology of statistics and probability.

The remainder of the chapter is organized as follows: In Section 12.2 we list current guidelines for reviewers from several journals in the fields of statistics and data mining in order to justify the need for a more guided paper reviewing process framework. We show that the existing guidance ranges from none to general, with different types of questions and approaches. How can a reviewer use the InfoQ framework? Firstly, by identifying the four InfoQ components in the paper (goal, data, analysis, and utility) and then assessing each of the eight InfoQ dimensions. Section 12.3 presents guidelines for reviewers based on the eight InfoQ dimensions. For each dimension, we list questions that a reviewer should ask while reviewing. The goal is to help reviewers,

---

[2] www.tandfonline.com/action/authorSubmission?journalCode=uasa20&page=instructions
[3] www.sciencemag.org/site/feature/contribinfo/RAinstr13.pdf

associate editors, and editors assess the contribution of a paper, its suitability to the journal, and judge the potential for improvement when requesting a revision by the paper's author/s. Reviewing of papers is mostly done by volunteers, so our proposal should be considered as a suggestion for those interested in achieving clarity in reviews. Associate editors and editors could consider these suggestions as a structured approach to increase homogeneity in the quality of the review process.

12.2   Current guidelines in applied journals

Tables 12.1–12.7 summarize results from a search of the websites of applied journals published by the leading statistical societies (American Statistical Association (ASA), Institute of Mathematical Statistics (IMS), the Royal Statistical Society (RSS)) leading scientific journals (Science and Nature), and top machine learning journals and conferences (JMLR, ML, KDD) Table 12.1. The tables are designed to emphasize the lack of publicly available guidelines for reviewers. While journal guidelines range in terms of detail provided to reviewers, the key criteria for judging novelty and importance are typically generally defined (e.g., "Interest and importance and novelty as a scientific contribution" and "Is the paper of practical importance?"). For example, *Science* requests reviewers to comment on two aspects:

1. **Technical rigor**: Evaluate whether, or to what extent, the data and methods substantiate the conclusions and interpretations. If appropriate, indicate what additional data and information are needed to do so.

2. **Novelty**: Indicate in your review if the conclusions are novel or similar to work already published.

The *Journal of the Royal Statistical Society (JRSS)-C (Applied Statistics)* asks reviewers to consider the following point: "Are the facts, arguments, and conclusions in the paper technically valid and accurate?"

These types of guidelines, while helpful at a theoretical level, need to be operationalized for actual reviewing. The eight InfoQ dimensions presented in Section 12.3 provide a way to do that.

As a final example, consider the statements on the website of *The Statistical Journal of the International Association of Official Statistics* (IAOS):

> The main aim of the journal is to support the IAOS mission by publishing articles to promote the understanding and advancement of official statistics and to foster the development of effective and efficient official statistical services on a global basis.

This stated goal poses a unique challenge to publications of official statistical services in general, including the review process of such publications. For more on InfoQ applications in official statistics, see Chapter 10 and Kenett and Shmueli (2016).

Table 12.1   List of journals published by American Statistical Association (ASA) Referee guidelines web pages were not found for any of these journals.

Journal of the American Statistical Association
The American Statistician
Journal of Business & Economic Statistics
Statistics in Biopharmaceutical Research
Journal of Agricultural, Biological, and Environmental Statistics
Journal of Computational and Graphical Statistics
Technometrics
Statistics and Public Policy
Journal of Educational and Behavioral Statistics
Journal of Statistics Education
Journal of Statistical Software
Statistical Analysis and Data Mining
Statistics Surveys
Journal of Nonparametric Statistics
Journal of Quantitative Analysis in Sports
SIAM/ASA Journal on Uncertainty Quantification
Journal of Survey Statistics and Methodology

12.3   InfoQ guidelines for reviewers

We now revisit the eight InfoQ dimensions that were introduced in Chapter 3. We propose specific guiding questions for reviewers interested in assessing the level of InfoQ of a manuscript submitted for publication that involves data analysis and applied research.

12.3.1   Data resolution

Data resolution refers to the measurement scale and aggregation level of the data. The measurement scale of the data should be carefully evaluated in terms of its suitability to the stated goal, the analysis methods used, and the required resolution of the research utility. Questions that a reviewer should ask to figure out the strength of this dimension:

- Is the data scale aligned with the stated goal?

- How reliable and precise are the measuring devices or data sources?

- Is the data analysis suitable for the data aggregation level?

A low rating on data resolution can be indicative of low trust in the usefulness of the study's findings. As an example, consider Google's ability to predict the prevalence of flu on the basis of the type and extent of internet searches.[4] These predictions match quite well with the official figures published by the Centers for Disease Control and Prevention (CDC). The point is that Google's tracking has only one-day delay compared with the week or more it takes for the CDC to assemble a picture based on doctors' reports. Google is faster because it tracks the outbreak by finding a correlation between what people search for online and whether they have flu symptoms.

Another example is provided by the RISCOSS project (www.riscoss.eu) that has developed a risk management methodology for adopters of open-source software (OSS). The data used to generate risk indicators combines online data of the OSS community, such as time to fix bugs, social network analysis of the OSS community, and expert opinions. The community data can be collected online with continuous updates. Risk management is based on evaluating risk indicators on a weekly or monthly basis so that the OSS data in RISCOSS is aggregated on a weekly or monthly basis to match the needs of the adopter's risk management activity (Kenett et al., 2014). Data collected on a minute by minute basis or on a yearly basis would not have the proper resolution.

12.3.2   Data structure

Data structure relates to the type(s) of data and data characteristics, such as corrupted and missing values, due to the study design or data collection mechanism. Data types include structured, numerical data in different forms (e.g., cross-sectional, time series, and network data) as well as unstructured, nonnumerical data (e.g., text, text with hyperlinks, audio, video, and semantic data). The InfoQ level of a certain data type depends on the goal at hand. Questions that a reviewer should ask to figure out the strength of this dimension:

- Is the type of data used aligned with the stated goal?

- Are data integrity details (corrupted/missing values) described and handled appropriately?

- Are the analysis methods suitable for the data structure?

A low rating on data structure can be indicative of poor data coverage in terms of the project goals. For example, using a cross-sectional analysis method to analyze a time series warrants special attention when the goal is parameter inference but is of less concern if the goal is forecasting future values. Another example is removing records with missing data when missingness might not be random. A paper analyzing online transactions with the objective of evaluating actual behavior versus declared behavior also needs data on declared behavior through focused queries or questionnaires. Without that, the structure of the data will not provide adequate information quality.

---

[4]http://www.ft.com/cms/s/2/21a6e7d8-b479-11e3-a09a-00144feabdc0.html#axzz2y6ASfagk

12.3.3 Data integration

With the variety of data source and data types available today, studies sometimes integrate data from multiple sources and/or types to create new knowledge regarding the goal at hand. Such integration can increase InfoQ, but in other cases it can reduce InfoQ, for example, by creating privacy breaches. A low rating on data integration can be indicative of missed potential in data analysis. Questions that a reviewer should ask to determine the strength of this dimension:

- Is the data integrated from multiple sources? If so, what is the credibility of each source?

- How is the integration done? Are there linkage issues that lead to dropped crucial information?

- Does the data integration add value in terms of the stated goal?

- Does the data integration cause any privacy or confidentiality concerns?

A prime example of data integration is the fusion feature in Google.[5] In the RISCOSS methodology, aggregated quantitative data captured from OSS communities is integrated with qualitative expert opinion through an assessment of risk scenarios to derive risk indicators using Bayesian networks (Kenett et al., 2014).

Other examples of data integration include the combination of structured and unstructured semantic data using ETL methods (Kenett and Raanan, 2010) and the calibration of organizational data with official statistics data using copulas (Dalla Valle, 2014).

12.3.4 Temporal relevance

The process of deriving knowledge from data can be arranged on a timeline that includes data collection, data analysis, and results' usage periods as well as temporal gaps between these three stages. The different durations and gaps can each affect InfoQ. The data collection duration can increase or decrease InfoQ, depending on the study goal, for example, studying longitudinal effects versus a cross-sectional goal. Similarly, if the collection period includes uncontrollable transitions, this can be useful or disruptive, depending on the study goal. Questions that a reviewer should ask to determine the strength of this dimension:

- Are any of the data collection, data analysis, and deployment stages time sensitive?

- Does the time gap between data collection and analysis cause any concern?

- Is the time gap between the data collection and analysis and the intended use of the model (e.g., in terms of policy recommendations) of any concern?

---

[5] support.google.com/fusiontables/answer/2571232?hl=en

A low rating on temporal relevance can be indicative of an analysis with low relevance to decision makers due to the data collected in a different contextual condition. This can happen in economic studies with policy implications that are based on an old or outmoded data.

12.3.5   Chronology of data and goal

InfoQ is affected by the choice of variables to collect, the temporal relationship between them, and their meaning in the context of the goal at hand.

Questions that a reviewer should ask to determine the strength of this dimension:

- If the stated goal is predictive, are all the predictor variables available at the time of prediction?

- If the stated goal is causal, do the causal variables precede the effects?

- In a causal study, are there issues of endogeneity (reverse causation)?

A low rating on chronology of data and goal can be indicative of low relevance of a specific data analysis due to misaligned timing. A customer satisfaction survey, designed to be used as input for the annual budget planning cycle, becomes irrelevant if its results are communicated after the annual budget is finalized (Kenett and Salini, 2012). Reporting of air quality indicators to help allergic patients take proactive actions needs to be reported before the potentially dangerous conditions arise (for such an indicator, see the environmental protection agency's air quality index).[6]

In another example, in the context of online auctions, classic auction theory dictates that the number of bidders is an important driver of auction price. Models based on this theory are useful for explaining the effect of the number of bidders on price. However, for the purpose of predicting the price of ongoing online auctions, where the number of bidders is unknown until the auction end, the variable "number of bidders," even if available in the data, cannot be used as a predictor. Hence, the level of InfoQ contained in number of bidders for models of auction price depends on the goal at hand.

12.3.6   Generalizability

The utility of $f(X \mid g)$ is dependent on the ability to generalize f to the appropriate population. Two types of generalizability are *statistical generalizability* and *scientific generalizability*. Statistical generalizability refers to inferring from a sample to a target population. Scientific generalizability refers to applying a model based on a particular target population to other populations. This can mean either generalizing an estimated population pattern/model f to other populations or applying f estimated from one population to predict individual observations in other populations. Determining the level of generalizability requires careful characterization of g.

---

[6]www.airnow.gov/?action=aqibasics.aqi

Generalizability is related to the concepts of *reproducibility*, *repeatability*, and *replicability*. Some define reproducibility as the ability to recreate the scientific conclusions and insights, while repeatability is the ability to recreate the exact same numerical results (Drummond, 2009; Kenett and Shmueli, 2015). Replicability (used mainly in biostatistics) refers to replicating results under different conditions, and therefore it relates to scientific generalization. Generalizability is also a potent enabler to improve communication of results. With growing specialization, practitioners, such as clinical physicians, are finding it more and more difficult to extract useful information from highly specialized journals (Chariklia Tziraki-Amiel, 2015, personal communication to the first author). For a discussion of these terms and their relationship with generalizability, see Chapter 11.

Other aspects of questionable research practices are described by John et al. (2012). They show that in psychology, it is common to find papers that do not report subsidiary results in postexperimental interviews but do report irrelevant extra analyses broken down by demographic variables. Even more disturbing is a prevalent response by authors that they selectively reported studies with positive outcomes. Experienced reviewers will know how to provide feedback to authors to minimize such questionable research practices.

Questions that a reviewer should ask to determine the strength of the generalizability of a report:

- Is the stated goal statistically or scientifically generalizable?

- For statistical generalizability in the case of inference, does the paper answer the question "What population does the sample represent?"

- For generalizability in the case of a stated predictive goal (predicting values of new observations; forecasting future values), are the results generalizable to the to-be-predicted data?

- Does the paper provide sufficient detail for the type of needed reproducibility and/or repeatability and/or replicability?

A low rating on generalizability reflects a study with relatively low impact. Pearl (2013) states: "Science is about generalization, and generalization requires transportability. Conclusions that are obtained in a laboratory setting are transported and applied elsewhere, in an environment that differs in many aspects from that of the laboratory." Rasch (1961, 1977) used the term *specific objectivity* to describe that case essential to the measurement in which comparisons between individuals become independent of which particular instruments—tests or items or other stimuli—have been used.

Symmetrically, it is thought to be possible to compare stimuli belonging to the same class—measuring the same thing—independent of which particular individuals, within a class considered, were instrumental for comparison. The term general objectivity is reserved for the case in which absolute measures (i.e., amounts) are independent of which instrument (within a considered class) is employed, and no other object is required. By "absolute" we mean the measure "is not dependent on, or without reference to, anything else—not relative." In reviewing a paper, one should assess the contribution of the paper also in terms of its generalization.

12.3.7   Operationalization

Two types of operationalization are considered, construct operationalization and action operationalization, of the analysis results. Constructs are abstractions that describe a phenomenon of theoretical interest. Measurable data are an operationalization of underlying constructs. The relationship between an underlying construct and its operationalization can vary, and its level relative to the goal is another important aspect of InfoQ. The role of construct operationalization is dependent on the goal, and especially on whether the goal is explanatory, predictive, or descriptive. In explanatory models, based on underlying causal theories, multiple operationalizations might be acceptable for representing the construct of interest. As long as the data is assumed to measure the construct, the variable is considered adequate. In contrast, in a predictive task, where the goal is to create sufficiently accurate predictions of a certain measurable variable, the choice of operationalized variable is critical.

Action operationalization of results refers to three questions posed by Deming[7] (1982):

1. What do you want to accomplish?

2. By what method will you accomplish it?

3. How will you know when you have accomplished it?

Questions that a reviewer should ask to determine the strength of construct operationalization:

- Are the measured variables themselves of interest to the study goal or is their underlying construct the focus of the study?

- What are the justifications for the choice of variables?

Questions that a reviewer should ask to determine the strength of action operationalization:

- Who will be affected (positively or negatively) by the research findings?

- What can the affected parties do about it? What are the practical options of the affected parties?

A low rating on operationalization indicates that the research may have academic value but little or no practical impact.

---

[7] www.spcpress.com/pdf/DJW187.pdf

12.3.8   Communication

Effective communication of the analysis and its utility directly impacts InfoQ. There are plenty of examples where miscommunication of valid results has led to disasters, such as the NASA shuttle Challenger disaster. For a description of the communication issues in the NASA shuttle program, see Kenett and Thyregod (2006).

This is the dimension that is typically covered in the greatest detail in reviewer guidelines. Questions that a reviewer should ask to determine the strength of this dimension:

- Is the exposition of the goal, data, analysis, and utility clear?

- Is the exposition level appropriate for the readership of the journal?

- Are there any confusing details or statements that could lead to confusion or misunderstanding?

A low rating on communication can indicate that poor communication may blur or obfuscate the true value of the analysis and destroy the value of the information provided by the analysis.

12.4   Summary

The role of editors and associate editors is to aggregate information from reviewers in order to form an opinion regarding a specific submission. An unstructured approach in the review process leads to inconsistencies and high variability in the depth and breadth of reviews. Another aspect of this condition is the range of feedback provided to authors, from very brief and minimally informative comments to overly detailed reviews. This leads to a sense of arbitrariness in the process, sometimes justified but usually not.

In considering official guidelines for review (Tables 12.2–12.7), we found a general lack of clarity in guidelines for reviewers. With the proliferation of journals and communication channels, good journals can distinguish themselves by establishing structured guidelines for reviewers and as training methods for new reviewers. The general assumption, that experienced authors can instantly

Table 12.2   Partial list of journals published by American Society for Quality (ASQ) Referee guidelines web pages were not found for any of these journals. The same lack of guidelines applies to all other ASQ journals (http://asq.org/pub/).

Journal of Quality Technology
Quality Progress

Table 12.3   List of journals published by the Institute of Mathematical Statistics (IMS) and URLs for referee guidelines (accessed July 7, 2014).

Journal	Reviewer guidelines URL	Main criteria
Annals of Statistics	http://www.imstat.org/aos/referee.html	Interest and importance and novelty as a scientific contribution Quality of writing and presentation Technical correctness
Annals of Applied Statistics	http://www.imstat.org/aoas/referee.html	Does the paper genuinely concern applied statistics? Is the paper clearly written? Is the paper correct? Is the paper interesting?
Annals of Probability	http://www.imstat.org/aop/referee.htm	Is the presentation clear and well organized? Are the results new and interesting? Are the proofs correct and given in adequate detail, or can they be substantially simplified? Do the introduction and abstract give an adequate summary?
Annals of Applied Probability	www.imstat.org/aap/referee.html	Is the presentation clear and well organized? Are the results new and interesting? Are the proofs correct and given in adequate detail, or can they be substantially simplified? Do the introduction and abstract give an adequate summary?
Statistical Science	www.imstat.org/sts/referee.html	Is the presentation clear and well organized? Are the results new and interesting? Are the proofs correct and given in adequate detail, or can they be substantially simplified? Do the introduction and abstract give an adequate summary? Can the paper be streamlined?

Table 12.4   List of journals published by the Royal Statistical Society (RSS) and URLs for referee guidelines (accessed July 7, 2014).

Journal	Reviewer guidelines URL	Main criteria
JRSS Series A: Statistics in Society	http://www.rss.org.uk/ Images/PDF/publications/ rss-journal-notes-for- authors-2014.pdf	Suitability for Series A Importance Interest Originality Correctness
JRSS Series B: Statistical Methodology	http://www.rss.org.uk/ Images/PDF/publications/ rss-journal-notes-for- authors-2014.pdf	Suitability for Series B Importance Interest Originality Correctness
JRSS Series C: Applied Statistics	http://www.rss.org.uk/ Images/PDF/publications/ rss-journal-notes-for- authors-2014.pdf	Contents Are the facts, arguments, and conclusions in the paper technically valid and accurate? Is previous work adequately referenced and integrated with the new results? Is the paper of practical importance? Structure Exposition

become good reviewers, is not necessarily valid. We suggest that InfoQ can provide a structured framework for the review process of applied research papers. Table 12.8 provides a questionnaire to help guide the review of a submitted manuscript of applied research by pointing out key questions that relate to each of the InfoQ dimensions. We suggest incorporating such a questionnaire, either formally or informally, into the reviewing process of journals that publish empirical analyses.

Given the substantial effort and budgets dedicated to research, improving the review process appears as a necessary step. We provided here some elements that can be used to structure the review process without limiting critical thinking, independent

Table 12.5   List of journals in machine learning and URLs for referee guidelines (accessed July 7, 2014).

Journal	Reviewer guidelines URL	Main criteria
Journal of Machine Learning Research	jmlr.org/reviewer-guide.html	**Goals**: What are the research goals and learning tasks? **Description**: Is the description adequately detailed for others to replicate the work? **Evaluation**: Do the authors evaluate their work in an adequate way (theoretically and/or empirically)? **Significance**: Does the paper constitute a significant, technically correct contribution to the field that is appropriate for JMLR? **Related work and discussion**: Are strengths and limitations and generality of the research adequately discussed? Clarity: Is it written in a way such that an interested reader with a background in machine learning, but no special knowledge of the paper's subject, could understand and appreciate the paper's results?
Machine Learning	Instructions for authors: www.springer.com/ computer/ai/journal/10994	What is the main claim of the paper? Why is this an important contribution to the machine learning literature? What is the evidence you provide to support your claim? What papers by other authors make the most closely related contributions, and how is your paper related to them? Have you published parts of your paper before?

Table 12.6   Reviewing guidelines for major data mining conference (accessed July 7, 2014).

Conference	Reviewer guidelines URL	Main criteria
KDD 2012	http://www. kddcup2012. org/c/ kddcup2012- track2/details/ Evaluation	**Novelty**: This is arguably the single most important criterion for selecting papers for the conference **Technical quality**: Are the results sound? **Potential impact and significance**: Is this really a significant advance in the state of the art? Clarity of Writing

Table 12.7   List of top scientific journals and URLs for referee guidelines (accessed July 7, 2014).

Journal	Reviewer guidelines URL	Main criteria
Science Magazine	www.sciencemag. org/site/feature/ contribinfo/ review.xhtml and www.sciencemag. org/site/feature/ contribinfo/ RAinstr13.pdf	**Technical rigor:** Evaluate whether, or to what extent, the data and methods substantiate the conclusions and interpretations. If appropriate, indicate what additional data and information are needed to do so **Novelty:** Indicate in your review if the conclusions are novel or are too similar to works already published
Nature Journals	www.nature.com/ authors/policies/ peer_review.html	Who will be interested in reading the paper and why? What are the main claims of the paper and how significant are they? Is the paper likely to be one of the five most significant papers published in the discipline this year? How does the paper stand out from others in its field? Are the claims novel? If not, which published papers compromise novelty? Are the claims convincing? If not, what further evidence is needed? Are there other experiments or work that would strengthen the paper further? How much would further work improve it, and how difficult would this be? Would it take a long time? Are the claims appropriately discussed in the context of previous literature? If the manuscript is unacceptable, is the study sufficiently promising to encourage the authors to resubmit? If the manuscript is unacceptable but promising, what specific work is needed to make it acceptable?

Table 12.8   Questionnaire for reviewers of applied research submission.

Dimension	Questions
1. Data resolution	1.1 Is the data scale aligned with the stated goal?
	1.2 How reliable and precise are the measuring devices or data sources?
	1.3 Is the data analysis suitable for the data aggregation level?
2. Data structure	2.1 Is the type of data used aligned with the stated goal?
	2.2 Are data integrity details (corrupted/missing values) described and handled appropriately?
	2.3 Are the analysis methods suitable for the data structure?
3. Data integration	3.1 Is the data integrated from multiple sources? If so, what is the credibility of each source?
	3.2 How is the integration done? Are there linkage issues that lead to dropped crucial information?
	3.3 Does the data integration add value in terms of the stated goal?
	3.4 Does the data integration cause any privacy or confidentiality concerns?
4. Temporal elevance	4.1 Considering the data collection, data analysis, and deployment stages, are any of them time sensitive?
	4.2 Does the time gap between data collection and analysis cause any concern?
	4.3 Is the time gap between the data collection and analysis and the intended use of the model (e.g., in terms of policy recommendations) of any concern?
5. Chronology of data and goal	5.1 If the stated goal is predictive, are all the predictor variables expected to be available at the time of prediction?
	5.2 If the stated goal is causal, do the causal variables precede the effects?
	5.3 In a causal study, are there issues of endogeneity (reverse causation)?
6. Generalizability	6.1 Is the stated goal statistical or scientific generalizability?
	6.2 For statistical generalizability in the case of inference, does the paper answer the question "What population does the sample represent?"

(*Continued*)

Table 12.8   (*Continued*)

Dimension	Questions
	6.3 For generalizability, in the case of a stated predictive goal (predicting the values of new observations and forecasting future values), are the results generalizable to the to-be-predicted data?
	6.4 Does the paper provide sufficient detail for the type of needed reproducibility and/or repeatability and/or replicability?
7. Operationalization	Construct operationalization:
	7.1 Are the measured variables themselves of interest to the study goal or is the focus on their underlying construct?
	7.2 What are the justifications for the choice of variables?
	Strength of operationalizing results:
	7.3 Who can be affected (positively or negatively) by the research findings?
	7.4 What can the affected parties do about it?
8. Communication	8.1 Is the exposition of the goal, data, and analysis clear?
	8.2 Is the exposition level appropriate for the readership of the journal?
	8.3 Are there any confusing details or statements that might lead to confusion or misunderstanding?

opinion making, and reviewing approaches. Adapting an InfoQ infrastructure to reviews would provide both flexibility and efficiencies to reviews, without affecting the basic reviewing elements of impartiality, objectivity, and the upholding of high application standards.

References

DallaValle, L. (2014) Official statistics data integration using vines and non parametric Bayesian belief nets. *Quality Technology and Quantitative Management*, 11(1), pp. 111–131.

Deming, W.E. (1982) *Quality, Productivity, and Competitive Position, Massachusetts Institute of Technology*. Massachusetts Institute of Technology, Center for Advanced Engineering Study, Cambridge, MA.

Drummond, C. (2009) Replicability Is Not Reproducibility: Nor Is It Good Science. *Proceedings of the Evaluation Methods for Machine Learning Workshop at the 26th ICML*, Montreal, Canada.

Francois, O. (2015) Arbitrariness of Peer Review: A Bayesian Analysis of the NIPS Experiment, http://arxiv.org/abs/1507.06411 (accessed May 19, 2015).

Gewin, V. (2011) What the novice peer reviewer needs to know before combing through a submission. *Nature*, 478, pp. 275–277.

John, L.K., Loewenstein, G. and Prelec, D. (2012) Measuring the prevalence of questionable research practices with incentives for truth-telling. *Psychological Science*, 23, pp. 524–532.

Kenett, R.S. and Raanan, Y. (2010) *Operational Risk Management: A Practical Approach to Intelligent Data Analysis*. John Wiley & Sons, Ltd, Chichester, UK.

Kenett, R.S. and Salini, S. (2012) *Modern Analysis of Customer Surveys: With Applications Using R*. John Wiley & Sons, Ltd, Chichester, UK.

Kenett, R.S. and Shmueli, G. (2015) Clarifying the terminology that describes scientific reproducibility. *Nature Methods*, 12(8), p. 699.

Kenett, R.S. and Shmueli, G. (2016) Helping reviewers ask the right questions: the InfoQ framework for reviewing applied research. *Journal of the International Association for Official Statistics*, 32(1), pp. 11–19.

Kenett, R.S. and Thyregod, P. (2006) Aspects of statistical consulting not taught by academia. *Statistica Neerlandica*, 60(3), pp. 396–412.

Kenett, R.S., Franch, X., Susi, A. and Galanis, N. (2014) Adoption of Free Libre Open Source Software (FLOSS): A Risk Management Perspective. *Proceedings of the 38th Annual IEEE International Computer Software and Applications Conference (COMPSAC)*, Västerås, Sweden.

Pearl, J. (2013) Transportability Across Studies: A Formal Approach. UCLA Cognitive Science Laboratory, Technical Report (R-372), San Francisco.

Rasch, G. (1961) On General Laws and the Meaning of Measurement in Psychology, in *Proceedings of the Fourth Berkeley Symposium on Mathematical Statistics and Probability, IV*, pp. 321–334. University of Chicago Press, Berkeley.

Rasch, G. (1977) On specific objectivity: an attempt at formalizing the request for generality and validity of scientific statements. *The Danish Yearbook of Philosophy*, 14, pp. 58–93.

13

Integrating InfoQ into data science analytics programs, research methods courses, and more

13.1 Introduction

The last several years have seen an incredible growth in the number of courses and new programs in "data science," "business analytics," "predictive analytics," "big data analytics," and related titles. Let us call them "analytics," for short. Figure 13.1, derived from Google Trends, shows the growth in popularity of the term "data science course" in search queries over time. Similar trends can be seen for "business analytics course" or, simply, "analytics course." Different programs have a different emphasis depending on whether they are housed in a business school, a computer science department, or a cross-departmental program. What is however common to all of them is their focus on data (structured and unstructured) and, specifically, on data analysis.

Many statistics and operations research programs and departments have been restructuring, revising, and rebranding their courses and programs to match the high demand for people skilled in data analysis. This change represents an evolution in quantitative thinking education where emphasis is increasingly on linking academic programs to real-world applications and facilitating student engagement. A pioneering paper in this direction, presenting the involvement of students in the

Information Quality: The Potential of Data and Analytics to Generate Knowledge,
First Edition. Ron S. Kenett and Galit Shmueli.
© 2017 John Wiley & Sons, Ltd. Published 2017 by John Wiley & Sons, Ltd.
Companion website: www.wiley.com/go/information_quality

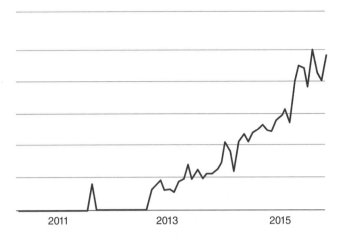

Figure 13.1   Google Trends data on "data science course." Source: © 2015 Google Inc, is used with permission. Google and the Google logo are registered trademarks of Google Inc.

design and analysis of practical experiments, is Hunter (1977). This approach leads naturally to a discussion on the goal of a study, the type of data that needs to be collected and the utility of the findings. Kenett and Steinberg (1987) present their experience in applying this approach in various introduction to statistics academic courses with physical designed experiments. With the ubiquity of computer technology, class experiments using simulations has become popular (see Froelich and Duckworth, 2010). An expansion of this idea to a life cycle view of statistics as a discipline is discussed in Kenett and Thyregod (2006) and Kenett (2015), with a retrospective assessment on experience gained by this approach provided in Kenett and Steinberg (2014).

Designing a new analytics program requires not only identifying needed courses but also tying the courses together into a cohesive curriculum with an overriding theme. Such a theme helps determine the proper sequencing of courses and creates a coherent linkage between different courses often taught by faculty from different domains. It is common to see a program with some courses taught by computer science faculty, other courses taught by faculty from the statistics department, and others from operations research, economics, information systems, marketing, or other disciplines. Applying an overriding theme not only helps students organize their learning and course planning but it also helps the teaching faculty in designing their materials and choosing terminology. A theme like this also helps in designing courses to be cotaught by faculty with different background and expertise.

In the previous chapters, we presented the principles, components and dimensions of InfoQ and many examples of application areas. In this chapter, we show how InfoQ can be used as an infrastructure for designing, conducting, and evaluating academic programs, workshops, and industry courses in analytics.

The InfoQ framework provides a theme that focuses the attention of faculty and students on the important question of the value of data and its analysis with flexibility that accommodates a wide range of data analysis topics.

By adopting an InfoQ framework, the instructor and students focus on the goals of a study, the data used in the analysis, the analysis methods, and the utility measures learned in a certain course. This focus helps compare and contrast knowledge provided across courses, creating richness and cohesion. The eight InfoQ dimensions can be used in structuring any empirical data analysis course—be it predictive analytics, econometrics, optimization, advanced regression analysis, database management, or any other course that touches on data and data analysis.

In the following, we present how the InfoQ framework, as a theme, can be used for:

- Designing a new program

- Redesigning an existing program

- Designing a new course

- Redesigning an existing course

The InfoQ framework can be integrated through a variety of media and activities. Examples include the following:

- Instructors can dedicate an in-class session to introduce the InfoQ framework, through presentation, short videos, discussion, examples, and in-class assignments.

- Instructors can assign individual and/or group assignments that analyze one or more empirical studies using the InfoQ framework and then present them to class and/or submit a written report.

- Faculty advisors and instructors of research methods courses can analyze and let students analyze their research proposals, dissertation, and papers.

- Instructors of seminar courses can use the InfoQ framework for analyzing the papers that students are assigned in the course (in seminar courses, students are often assigned reading and presenting papers).

- In programs with speaker series such as a weekly seminar with talks by visitors, students can be instructed to analyze the paper presented by using the InfoQ framework. Such an analysis can be assigned prior to the talk, to facilitate questions and interaction, and after the talk, to evaluate comprehension.

In Section 13.2, we present some experience with these activities.

13.2   Experience from InfoQ integrations in existing courses

This section describes three case studies that demonstrate the use of the InfoQ framework for developing a plan and for evaluating an empirical study. These investigations focused on in-class discussions and ratings of InfoQ elements but did not assess the reliability of ratings across raters. Future research is needed in this area.

Case 1:   Evaluating research proposals

Graduate students at the Faculty of Economics of the University of Ljubljana (FELU) undergo a three-day workshop on research methods, to equip them with methodology for developing and presenting a research proposal that is the basis for their doctorate thesis. One of the first milestones for students is to defend their research proposal in front of a committee.

The class experience presented here is based on 50 graduate FELU students from a wide range of areas, including organizational behavior, operations research, marketing, and economics. In 2009, InfoQ was integrated into the research methods workshop. The goal was to help students (and their advisors) figure out whether their proposed research is properly defined as to potentially generate effective knowledge, that is, whether their proposal can potentially generate information of sufficient quality.

Students worked in small teams and discussed the InfoQ dimensions of their draft research proposals. Each student gave a 15-minute presentation on his/her proposal to the entire team. Details of the workshop and preworkshop assignments are available at goo.gl/f6bIA. Students' grades in the workshop were derived from an InfoQ score of the research proposal they submitted, which consisted of a PowerPoint presentation and a written document. This approach was designed to make their research journey more efficient and more effective. Feedback by students and faculty indicated that the InfoQ-based research methods workshop has indeed met this goal (Kenett et al., 2010).

Case 2:   Ex post evaluation of empirical studies

This case study is based on an assignment that requires participants to evaluate five empirical studies presented in the form of PowerPoint slides and written reports. After evaluating the reports and a brief introduction to InfoQ, participants were asked to:

1.   Give a brief description of the goal, data, analysis and utility measure for each study

2.   Rate the study on each of the eight InfoQ dimensions

The evaluation form with information on InfoQ, the five studies, and the InfoQ questions and ratings are available at goo.gl/erNPF. Figure 13.2 shows the questions asked in one of the studies.

Evaluating Information Quality (InfoQ)

Study #1: Predicting days with unhealthy air quality in Washington DC

Several tour companies' revenues depend heavily on favorable weather conditions. This study looks at air quality advisories, during which people are advised to stay indoors, within the context of a tour company in Washington DC.

You will find the study report and presentation here (copy and paste the address into your browser): http://galitshmueli.com/content/tourism-insurance-predicting-days-unhealthy-air-quality-washington-dc

Stated objective of study
Is the stated objective to explain, predict, or to describe? If more than one objective is stated, choose the highest priority objective

○ Explain (how or which factors affect air quality)

○ Predict (air quality on new or future days)

○ Describe (the relationship between air quality and other variables)

Data used and its origin
Briefly describe the data used in the analysis

Analysis methods used in study
List all the methods mentioned (for example: histogram, logistic regression)

Utility of findings
What is the potential value of the study findings? To whom? When?

Figure 13.2   InfoQ evaluation form for an empirical study on air quality. The complete information and additional studies for evaluation are available at goo.gl/erNPF.

Kindly rate the project on the 8 InfoQ dimensions
Information about the 8 dimensions is available at http://tinyurl.com/6pmrwcx

	completely inadequate	inadequate	reasonable	achieved	fully achieved
Data resolution	○	○	○	○	○
Data structure	○	○	○	○	○
Data integration	○	○	○	○	○
Temporal relevance	○	○	○	○	○
Generalizability	○	○	○	○	○
Chronology of data and goal	○	○	○	○	○
Operationalization	○	○	○	○	○
Communication	○	○	○	○	○

Additional comments
Feel free to write any additional comments that you may have regarding this study

Figure 13.2   (Continued)

In 2012, the InfoQ assignment was integrated into a course in the statistics practice master's program at Carnegie Mellon University's Statistics Department, in Pittsburgh PA, USA. Each of the 16 students spent about 60–90 minutes reading the five studies and evaluating the eight dimensions for each study.

Comparing the responses of the 16 participants on each of the five studies revealed variability in respondents' ratings of the InfoQ dimensions. This variability indicates a need to further streamline the process of quantifying each dimension rating. An important result of this study was the feedback regarding the value added by going through the evaluation process. Participants reported that using this approach helped them "sort out all of the information," and several participants reported that they will adopt this evaluation approach for future studies.

Case 3:   Retrospective evaluation of completed research

The International PhD program at National Tsing Hua University's Institute of Service Science in Taiwan brings together students from a variety of masters' programs internationally—from engineering to statistics to business. The research methods course is designed to provide students with a broad array of methods for collecting and analyzing data, especially in the social sciences, which is relevant to research in their fields. The course covers both traditional research and state-of-the-art research methods useful for big data.

In fall 2015, one of the first three-hour sessions in the research methods course was devoted to introducing InfoQ. As a preparation, students were tasked with three preparations:

1. Read the paper "On Information Quality" (Kenett and Shmueli, 2014)

2. Watch a 30-minute video introducing the topic (https://youtu.be/ LeKn5zbs-cU)—this is a webinar that one of the authors presented at the Royal Statistical Society's Journal Club

3. Analyze their previous research work (masters' dissertation work) using the InfoQ framework

Specifically, in #3, students were instructed to:

• Identify the four components in the study: goal, data, analysis, and utility

• Evaluate each of the eight InfoQ dimensions and give a rating (on a 1–5 scale) to each dimension

During the class session, each student presented his/her research and the InfoQ analysis that they performed. The presentation ranged widely in research areas—from engineering to theoretical statistics, to social science, reflecting the background of the students.

Evaluating one's own research retrospectively, in a group format, provides an opportunity to discuss work in a structured way. With proper facilitation, this generates in-depth feedback and insights with impact on the research being evaluated and future work of the group involved. In a sense, the InfoQ assessment provides a strong pedagogical support with short-term and long-term implications.

In these case studies, the InfoQ framework proved useful in three main aspects. First, it helped students better communicate their empirical work, by laying out the four InfoQ components (goal, data, analysis, and utility). Second, students were able to learn about different goals, data, analysis methods, and utilities used outside of their own specific areas. Third, the InfoQ framework allowed students to organize their own thoughts about their previous research in a structured way. Students were especially appreciative of the latter. Examination, through the InfoQ framework and specifically through the eight dimensions, has led students realize flaws in their work that went unnoticed in casual examination.

Finally, when the InfoQ session is scheduled early in the course, the InfoQ terminology can be carried out throughout the course, helping to tie the value of disparate data collection methods such as survey sampling, design experiments, observational data from the Internet, and analysis methods (including causal explanatory modeling and predictive modeling). In this way, InfoQ adds to the cohesiveness and integration of a program on analytics.

13.3   InfoQ as an integrating theme in analytics programs

The number of new programs in analytics has been growing quickly in recent years, thanks to demand from industry. Two popular locations for new analytics programs are within a business school and as a cross-department university program. At the time of writing, multiple websites offer lists of such programs; see, for example, www.dnuggets.com/education/index.html. Unsurprisingly, rankings of such programs have also started to appear (e.g., https://tfetimes.com/msba-rankings/). The great majority of programs are currently a dedicated masters' program, typically one- to two-year-long programs. A typical program will include a set of courses on different analytics topics and one or more integrative activities such as a capstone course or an internship.

The first large business analytics program in India, at the Indian School of Business (ISB) in Hyderabad, India, in 2013, was designed by the second author. ISB is well known for its MBA-type program (Post Graduate Programme in Management, or PGP) and has been offering several analytics courses as elective courses within the PGP. The new business analytics program was designed as a separate 12-month certificate program for working professionals.

Important pedagogical considerations in designing a new program are the choice of courses and other learning activities to include, their sequencing and the integration of knowledge and experiences. Another consideration is the types of teaching, learning, and evaluation formats. These are all subject to constraints from the offering institution's side, in terms of faculty knowledge and expertise, faculty teaching schedules and teaching styles, and availability of supporting staff, classrooms, etc., as well as constraints from the students' side: background knowledge and experience, working schedules, and so on. Furthermore, many programs emphasize group work, as is commonly the case for analytics professionals in industry, which requires further consideration of learning and evaluation topics and styles. Lastly, analytics programs, even when housed within a business school, will have instructors from different disciplines, for example, marketing, operations research, statistics, and information systems, each with its different teaching styles and world view. Thus, analytics programs are of interdisciplinary nature. These settings are not new and have been the context for various new programs over the years. For example, Lee et al. (2007) describe the design of a new masters' program focused on data quality,[1] which, like analytics programs, aims to achieve an interdisciplinary curriculum design while maintaining a balance between theoretical rigor and practical relevance.

Due to the nature of such programs and given the considerations and challenges described earlier, it is extremely useful to have a unifying theme for the program, which is understood both by the faculty and the students. Business schools that use the case study method (e.g., Harvard Business School) create such an overarching theme for learning. In analytics programs, InfoQ can provide a more natural

---

[1]Although the official program name is of Masters of Science program in Information Quality (MSIQ), their definition of information quality is different from InfoQ and is closer to data quality.

alternative. It does not require much restructuring of existing courses that faculty might offer in other programs and provides a framework which can help students and faculty make connections across courses and other learning activities.

The InfoQ framework can help choose courses suitable for a program, by considering the different *analysis goals* that students should learn (e.g., predicting new records, quantifying causal relationships, visualizing patterns in text, and optimizing complex processes), *domain goals* (e.g., customer retention, risk management, and political campaign prediction), *data types* (e.g., cross-sectional, time series, text data, and network data), *analysis methods* (e.g., statistical models, machine learning algorithms, simulation and optimization methods, visualization methods, and econometric models), and *utility metrics* (e.g., out-of-sample predictive performance, statistical significance, and costs). A set of courses can then be selected to diversify the students' knowledge in terms of these components. Mapping courses along these components can also help with sequencing decisions and creating subspecialty tracks.

The inclusion of an integrative component such as a capstone project or internship is typically intended to help students cement the learnings from the different courses and gain some experience with analytics in practice. It aims to introduce students to real-world problem solving using analytics. However, these components tend to be the most challenging in terms of structuring and evaluation. Specifically, the requirements can vary from one internship to the other and from one project supervisor to another. InfoQ is a practical solution for this diversity, as it offers two anchors that are sufficiently flexible for a range of applications: identification of the four InfoQ components and evaluation of the eight InfoQ dimensions. The InfoQ framework can also help students provide immediate value to companies they are working with, by introducing them to a framework that will facilitate their own understanding of analytics in their context.

Finally, the InfoQ framework can be helpful in creating a shared language and discourse within a program, between faculty and students with diverse backgrounds. Section 13.4 discusses the design and redesign of analytic courses.

13.4   Designing a new analytics course (or redesigning an existing course)

With the growth in popularity of analytics courses, faculty—both research faculty and teaching faculty—are engaged in the design of new analytics courses and redesign of existing ones in order to expand education programs with new topics, new audiences, and new programs.

Redesigning an existing course for a new audience—for example, modifying an existing graduate-level data mining course for an undergraduate audience—requires considering the background, needs, and attitudes of the new audience. The important question is what parts should be modified and which components can remain as is. Using the InfoQ framework helps maintain the key structure of the course: there is always a need to focus on goal, data, analysis methods, and utility metrics. In this context, the instructor can ask in preparation for his/her lectures:

1. **Goal**: How do I explain the main business goals to the new audience? How will the new audience be able to learn about goals of interest in practice? How will students learn to connect an analytics goal to a domain goal?

2. **Data**: What types of datasets will the new audience be enthusiastic about? What can they handle technically?

3. **Analysis**: Which sets of analysis methods are most needed for the new audience? What related methods have they already learned? What type of software can be used by these students?

4. **Utility**: What types of utility metrics must students learn? How will they learn to connect the overall utility of the analysis with the utility metrics taught?

5. **InfoQ**: How will students learn to integrate the four components? Is this new audience comfortable working in groups? How will they communicate their integrative ability?

Based on these answers, the instructor can redesign the learning materials and the evaluation activities and testing criteria. In some cases, instructors might find of interest the addition of formative assessment methods, such as MERLO, for assessing the conceptual understanding levels of students as discussed in Section 6.4 (see also Etkind et al., 2010).

Designing a new course requires making further decisions about which topics to cover, what textbook and software to use, and what teaching style to use. While many instructors make choices based on a popular textbook, it can be useful to evaluate their options by considering the four InfoQ components and their integration:

1. **Goal**: What types of analysis and business or other domain goals are covered in the textbook?

2. **Data**: What datasets are provided with the textbook and/or software? What other sources of data can be used for this audience?

3. **Analysis**: Which analysis methods are covered in the textbook, and is the level of technicality (mathematical, computational, etc.) appropriate for the audience?

4. **Utility**: Which utility measures are presented in the textbook? Are they tied to the domain utility?

Following such an analysis, the instructor can map the methods and examples used in the textbook to the eight InfoQ dimensions: (i) data resolution, (ii) data structure, (iii) data integration, (iv) temporal relevance, (v) generalizability, (vi) chronology of data and goal, (vii) operationalization, and (viii) communication. In fact, these dimensions can be used as a checklist to assess any empirical research, as shown in the examples presented in Part II.

As an example, the instructor can evaluate any reference to semantic data under data structure, integration methods like ETL under data integration, studies designed to consider applicability of findings in different contexts as generalizability, and the approach to the visualization of results as communication.

In redesigning a course, the instructor can add a session on InfoQ early in the course. Find a paper or case study relevant to the material taught in the course (perhaps one that is already used in the course), and analyze it using the InfoQ framework.

In future following sessions, the instructor can start with an overview of the method or case to be taught using the InfoQ terminology. This requires minimal adjustment to existing materials—a couple of slides at the beginning and slightly reorganizing or highlighting the four InfoQ components and eight InfoQ dimensions in the existing materials. For example, in a course on time series forecasting, before moving into teaching forecasting methods, the instructor can highlight:

1. **Goal**: forecasting of future values of a series for different purposes (e.g., weekly demand forecasting for stocking decisions, daily air quality forecasting for issuing health advisory warnings; daily emergency department traffic for scheduling staff, and minute-by-minute web server traffic forecasting for detecting cyber attacks)

2. **Data**: time series (typically treated in univariate mode but possibly multivariate) and how it differs from cross-sectional data

3. **Analysis methods**: an overview of the methods to be learned in the course (e.g., extrapolation methods and regression-based methods)

4. **Utility**: the type of performance metrics and performance evaluation approaches to be used, such as partitioning the series into training and holdout period and evaluating out-of-sample forecasts

In Section 13.5, we present the structure of an InfoQ workshop on analytic methods conducted within an organization with teams of analysts and experts with statistical competencies.

13.5   A one-day InfoQ workshop

We present here a one-day InfoQ workshop that can be conducted by organizations in the industrial and service sectors. This workshop was designed to provide a methodology and tools for designing and evaluating empirical studies using the eight InfoQ dimensions. It is aimed at analysts involved in practical analytic work using statistical software tools such as JMP, Minitab, or R. The goal is to help participants improve the design and evaluation of empirical studies that they conduct by providing tools and methodology based on InfoQ components and dimensions.

The workshop described in the following took place internally in an organization developing web-based applications. The workshop was aimed at the company's analyst in charge of system optimization and customer experience enhancements and consisted of two phases. In phase one, InfoQ concepts and dimensions were presented and illustrated with examples. In the second phase, participants discussed case studies in small groups, followed by individual participants' sharing of past or current projects through the InfoQ lens.

The workshop agenda, which can be used as a general agenda, is listed as follows:

8:30–9:30	Background on InfoQ: The Pokemon case study
9:30–10:00	InfoQ at the study design phase and at the postdata collection phase
10:00–10:15	Break
10:15–12:00	Group evaluation of case studies in small groups followed by a class presentation
12:00–13:00	Lunch
13:00–17:00	Presentation by participants of work-related projects
17:00–17:30	Recap and lessons learned

Participants included machine learning experts, statisticians, and industrial engineers with undergraduate and graduate degrees. All participants had experience in R, and some had experience in SQL, Minitab, JMP, and MATLAB. The level of experience ranged from one to ten years.

The unanimous feedback was that InfoQ provided the missing link between what they learned at university and what their day-to-day work required. As a particularly gratifying feedback was an input from management that InfoQ components and dimensions will be used to assess reports and presentations based on empirical research.

13.6   Summary

This chapter integrates InfoQ in educational programs in topics such as data science, business analytics, and applied statistics. It shows how such programs can be designed or redesigned with InfoQ as a unifying theme. Providing an InfoQ-based integrating approach can make the difference between successful programs that provide participants with both the theoretical and practical exposure necessary for effective and efficient analytic work. Specifically, consideration of the eight InfoQ dimensions (data resolution, data structure, data integration, temporal relevance, generalizability, chronology of data and goal, operationalization, and communication) provides a skeleton for presenting various analytic methods handling structured (numerical) and unstructured (semantic) data. We discuss, in this chapter, how to design or redesign courses in analytics by focusing on the four InfoQ components: g (goal), X (data), f (data analysis), and U (utility) as well as by the relationships between X, f, g, and U, that form InfoQ(f, X, g, U) = $U(f(X \mid g))$. The chapter also provides a brief description and an agenda for an InfoQ workshop designed to be conducted within companies and organizations. Programs in specific domain areas such as healthcare, risk management, and official statistics can use the examples in Part II, which are already presented with the underlying theme of InfoQ. In addition, the following three chapters (Chapters 14, 15, and 16) present specific examples of InfoQ analysis in three popular statistical software platforms: R, Minitab, and JMP. The combination of a unifying theme, real-life examples, and supporting software provides a solid basis for the design (or redesign) of programs on analytics such as data science, business analytics, and applied statistics.

Acknowledgements

We thank Professors Joel Greenhouse (Carnegie Mellon University), Shirley Coleman (Newcastle University), and Irena Ograjensek (University of Ljubljana) for their support of integrating InfoQ into graduate courses at CMU and University of Ljubljana and helping assess its impact.

References

Etkind, M., Kenett, R.S. and Shafrir, U. (2010) The Evidence Based Management of Learning: Diagnosis and Development of Conceptual Thinking with Meaning Equivalence Reusable Learning Objects (MERLO). *Proceedings of the 8th International Conference on Teaching Statistics (ICOTS)*, Ljubljana, Slovenia.

Froelich, A. and Duckworth, W. (2010) Using New JMP® Interactive Modules to Teach Concepts in Introductory Statistics, SAS Global Forum 2010 Posters Paper 289-2010, Seattle, Washington http://support.sas.com/events/sasglobalforum/2010/ (accessed June 3, 2016).

Hunter, W. (1977) Some ideas about teaching design of experiments, with 25 examples of experiments conducted by students. *The American Statistician*, 31(1), pp. 12–17.

Kenett, R.S. (2015) Statistics: a life cycle view (with discussion). *Quality Engineering*, 27(1), pp. 111–129.

Kenett, R.S. and Shmueli, G. (2014) On information quality. *Journal of the Royal Statistical Society, Series A*, 177(1), pp. 3–38.

Kenett, R.S. and Steinberg, D. (1987) Some experiences teaching factorial designs in introductory statistics courses. *Journal of Applied Statistics*, 14, pp. 219–228.

Kenett, R.S. and Steinberg, D. (2014) Teaching Design of Experiments: A 25 Years Retrospective. *Proceedings of the European Network for Business and Industrial Statistics (ENBIS) 14th Annual Conference on Business and Industrial Statistics*, Linz, Austria.

Kenett, R.S. and Thyregod, P. (2006) Aspects of statistical consulting not taught by academia. *Statistica Neerlandica*, 60(3), pp. 396–412.

Kenett, R.S., Coleman, S. and Ograjenek, I. (2010) On Quality Research: An Application of Infoq to the phd Research Process. *Proceedings of the European Network for Business and Industrial Statistics (ENBIS) Tenth Annual Conference on Business and Industrial Statistics*, Antwerp, Belgium.

Lee, Y., Pierce, E., Talburt, J., Wang, R. and Zu, H. (2007) A curriculum for a master of science in information quality. *Journal of Information Systems Education*, 18(2), pp. 233–240.

14

InfoQ support with R

Federica Cugnata[1], Silvia Salini[2] and Elena Siletti[2]

[1] *University Centre of Statistics for Biomedical Sciences (CUSSB), Vita-Salute San Raffaele University, Milano, Italy*
[2] *Department of Economics, Management and Quantitative Methods, Università degli Studi di Milano, Milano, Italy*

14.1 Introduction

R is a free software environment for statistical computing and graphics. It runs on a wide variety of platforms such as MS-Windows, Mac OS X, and Linux. It is available as free software under the terms of the Free Software Foundation's (FSF) GNU General Public License in source code form. It can be freely downloaded from the main CRAN repository at the URL: http://cran.r-project.org or one of its mirrors.

Open-source software is a term used to describe computer software which publishes for free the source code and makes it available to any user. The R code and its applications are protected under GNU General Public License, version 2 or 3, that provides rights to study, change, and distribute the software for any purpose. This enables a collaborative environment where many people around the globe can tweak, modify, and improve the software in question. As a result, an open software like R is upgraded at a faster pace than many other commercial alternatives.

R is similar to the S language and environment which was developed in the late 1970s by John Chambers and colleagues at Bell Laboratories (formerly AT&T, now

Information Quality: The Potential of Data and Analytics to Generate Knowledge,
First Edition. Ron S. Kenett and Galit Shmueli.
© 2017 John Wiley & Sons, Ltd. Published 2017 by John Wiley & Sons, Ltd.
Companion website: www.wiley.com/go/information_quality

Lucent Technologies). The S software was however not an open-source project and failed to generate broad interest. Ultimately, the copyright was sold to TIBCO Software. Nowadays, R is surpassing what was originally imagined possible with S.

R was initially written by Ross Ihaka and Robert Gentleman (1996), also known as "R & R" of the Statistics Department at the University of Auckland. The current R is the result of a collaborative effort with contributions from all over the world. The open-source nature of R was a key component of its success. As it is, any user can contribute "packages" that extend R's capabilities. You can find community-created tools for R that perform many modern statistical tools, some of them not available in other commercial software. The extensibility of R has grown even beyond that. There are now packages that allow you to create interactive visualizations, maps, or Web-based applications, all within R. Ihaka himself attributes the accomplishments of R to its collaborative nature: "R is a real demonstration of the power of collaboration, and I don't think you could construct something like this any other way. We could have chosen to be commercial, and we would have sold five copies of the software."

While it is difficult to know exactly how many people are using R, in a *New York Times* article, it was estimated that close to 250 000 people work with it regularly (Vance, 2009). Revolution Analytics (http://www.revolutionanalytics.com/community), creators of an enhanced version of R, claims that the number of users surpasses two million.

Many users think of R as a statistics system; however, developers prefer to think of it as an environment within which statistical techniques are implemented. Its main environment properties are as follows:

- It allows for effective data management and storage.

- It is a suite of operators for calculations on objects (arrays, matrices, and vectors).

- It is a collection of tools for data analyses.

- It incorporates graphical facilities for data analysis and allows viewing on different media.

- It is made by a simple, effective, and well-developed language which includes conditionals, loops, user-defined recursive functions, and input and output facilities.

Being designed as a computer language, R allows users to add additional functionality by defining new functions. Furthermore, R can be extended using "packages." Packages are collections of R functions, data, and compiled code in a well-defined format. The directory where packages are stored is called a *library*. There are eight packages supplied with R, and many more are available through the CRAN covering a very wide range of modern statistics. On November 2015, the CRAN package repository featured 7531 available packages. Anyone can create an R package following the R development guidelines, support the R-project sharing code, or simply save time by automating work. When a package is available through CRAN, it can be directly uploaded using the graphical user interface (GUI) functionality or by typing in the R console (`install.packages("package.name")`).

R has its own LaTeX-like documentation format which is used to provide comprehensive documentation both online in a number of formats and in hard copy.

Another way to use the R environment is with RStudio (http://www.rstudio.com). Specifically, RStudio is an integrated development environment (IDE) for R that includes a console, a syntax-highlighting editor that supports direct code execution, and tools for plotting, history, debugging, and workspace management (see Figure 14.1). RStudio offers a set of tools for the R computing environment. It is available both as open source, with limited features, and in a commercial version. It runs on a desktop (Windows, Mac, and Linux) and with a browser connected to RStudio Server or RStudio Server Pro (Debian/Ubuntu, Red Hat/CentOS and SUSE Linux). RStudio includes powerful coding tools designed to enhance productivity, enable rapid navigation to files and functions, and make it easy to start new or find existing projects; integrates support for Git and Subversion; supports authoring HTML, PDF, Word documents, and slide shows; and supports interactive graphics with Shiny and ggvis. Furthermore, it integrates R help and documentation and makes it easy to manage multiple working directories using projects. It also includes an interactive debugger

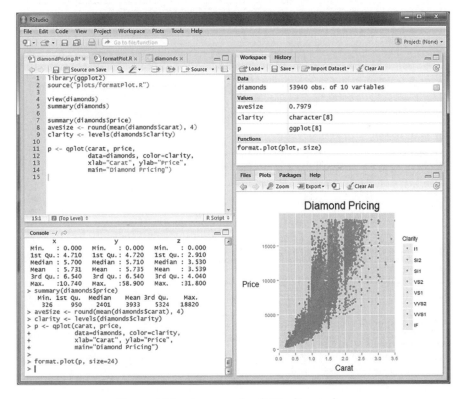

Figure 14.1   An example of RStudio window.

to quickly diagnose and fix errors. Finally, it is an extensive package development tool, and the RStudio team contributes code to many R packages and projects. For a comprehensive treatment of modern industrial statistics and the mistat package available in CRAN see Kenett and Zacks, 2014.

Besides the traditional R command prompt, there are many GUIs which enable the analyst to use click-and-point methods to analyze data without getting into the details of learning the R syntax. R GUIs are very popular both as mode of instruction in academia and in industry applications as it cuts down considerably on time taken to adapt to the language. As with all command line and GUI software, for advanced tweaks and techniques, command prompts also come in handy.

One of the most widely used R GUI is the R Commander (Fox, 2005), commonly known as *Rcmdr* (see Figure 14.2). Basically, R provides the engine that carries out

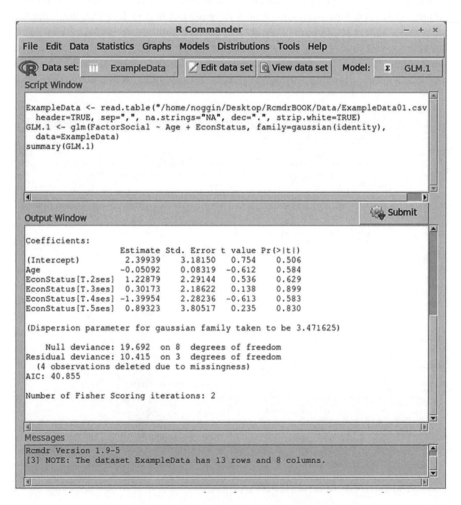

Figure 14.2   An example of R Commander window.

the analyses, and *Rcmdr* provides a convenient way for users to input commands. The *Rcmdr* program enables analysts to access a selection of commonly used R commands using a simple interface that is familiar to most computer users. It was created by John Fox to cover the content of a basic statistics course. However it is extensible and many other packages can be added in the form of R Commander Plugins.

The R Commander GUI consists of a window containing several menus, buttons, and information fields. In addition, the Commander window contains script and output text windows. The menus are easily configurable through a text file or, preferably, through plug-in packages, of which many are now available on CRAN. The menus lead to simple dialog boxes, the general contents of which are more or less obvious from the names of the menu items. These boxes have a common structure, including a help button leading to the help page for a relevant function, and a reset button to reset the dialog to its original state. By default, commands generated in dialogs are posted to the output window, along with printed output, and to the script window. Lines in the script window can be edited and submitted for execution. In other words, using the script window, you can write R code in combination with the menus. As you point and click a particular menu item, the corresponding R code is automatically generated in the log window and executed. Traditionally error messages, warnings, and "notes" appear in a messages window.

All these features, in addition to making things simple, have the important role of helping users implement R commands and develop their knowledge and expertise in using the traditional command line that is needed by those wishing to exploit the full power of R. R Commander is also integrated with RStudio.

Nowadays, R provides a wide variety of statistical, linear and nonlinear modeling, classical statistical tests, time series analysis, classification, clustering, and graphical techniques, always being highly extensible. It is not only an application for standard statistical procedures, it is also the de facto standard among statisticians for the development of new statistical procedures. R's advantages include the ease with which we can import data and export data or output and well-designed quality plots that can be produced, including mathematical symbols and formulas.

Importing data into R is fairly simple, and there are many ways to import data: getting data directly from different file types or directly from the Web. This means that data not only can be saved in a file on a personal computer (e.g., a local Excel, SPSS, or some other type of file) but can also be saved on the Internet or be obtained through other sources. Munzert and colleagues (2015) present a very exhaustive collection of techniques enabling the creation of powerful collections of unstructured or unsorted data, focused on Web scraping and text mining with R.

Using the JMP app (see Chapter 16), one can interact with R using JSL, that is, submitting statements to R from within a JSL script, exchanging data between JMP, and R and displaying graphics produced by R.

Usually we obtain a data frame by importing it from SAS, SPSS, Excel, Stata, Minitab, a database, or an ASCII file. Such a file can be also created interactively.

The `read.table()` function is the most important and commonly used function to import simple data files into R (.csv or .txt). It is easy and flexible. One of the best ways to read an Excel file is to export it to a comma delimited file (.csv) and import it using the aforementioned method. Alternatively you can use the xlsx package to access Excel files. For SPSS and SAS, we recommend the Hmisc package for ease and functionality. For Stata and Minitab, we recommend the foreign package. The RODBC package provides access to databases, including Microsoft Access and Microsoft SQL Server, through an ODBC interface. It is also possible to import data from other database interfaces, simply changing package; some examples include the RMySQL package that provides an interface to MySQL, the ROracle package that provides an interface for Oracle, and the RJDBC package that provides access to databases through a JDBC interface.

Getting data from HTML tables into R is pretty straightforward too, with a simple command or using httr package.

Using quantmod package we can easily import financial data from an Internet source. This package is closely related to the packages "xts" and "zoo" to analyze time series. The function that we use to get data into R is "getSymbols()," like in the following example, where we download the time series of daily values of FTSEMIB from 01/01/2010 to 31/12/2015 from the website http://it.finance.yahoo.com:

```
library(quantmod)
getSymbols("FTSEMIB.MI",from="2010-01-01", to="2015-12-31")
to.monthly(FTSEMIB.MI, indexAt='lastof')
```

In the case presented here, the prices are converted from daily to monthly, presenting for each bar the opening, the max, the min, and the end of each month; you can switch to different time frames, such as weekly or yearly, but it is not possible to go down from longer to shorter ones (e.g., from months to days).

In recent years there is a growing interest in data from social media, like Twitter or Facebook. In R we can import and analyze this data very easily. The following sections present an application that uses data from Twitter.

There are numerous methods for exporting R objects into other formats. The simple way is with the function write.table() for .csv or .txt format. For SPSS, SAS, Stata, and Minitab, we need to load the foreign package. For Excel, we need the xlsReadWrite package.

Turning plain text output into well-formatted tables can be a repetitive task, especially when many tests or models are being incorporated into a paper. For R users, there are a lot of ways to make it easier, regardless of what typesetting system is used.

R is able to quickly export results and the xtable package produces LaTeX-formatted tables. Using specific kinds of R objects, such as linear model summaries, we can turn them into "xtables," which can in turn be the output to either LaTeX or HTML. An alternative to xtable is the R2HTML package, which is similar, but does not require "xtable" objects to be created in order to generate HTML output.

With the *Sweave* environment, it is possible to work by combining R and LaTeX: this means that the source of a text could be a plain LaTeX file with special R tags. R reads and processes the file, producing the correct LaTeX code, graphics, tables, and all the R output. The new file is passed on to a standard LaTeX compiler, and the result is the final text ready to publish. The *Sweave* document is a live document, where changing the data frame updates all the statistical analysis.

14.2   Examples of information quality with R

Following Hand's (2008) definition of statistics as "The technology of extracting meaning from data," we consider the utility of applying a technology f to a resource X for a given purpose g, with R.

14.2.1   Text mining of Twitter data

Twitter is an online social network that enables users to send and read short 140-character messages called "tweets." Registered users can read and post tweets, but unregistered users can only read them. Users access Twitter through the website interface, SMS, or mobile device app. This service was created in 2006 and rapidly gained worldwide popularity, with more than 100 million users who in 2012 posted 340 million tweets and handled 1.6 billion search queries per day. In 2015, Twitter had more than 500 million users, out of which more than 302 million were active users.[1]

Despite the potential, many researchers do not exploit big data from social network sites and do not know how to process it. In this section, we provide a basic introduction to text analysis of data from Twitter with R. Though data from Twitter is not different from data from other social media networks, Twitter has important characteristics that make it particularly interesting for text mining: on the one hand, weak-tie connection among people on Twitter is stronger than other networks, which greatly increases information exposure, and on the other hand, it has word limits on each tweet, and for this reason users tend to be more precise than in conversational language.

As well as being a great tool of communication, Twitter can also be considered an open mine for text and social web analyses. Among the different applications that can be used to analyze Twitter's data, R offers a wide variety of options. We don't cover here all the topics associated with Twitter data analysis, but we offer some examples.

People gather data from Twitter for goals such as to discover useful, valid, unexpected, and understandable knowledge from data. We analyze information because we have some particular questions in mind, such as "What are people talking about when they refer to Expo 2015?," "What is the average number of words per tweets on a specific topic?," or "What's the opinion of people about Expo 2015?."

Get data from Twitter

There are several ways to get data from Twitter, but, in our opinion, the simplest and quickest way is by using the R package `twitteR` (Gentry, 2015).

---

[1] https://en.wikipedia.org/wiki/Twitter#cite_note-12

Before using this package you need a developer account. To register go to https://dev.twitter.com/. Once you have it, you can obtain the API code creating a new application. In the bottom of the page, click "Manage Your Apps" under "Tools" and then on "Create New Application." On the application creation page, you have to fill all the questions except the *Callback*. When you finish the creation step, you can check the details of our application. The generated consumer keys and secrets (API) would be under the tab "Keys and Access Token."

This information is important to successfully connect to Twitter, and you need to run this procedure only once: with the same code you can connect forever:

```
install.packages(c("devtools", "rjson", "bit64", "httr"))
#RESTART R session!
library(devtools)
install_github("twitteR", username="geoffjentry")
#this is important to be sure to use the last version of the
package

library(twitteR)
api_key <- "YOUR API KEY"
api_secret <- "YOUR API SECRET"
access_token <- "YOUR ACCESS TOKEN"
access_token_secret <- "YOUR ACCESS TOKEN SECRET"
setup_twitter_oauth(api_key,api_secret,access_token,access_
token_secret)
```

Goal 1: Determine what people are talking about regarding "Expo 2015" and "Expo 2020"

In many cases, the main question that we want to answer from a twitter analysis is to know what people are talking about for a given topic. Usually the quickest way to know more about what people are talking about is by visualizing the most frequent words and terms contained in the tweets. We can do this by creating a plot—a bar plot with the most frequent terms—or we can use a more visually attractive way using a word cloud. A *wordcloud* is generally very helpful if we only want to take a quick and simple look at the data.

We download a maximum number of 1000 tweets in one week, containing Expo+2015+Milan and Expo+2020+Dubai:

```
expo2015 <- searchTwitter("Expo+2015+Milan",n=1000, lang="en")
expo2020 <- searchTwitter("Expo+2020+Dubai",n=1000, lang="en")
```

Before beginning our analysis, we load all the required packages for text mining (Feinerer, 2015) and for creating (Fellows, 2013) *wordcloud*:

```
library(tm)
library(wordcloud)
```

We extract the text from the tweets in two different vectors:

```
text_2015 <- sapply(expo2015, function(x) x$getText())
text_2020 <- sapply(expo2020, function(x) x$getText())
```

We then need to clean the data, but `tolower` doesn't always behave as expected and returns error messages. This is why we apply a modified version that skips such errors:

```
newTolower = function(x)
{
    y = NA
    try_error = tryCatch(tolower(x), error=function(e) e)
    if (!inherits(try_error, "error"))
        y = tolower(x)
    return(y)
}

text_2015 <- sapply(text_2015, newTolower)
names(text_2015) = NULL
text_2015 <- text_2015[text_2015 != ""]
text_2020 <- sapply(text_2020, newTolower)
names(text_2020) <- NULL
text_2020 <- text_2020[text_2020 != ""]

clean.text = function(x)
{
    x = tolower(x)
    # remove rt
    x = gsub("rt", "", x)
    # remove at
    x = gsub("@\\w+", "", x)
    # remove punctuation
    x = gsub("[[:punct:]]", "", x)
    # remove numbers
    x = gsub("[[:digit:]]", "", x)
    # remove links http
    x = gsub("http\\w+", "", x)
    # remove tabs
    x = gsub("[ |\t]{2,}", "", x)
    # remove blank spaces at the beginning
    x = gsub("^ ", "", x)
    # remove blank spaces at the end
    x = gsub(" $", "", x)
    return(x)
}

text_2015 = clean.text(text_2015)
text_2020 = clean.text(text_2020)
```

We then construct the lexical corpus. Furthermore, we consider some stop words and create the term document matrix:

```
corpus_2015 <- Corpus(VectorSource(text_2015))
corpus_2020 <- Corpus(VectorSource(text_2020))

skipwords <-c("and","with","than","expo","the","for","2015",
  "2020","from","will",
        "you","not","this","these","those","has","are","our",
  "in","into",
        "your","which", "it","its")

corpus_2015 <- tm_map(corpus_2015,removeWords,skipwords)
corpus_2020 <- tm_map(corpus_2020,removeWords,skipwords)

tdm_2015 <- TermDocumentMatrix(corpus_2015)
tdm_2020 <- TermDocumentMatrix(corpus_2020)
```

Now we can obtain words and their frequencies:

```
milan = as.matrix(tdm_2015)
word_freqs_milan = sort(rowSums(milan), decreasing=TRUE)
dmilan = data.frame(word=names(word_freqs_milan),
  freq=word_freqs_milan)
dubai = as.matrix(tdm_2020)
word_freqs_dubai = sort(rowSums(dubai), decreasing=TRUE)
ddubai = data.frame(word=names(word_freqs_dubai),
  freq=word_freqs_dubai)
```

Finally, we create the *Wordcloud* plots (Figure 14.3). *Wordclouds* are graphical representations of word frequency that give greater prominence to words that appear more

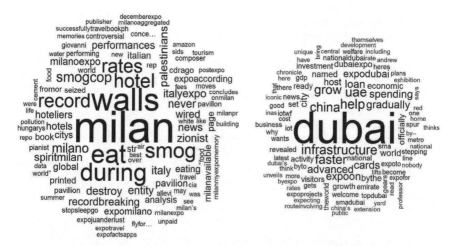

Figure 14.3   Wordclouds for the two datasets.

frequently in a source text. The larger the word in the visual, the more common the word was in the document. This type of visualization can assist evaluators with exploratory textual analysis by identifying frequent words. It can also be used for communicating the most salient points or themes in the reporting stage:

```
pdf("expo_2015.pdf")
wordcloud(dmilan$word,dmilan$freq,
          random.order=FALSE,colors=brewer.pal(8,"Dark2"),
          max.words=100)
dev.off()
pdf("expo2020.pdf")
wordcloud(ddubai$word,ddubai$freq,
          random.order=FALSE,colors=brewer.pal(8,"Dark2"),
          max.words=100)
dev.off()
```

Goal 2: Compare what people are talking about regarding "Expo 2015" and "Expo 2020"

When we work on a project comparing two text files, we can generate a unique *wordcloud* that can provide more insights. For this purpose, we have to make some changes in the script.

First, we join texts in a vector for each event:

```
m <- paste(text_2015,collapse="")
d <- paste(text_2020,collapse="")
all <- c(m, d)
```

We create the corpus, clean it, and create the term document matrix:

```
corpus <- Corpus(VectorSource(all))
corpus <- tm_map(corpus,removeWords,skipwords)
tdm <- TermDocumentMatrix(corpus)
tdm <- as.matrix(tdm)
colnames(tdm) <- c("Expo 2015", "Expo 2020")
```

For this goal, we can use the other type of graphic in the wordcloud package: the comparison and the commonality *wordcloud* before we plot a cloud, and we can compare the frequencies of words across documents:

```
pdf("comp_expo.pdf")
comparison.cloud(tdm,random.order=FALSE,colors=c("green",
  "red"), max.words=100)
dev.off()
```

Let p_{ij} be the rate at which word i occurs in document j and p_j be the average across documents: $p_j = \sum_i p_{ij} / (\text{number of documents})$. The size of each word is mapped to

Figure 14.4   Comparison (Expo 2015 = dark, Expo 2020 = light) and commonality clouds.

its maximum deviation ($\max_i(p_{ij} - p_i)$), and its angular position is determined by the document where that maximum occurs. The commonality cloud plot permits to plot a cloud of words shared across the two searches:

```
pdf("comm_expo.pdf")
commonality.cloud(tdm,random.order=FALSE,colors=c("green",
  "red"), max.words=100)
dev.off()
```

The *wordclouds* in Figure 14.4 are only examples used to analyze Twitter data with R.

Wordclouds are an easy to use and inexpensive option for visualizing text data, but we have to remember that the right way to interpret *wordclouds* is that the display emphasizes frequency of words, not necessarily their importance. *Wordclouds* will not accurately reflect the content of text if slightly different words are used for the same idea. They also do not provide the context, so the meaning of individual words may be lost. Because of these limitations, *wordclouds* are best suited for exploratory qualitative analysis.

An interesting analysis about Expo 2015 using R was conducted by the Voices from the Blogs.[2] This application with ExpoBarometro[3] presents an indicator that measures the daily social experience of Expo 2015, meaning the explicit reactions of the visitors of the event in Milan as they appear from posts published on Twitter, every day. ExpoBarometro presents an indicator of sentiment constructed as the ratio

---

[2] http://voicesfromtheblogs.com/
[3] http://dati.openexpo2015.it/en/expobarometro

between the percentage of posts making a positive reference to the Expo experience and the sum of the positive and negative posts, excluding from the calculation those of neutral content.

All ExpoBarometro analyses are carried out through iSA® technology by Voices from the Blogs, a spin-off of the University of Milan. In-depth quality is achieved with supervised statistical techniques that initially predict a classification of a training set by means of human coders. This is the key to retrieve meaningful information from texts published on the Web, avoiding problems of linguistic expression that appear in analysis based on semantic engines or on other fully automated techniques.

The data feed analyzed daily by iSA relates to all tweets written in Italian containing one of the following expressions: "expo, Expo 2015, Expo 2015, expomilan, expomilano." These tweets are obtained through Twitter's API streaming, and the analysis does not contain duplications and excludes posts by official Expo accounts. Figure 14.5 shows a screenshot of the results obtained and published on the official website of Expo 2015 organization team.

14.2.2   Sensory data analysis

Sensory analysis is a scientific discipline that applies principles of experimental design and statistical analysis to human senses (sight, smell, taste, touch, and hearing) for the purposes of evaluating consumer products. The discipline requires panels of human assessors, on whom the products are tested, and recordings of their responses. By applying statistical techniques to the results, it is possible to make inferences and generate insights about the products under test. Most large consumer goods companies have departments dedicated to sensory analysis. Sensory analysis uses human senses to characterize a product. The human being becomes a measurement instrument asked to quantify, compare, or score during tests what he/she felt when observing, tasting, smelling, touching, or listening to different stimuli. The tester is at the heart of the system that is used to identify and track the ideal product. Sensory analysis is a multidisciplinary discipline. Quality control combined with research and development, statistics, and marketing initiatives makes sensory data very powerful.

To derive precise results, this approach, based on the collection and computation of the sensations of a significant number of persons, needs a rigorous protocol and a suitable computer tool with proper session organization, data collection and storage, and result computation. Food tasting experiments have usually taken too long to get results, to reach good decisions, to make quick decisions, and to predict problems before they occur. Food tasting experiments have usually been conducted only in multinational food sector companies, with expensive and sophisticated software and technology.

Today, sensory analysis is a widely deployed technique. Various applications of sensory analysis exists not only in the food industry. The cosmetics industry and fashion and clothing companies can use sensory analysis to design better products. It is possible to measure the popularity of a theater play or a movie or the satisfaction

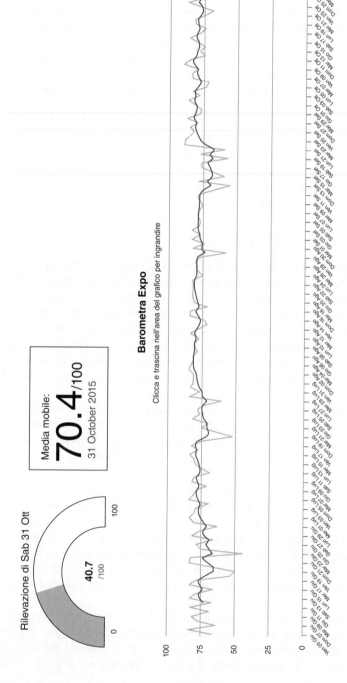

Figure 14.5   ExpoBarometro results.

from an event or an analysis of the characteristics of a sofa or a kitchen. Modern sensory data can be obtained in a cheaper way using tablets or smartphones, in real time, quickly and easily.

The problem is often how to analyze the collected data, that is, taking unstructured data, sometimes organized in many files produced by an app, and organizing and analyzing it with sophisticated statistical techniques.

In the foreword of the book *Analyzing Sensory Data with R* (Lê and Worch, 2014), Dr. Hal MacFie, a statistician with international reputation in the areas of product assessment and consumer research, wrote: "Back in the 1980s when I was still a consultant statistician supporting clients from sensory and other disciplines, I dreamed of the day when there would be free, easy-to-use statistical software for everyone to use. It seems that in sensory science that day is approaching fast with the arrival of SensoMineR [.] (Husson et al., 2015) Sensometrics is still a dynamic and fast-moving discipline, the beauty of the R software is that new methods can be quickly programmed and added to the R portfolio."

With this description, it is clear that R programs play a key role in approaching sensory data analysis following the InfoQ paradigm. In the next application, we show, with the four InfoQ components of Chapter 2, several examples of sensory data analysis. We then assess the application with the eight InfoQ dimensions of Chapter 3.

In this example, all the analyses are made using R and in particular the SensoMineR package http://sensominer.free.fr that was conceived and programmed in R language. The package is free and can be downloaded from CRAN.

SensoMineR includes several classic methods used when analyzing sensory data as well as advanced methods conceived and developed in a laboratory. It provides numerous graphical outputs that are easy to interpret, as well as syntheses of results derived from various analysis of variance models or from various factor analysis methods, accompanied with confidence ellipses. The package deals with classical profile data as well as with more specific data such as napping data as explained later.

SensoMineR is easy to use also for not advanced R users. Graphical interfaces are available using R commander, presented in Section 14.1 and SensoMineR Excel. Thanks to Heiberger and Neuwirth (2009) who developed RExcel, an Excel plug-in, the R software can be used via Excel (see Figure 14.6). We can easily add the Rcmdr menu to an Excel menu, and since SensoMineR can also be integrated in Rcmdr, one is able to use it from Excel. When we open RExcel, R is opened automatically.

In sensory analysis, different goals can be identified, and consequently, different types of data and different statistical techniques are applied.

In the InfoQ framework, in order to increase the information and to generate insights with action items, it is important to match and combine all the goals g_i, possibly using a flexible tool, like R, to integrate data with different structural form X_i, including a large portfolio of techniques f_i, producing different utilities U_i.

In this section we consider the example of perfumes taken from Lê and Worch (2014). All the datasets mentioned here can be downloaded from www.sensorywithr.org/.

The first dataset *perfumes_qda_experts.csv* consists of a test involving 12 perfumes, rated by 12 trained panelists twice, on 12 attributes.

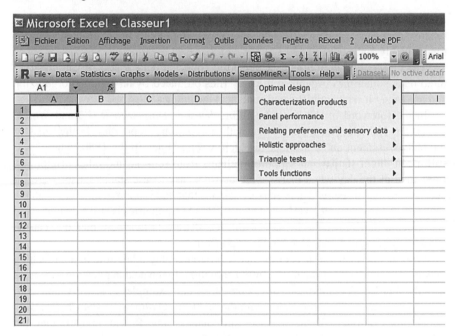

Figure 14.6   `SensoMineR` menu in Excel.

The second dataset *perfumes_qda_customers.csv* consists of a test involving 12 perfumes, rated by consumers on 21 attributes.

The third dataset *perfumes_linking.csv* consists of hedonic scores for the 12 perfumes, given by consumers.

Goal 1: Determine the performance of a sensory panel

For this goal, we consider the experts dataset. For each sensory attribute, it is possible to test the product effect, panelist effect, and session effect. Using R and in particular using the functions `panelperf()` and `coltable()` of `SensoMineR`, it is possible to get a useful plot that summarizes 12 different ANOVA models, one for each attribute, in which the cells are highlighted in yellow when p-values are lower than 0.05:

```
## Goal 1 ###
Experts <- read.table(file="perfumes_qda_experts.csv",
        header=TRUE, sep=",", quote="/", dec=".")
experts$Session <- as.factor(experts$Session)
experts$Panelist <- as.factor(experts$Panelist)
res.panelperf1 <- panelperf(experts,formul="~Product+Panelist
        +Session+Product:Panelist+Product:Session+Panelist:
  Session",firstvar=5)
coltable(res.panelperf$p.value[order(res.panelperf$p.
  value[,1]),],
        col.lower="yellow")
```

	Product	Panelist	Session	Product:Panelist	Product:Session	Panelist:Session
Heady	4.308e–31	1.703e–15	0.6443	1.438e–07	0.3529	0.1002
Greedy	1.147e–30	6.125e–05	0.4906	2.958e–07	0.6149	0.0009707
Vanilla	9.031e–23	8.553e–10	0.2305	7.708e–07	0.5682	0.02814
Spicy	1.996e–22	1.648e–12	0.8911	0.0006576	0.2128	0.1516
Floral	1.076e–19	4.326e–16	0.02697	0.0003138	0.0375	0.1845
Fruity	2.769e–13	3.704e–09	0.9368	8.842e–05	0.9033	0.7593
Oriental	2.682e–12	5.35e–18	0.1423	1.474e–12	0.9806	0.0009239
Wrapping	4.459e–11	3.472e–17	0.007783	1.795e–07	0.4035	0.4648
Green	2.586e–09	8.403e–09	0.06641	0.0001503	0.8727	0.7347
Woody	1.623e–06	6.315e–11	0.3825	0.03666	0.5552	0.05019
Marine	0.01519	2.843e–05	0.1156	0.5438	0.01695	2.458e–06
Citrus	0.08308	1.799e–09	0.1557	2.213e–06	0.5643	0.5178

Figure 14.7   Assessment of the performance of the panel with the `panelperf()` *and* `coltable()` *functions.*

Figure 14.7 shows that the panel discriminates between the products for all the sensory attributes, except *Citrus*. It also shows that the panelists have particularly well differentiated the products.

Goal 2: Represent the product space in a map

For this goal, we consider again the experts dataset. It is possible to define the sensory profile (adjusted mean) of the products using the `decat()` function in `SensoMineR` and then produce individual factor map and variable factor map using the `PCA()` function (see Figure 14.8):

```
## Goal 2 ###
res.decat<-decat(experts,formul="~Product+Panelist",
          firstvar=5,lastvar=ncol(experts))
res.pca<-PCA(res.decat$adjmean)
```

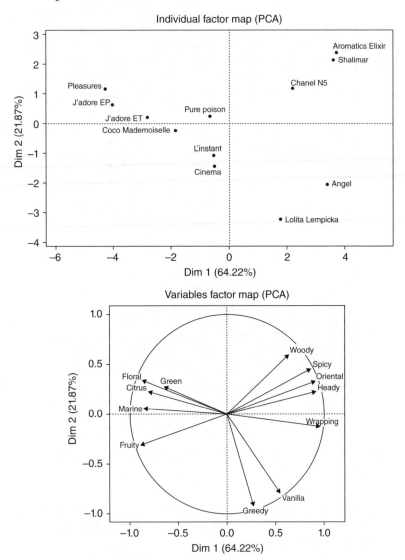

Figure 14.8   Representation of the perfumes and the sensory attributes on the first two dimensions resulting from PCA () on adjusted means of ANOVA models.

The main dimension of variability (i.e., the first component) contrasts products such as *Pleasures* with products such as *Angel*. In the second plot, for each dimension, only significant attributes are shown. *Wrapping* and *Heady* are most positively linked with the first dimension, whereas the sensory attributes that are the most negatively linked are *Fruity* and *Floral*.

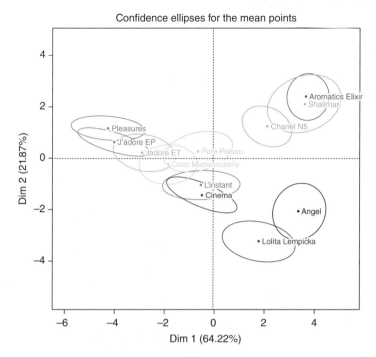

Figure 14.9   Representation of the perfumes on the first two dimensions resulting from PCA () in which each product is associated with a confidence ellipse.

A possible new question that arises looking at these plots can be: How would the positioning of the perfumes evolve if we would slightly change the composition of the panel? To answer this question, the developers of SensoMineR provide the panellipse() function:

```
res.panellipse <- panellipse(experts,col.p=4,col.
  j=1,firstvar=5,
                level.search.desc=1)
```

In Figure 14.9, the more one sees ellipse overlaps, the less distinctive (or the closer) the two products are. Hence it appears clearly that *Aromatics Elixir* and *Shalimar* are perceived as similar, whereas *Shalimar* and *Pleasure* are perceived as different.

Goal 3: Compare the sensory profile of a panel of experts and a panel of customers

For this goal, we consider the experts dataset and the consumers dataset. A multiple factor analysis can be applied using MFA() function of SensoMineR in order to produce maps including both consumer and expert sensory responses, which are based on different sets of attributes (see Figure 14.10):

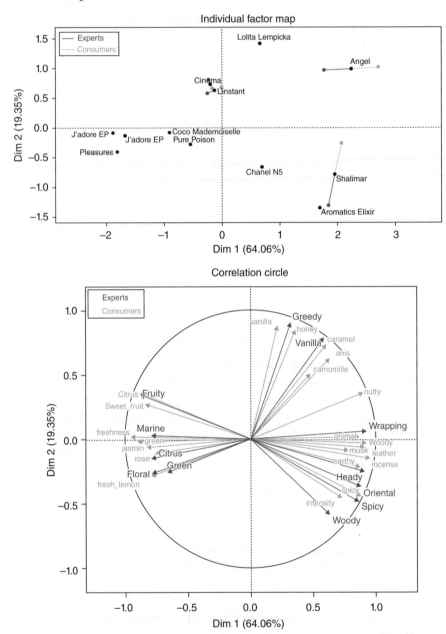

Figure 14.10   Representation of the perfumes and the sensory attributes on the first two dimensions resulting from MFA ( ) of both experts and consumers data.

```
experts.avg <- decat(experts,formul="~Product+Panelist+Session+
  Product:Panelist+
            Product:Session+Panelist:Session",firstvar=5,graph=
  FALSE)$adjmean
consumers <- read.table("perfumes_qda_consumers.
  csv",sep=",",header=TRUE,
            dec=".",quote="\"")
consumers$consumer <- as.factor(consumers$consumer)
consumers.avg <- decat(consumers,formul="~product+consumer",
  firstvar=3,
            graph=FALSE)$adjmean
data.mfa2 <- cbind(experts.avg,consumers.avg[rownames(experts.
  avg),])
res.mfa2 <- MFA(data.mfa2,group=c(12,21),type=c("s","s"),
            name.group=c("Experts","Consumers"))
```

From this representation, consumers and experts agree with the perfumes' similarity even if they use different sets of attributes. The correlation circle highlights also the correlation between the attributes evaluated by experts and the attributes evaluated by consumers.

Goal 4: What is the best product?

For this goal, we consider the liking dataset, that is, the hedonic score for the 12 perfumes given by consumers. In order to make the plot more readable, we consider a set of only two panelists, A and B.

The null hypothesis to be tested is that there are no differences between A and B. This can be tested using a t-test. Moreover, a useful plot implemented in SensoMineR is the graphinter() function:

```
Small <- read.table(file="perfumes_linking_small.csv",
  header=TRUE,
            sep=",", quote="/", dec=".")
graphinter(small,col.p=2,col.j=1,firstvar=3,numr=1,numc=1)
graphinter(small,col.p=1,col.j=2,firstvar=3,numr=1,numc=1)
```

As shown in Figure 14.11 (left panel), although the consumers provided different scores to the products (consumer A scoring higher than consumer B), both seem to evaluate the product in a similar way, at least in terms of ranking. Indeed, both consumers appreciate *J'adore EP* and *J'adore ET* the most and *Shalimar* the least. This finally reveals a strong product effect. Figure 14.11 (right panel) shows some inversion ranking for some perfumes, *J'adore EP* and *J'adore ET* and *Chanel N°5* and *Aromatics Elixir*. In their book, Lê and Worch (2014) consider the complete liking dataset with all the consumers; ANOVA is applied and Fisher's LSD is used. For a multivariate approach, the so-called internal preference mapping is considered.

Lê and Worch (2014) also present other advanced analysis methods combining different datasets. Other possible data structure is textual data, napping data, or just-about-right (JAR) data.

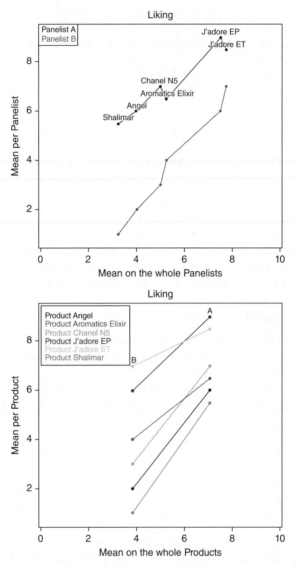

Figure 14.11   Visualization of the hedonic scores given by the panelists.

Textual data are textual comments on product given by consumers. Text mining techniques have to be applied in order to handle such unstructured data. In this case typical words can be considered as attributes and used to describe the product profile.

Napping data are issued from sessions where panelists arrange all the products on a blank paper sheet. Napping allows collecting directly in only one session the sensory distances panelists perceived between products. Each panelist discriminates products according to his/her own criteria. The closer the two products are, the more alike the panelist perceives them. The dataset has as many rows as there are products and twice the number of panelist columns, each pair of columns corresponding to the *X* and *Y* coordinates of the products for one panelist.

JAR scales are commonly used in consumer research to identify whether product attributes are perceived at levels that are too high, too low, or just about right for that product.

Many other goals can be identified, for example, how the relationship between each sensory attribute and the hedonic score can be evaluated at different levels, how to get homogeneous clusters of products or of consumers, etc. All these goals can be easily reached using R and his packages.

14.2.3   Explore European tourism

In the past decades, lowered transportation costs led to a sharp increase in the physical accessibility of tourist localities. As a consequence, the proportion of foreigner tourists rapidly increased all over the word. According to the United Nations World Tourism Organization, Europe was the most frequently visited region in the world in 2013, accounting for over half (52%) of all international tourist arrivals.

Goal 1: Based on data from the Eurostat database, what are European tourism patterns?

Using R, it is possible to import and manipulate the Eurostat data very easily. The `eurostat` package (Leo et al., 2015) provides tools to access the data directly from the Eurostat database in R with search and manipulation utilities.

We can search the available variables using `search_eurostat()` function. In order to efficiently represent the European tourism, we consider the number of overnight stays in tourist accommodation by NUTS 2 regions (table *tour_occ_nin2*, source Eurostat). This indicator represents both the length of stay and the number of visitors.

To download the data, we use the `get_eurostat()` function:

```
dt <- get_eurostat("tour_occ_nin2", time_format = "raw")
```

An effective way to visualize how a measurement varies across a geographic area is a choropleth map, that is, a thematic map in which areas are shaded in proportion to the value assumed by the variable of interest in that particular region.

R offers a wide variety of ways to create maps. We'll use principally two packages: `maptools` and `ggplot2`. The `maptools` package (Bivand and Lewin-Koh, 2015) provides tools for reading and manipulating geographic data. In particular we use it to read shapefiles of Europe published by Eurostat. We will use the shapefile in a 1:60 million scale from year 2010 and subset it at the level of NUTS2. The `ggplot2` package (Wickham and Chang, 2015) provides powerful graphics language for creating complex plots. In order to prepare the data for the visualization, we will use the `fortify()` function in `ggplot2` to convert the shapefile into a data frame, and we will join the data about the tourism and the spatial data. Finally, we will create the choropleth map using the `geom_polygon()` function. R is not the easiest way to generate maps, but it allows for full control of what the map looks like.

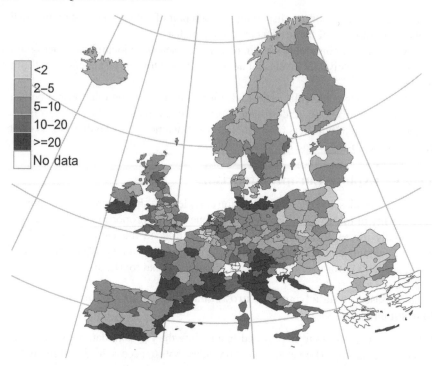

Figure 14.12   Nights spent in tourist accommodation establishments by NUTS level 2 region, 2013 (million nights spent by residents and nonresidents). Source: Eurostat (tour_occ_nin2). © European Union.

The map in Figure 14.12 represents the total number of nights spent in tourist accommodation in 2013 at regional level. Using this representation, it is easy to identify where tourism in the European Union is concentrated. The most attractive areas are the coastal regions, Alpine regions, and some major cities. Spain and Italy are the most common tourism destinations in the European Union.

Goal 2: Identify the determinants of overall tourists' satisfaction

The tourism phenomenon raised several policy implications for the efficient supply of tourist services and amenities to the visitors from abroad. Among other issues, some works were devoted to the analysis of the foreigner tourists' experience and their self-reported satisfaction.

This example is taken from Cugnata and Perucca (2015). We use a large sample of data from a survey conducted every year since 1996 on behalf of the Bank of Italy. In every survey study, a sample of foreign tourists is asked to report (among other things) their satisfaction with various aspects of their journey in Italy. At the same time, we include some variables for the socio-economic characteristics of the provinces visited by the respondent.

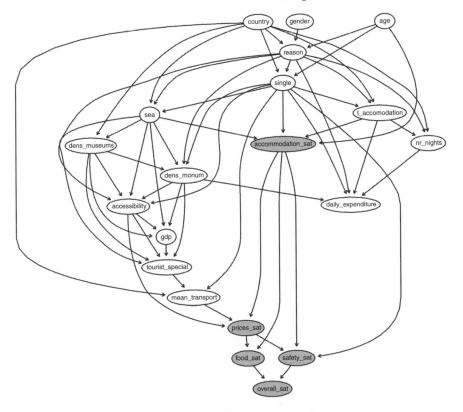

Figure 14.13   Bayesian network.

R offers efficient ways to integrate multiple data frames with different structures, for example, the merge function merges two data frames by common columns or row names or other versions of database *join* operations.

The goal of the application paper is to identify the determinants of overall tourists' satisfaction, among individual characteristics and some features of the tourist services supplied.

We use a Bayesian network (BN) approach to the study of tourist satisfaction. As discussed by Salini and Kenett (2009), BN presents several advantages compared with other statistical techniques. First, they are able to outline the complex set of links and interdependencies among the different components of satisfaction. Second, they are extremely useful for policy-design purposes, since they allow for the building of hypothetical scenarios, letting change the distribution of some parameters under the control of the policymakers and controlling the impact on the variables of interest.

Several R packages implement algorithms and models for constructing BN. We use the bnlearn package (Scutari, 2010, 2015). The hc() function is applied to learn the structure of the BN using the Hill-climbing algorithm and the graphviz.plot() function to plot the graph associated with the estimated BN. Figure 14.13 shows the BN obtained with the Hill-climbing algorithm with score functions such as AIC.

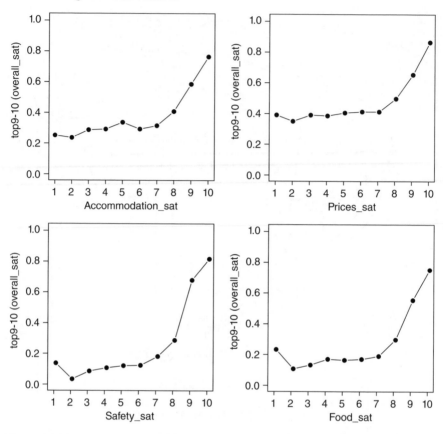

Figure 14.14   Distribution of the overall satisfaction for each level of each variable.

As expected, the three groups of variables are (within each category of data) joined together. However, both daily expenditure and the dimensions of satisfaction (in particular with the accommodation) are influenced by the individual characteristics and the socio-economic indicators.

One of the main benefits of BNs is that they provide an opportunity to conduct "what-if" sensitivity scenarios. Such scenarios allow us to set up goals and improvement targets. Comparing R to other BN software, an advantage of R is that it is a flexible tool.

Considering the dimensions of self-reported satisfaction, overall satisfaction is directly influenced by safety and food satisfaction and (indirectly) from prices and accommodation. If we fix the dimension of satisfaction to a certain level, the proportion of (overall) very satisfied respondents (9–10) changes. Figure 14.14 shows the distribution of the overall satisfaction for each level of each variable.

14.3   Components and dimensions of InfoQ and R

In the following table, the three applications presented in Section 14.2 are summarized according to the InfoQ components.

Example	g = goal	X = dataset	f = method	U = utility
14.2.1	1: Exploring what are people talking about "Expo 2015" and "Expo 2020" in these days	Two datasets created by info downloaded from Twitter	Function **search.Twitter** of `TwitterR` and packages `tm` and `wordcloud`	Wordle
	2: Comparing what are people talking about "Expo 2015" and "Expo 2020" in these days	One dataset created merging info download from Twitter	Function **search.Twitter** of `TwitterR` and packages `tm` and `wordcloud`	Wordle
14.2.2	1: How to assess the panel performance	Experts dataset	Functions **panelperf** and **coltable** of `SensoMineR`	F, p-values
	2: How to represent the product space in a map	Experts dataset	Functions **decat** and **PCA** of `SensoMineR`	Individuals and variables maps, confidence ellipse
	3: Do panel of experts and panel of customers produce the same sensory profile?	Experts and consumers datasets	Functions **decat** and **MFA** of `SensoMineR`	Multiple individuals and variables maps
	4: How can I identify the best product?	Liking data	Function **graphinter** of `SensoMineR`	Graphical representation of product ranking
14.2.3	1: Effectively represent the European tourism	Eurostat data	`eurostat`, `maptool` and `ggplot2` R packages	Choropleth map
	2: Identify the determinants of overall tourists' satisfaction	Bank of Italy survey data + ISAT data	`bnlearn` R package	Bayesian network

We discuss now how R supports InfoQ dimensions, keeping in mind the afore-mentioned examples.

As presented in the introduction, R is able to deal with any type of data. Moreover, most modeling functions in R offer options for dealing with missing values. You can go beyond pairwise of listwise deletion of missing values through methods such as multiple imputation. Good implementations that can be accessed through R include `Amelia II`, `Mice`, `ForImp`, `missForest`, `mitools`, and many others. Also the outlier detection techniques are developed in a lot of R packages, for example, `extremevalues`, `mvoutlier`, `outlier`, `outlierD`, and many others. Unlike other software, R provides all the estimators and approaches for both robust regression and multivariate analysis. So when the **data resolution** is not high, R can help to increase that score.

Furthermore, the previous examples, expressly chosen, highlight the extreme flexibility of R to various **data structure**. This is very important, especially in big data where data arrives quickly, with high frequency and sometimes in unstructured formats. R has packages able to structure the data collected in different ways from various sources. So **data integration** is another strength of R software. Not only the matching techniques but also the possibility to merge objects (models, plot, table, index, ranking, etc.) coming from different dataset and different sources in order to produce unified analyses is allowed.

As mentioned, with big data, one needs to be able to work with high-frequency data. The possibility to download updated data directly from the source (Eurostat, Twitter, finance.yahoo.com, sensory device, etc.) increases the score of the **temporal relevance** dimension and helps in the **chronology of data and goal** that obviously depends on the available data and on the specific goal.

The **generalizability** of the results depends on the context. R routines and R scripts, once written, are easily adaptable with other data sources and other application domains because of the versatility of the programming language. This definitely accelerates the process of generalization.

Operationalization is dependent on the emerged insights. Often the R functions are created with a precise operational value, with a clear objective. Every code inserted on CRAN needs to be motivated by an application example. This can help to follow a path that leads to maximum uptime.

Finally, **communication** is one of the greatest values of R community, intrinsically linked to the first absolute value, that is, to make statistics available to all. In R packages the graphics functions are often the most numerous. It is quite clear to all developers how important the end is to export any result, from the simplest to the most complex, in a format that anyone can easily understand. `SensoMineR` is a prime example of this philosophy.

14.4 Summary

Section 14.1 presents a brief introduction to R software environment, its origins, and its philosophy. The R graphical interfaces are mentioned, and the main data import and export functions are described. In Section 14.2 three non-standard data analysis applications are presented with the aim to show the R potential with respect to other classical statistical packages.

The first example starts by downloading, within the R environment, Twitter data and analyzing it with text mining techniques. The second example deals with sensory data and sensory analysis; it is based on SensoMineR packages and, following Lê and Worch (2014), presents different datasets and different goals. The third example considers official statistics of Eurostat on tourism, directly downloaded in the R environment, and produces a geographical map; afterwards it considers survey data and tries to determine the key driver of tourism satisfaction using an advanced technique, a BN, provided by R.

Finally, in Section 14.3, the three applications are revised according to the InfoQ components, and a discussion on how R supports the InfoQ dimensions is given.

References

Bivand, R. and Lewin-Koh, N. (2015) maptools Package. https://cran.r-project.org/web/packages/maptools/maptools.pdf (accessed May 2, 2016).

Cugnata, F. and Perucca, G. (2015) International Tourism in Italy: A Bayesian Network Approach, CLADAG 2015, Sardegna, Italy.

Feinerer, I. (2015) tm Package. https://cran.r-project.org/web/packages/tm/vignettes/tm.pdf (accessed May 2, 2016).

Fellows, I. (2013) wordcloud Package. https://cran.r-project.org/web/packages/wordcloud/wordcloud.pdf (accessed May 2, 2016).

Fox, J. (2005) The R commander: a basic-statistics graphical user interface to R. *Journal of Statistical Software*, 14(9), pp. 1–42.

Gentry, J. (2015) twitteR Package. https://cran.r-project.org/web/packages/twitteR/twitteR.pdf (accessed May 2, 2016).

Hand, D.J. (2008) *Statistics: A Very Short Introduction*. Oxford University Press, Oxford.

Heiberger, R.M. and Neuwirth, E. (2009) *R Through Excel. A Spreadsheet Interface for Statistics, Data Analysis, and Graphics*. Springer, New York, NY.

Husson, F., Le, S. and Cadoret, M. (2015) SensoMineR Package. https://cran.r-project.org/web/packages/SensoMineR/SensoMineR.pdf (accessed May 2, 2016).

Ihaka, R. and Gentleman, R. (1996) R: A language for data analysis and graphics. *Journal of Computational and Graphical Statistic*, 5(3), pp. 299–314.

Kenett, R.S. and Zacks, S. (2014) *Modern Industrial Statistics: With Applications in R, MINITAB and JMP*, 2nd edition. John Wiley & Sons, Ltd, Chichester, UK.

Lê, S. and Worch, T. (2014) *Analyzing Sensory Data with R*. CRC Press. Chapman & Hall/CRC The R Series, Boca Raton, FL.

Leo, L., Przemyslaw, B., Markus, K. and Janne, H. (2015) eurostat Package. https://cran.r-project.org/web/packages/eurostat/eurostat.pdf (accessed May 2, 2016).

Munzert, S., Rubba, C., Meißner, P. and Nyhuis, D. (2015) *Automated Data Collection with R: A Practical Guide to Web Scraping and Text Mining*. John Wiley & Sons, Ltd, Chichester, UK.

Salini, S. and Kenett, R.S. (2009) Bayesian networks of customer satisfaction survey data. *Journal of Applied Statistics*, 36(11), pp. 1177–1189.

Scutari, M. (2010) Learning Bayesian networks with the bnlearn R package. *Journal of Statistical Software*, 35(3), pp. 1–22.

Scutari, M. (2015) bnlearn Package. https://cran.r-project.org/web/packages/bnlearn/bnlearn.pdf (accessed May 2, 2016).

Vance, A. (2009) Data Analysts Captivated by R's Power. *The New York Times*. http://www.nytimes.com/2009/01/07/technology/business-computing/07program.html?pagewanted=all&_r=0 (accessed May 2, 2016).

Wickham, H. and Chang, W. (2015) ggplot2 Package. https://cran.r-project.org/web/packages/ggplot2/ggplot2.pdf (accessed May 2, 2016).

15

InfoQ support with Minitab

Pere Grima, Lluis Marco-Almagro and Xavier Tort-Martorell

Department of Statistics and Operations Research, Universitat Politècnica de Catalunya, BarcelonaTech, Barcelona, Spain

15.1 Introduction

Minitab is a statistical software package available from www.minitab.com that combines ease of use with the ability to handle large volumes of data and perform a wide range of statistical analyses. Simplicity and the feeling of always having everything "under control," without the software making unclear assumptions, are strengths of Minitab that differentiate it from other statistical packages.

Data are placed in Minitab in Excel-like datasheets, so that they are easy to view and manage. Furthermore, analysis of these data can be conducted using clear and well-structured menus. In fact, you can start using Minitab without devoting a lot of time to learn how it works. This fact makes it very popular among those who occasionally need to perform statistical analyses with rigor and versatility but without mastering the intricacies of highly technical or specialized software. Perhaps for this reason, Minitab is widely used in industrial environments and in the field of quality improvement (often within the framework of a Six Sigma improvement project).

Information Quality: The Potential of Data and Analytics to Generate Knowledge,
First Edition. Ron S. Kenett and Galit Shmueli.
© 2017 John Wiley & Sons, Ltd. Published 2017 by John Wiley & Sons, Ltd.
Companion website: www.wiley.com/go/information_quality

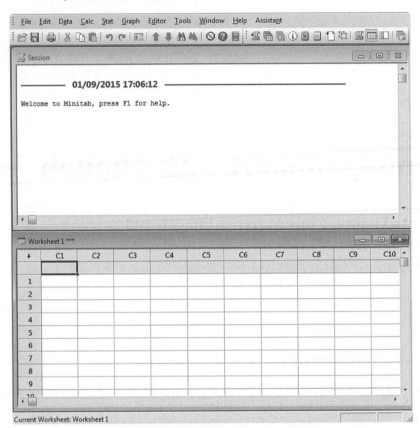

Figure 15.1   Minitab user interface, with session and worksheet windows.

Minitab's initial screen is shown in Figure 15.1. In addition to the main menu at the top, there is the session window (where the results of conducted analyses appear) and the worksheet, the Excel-like datasheet where data are placed (each column being a variable).

15.1.1   Data distribution: Differences between Minitab and Excel

When the worksheet is maximized, we can see a user interface very similar to an Excel spreadsheet, with which almost all of us are familiar. Data in a worksheet can be saved in Minitab format or Excel format, among others. Excel sheets can also be directly opened in Minitab, so if preferred, data manipulation and organization can be done in Excel and Minitab used only for the analysis.

The worksheet can contain up to 4000 columns and up to 10 000 000 rows, depending on the available memory in the computer. Unlike Excel, you cannot assign a specific formula to a cell but only to a column. After creating a column that is the result of a formula based on other columns, they may be linked (the values of the new column change when changing some of the columns involved in the formula, as in Excel) or fixed (default behavior in Minitab).

Similarly, when a chart is created from data, the graph can be updated as data changes (Excel behavior) or not (Minitab's default behavior).

Some details make working with Minitab spreadsheets easier than working with Excel. For example, columns in Minitab are named Cx, where x is the column number. So, what in Excel is column BJ, in Minitab is column C62, a much clearer denomination. An appropriate habit when working with data is giving each column (each variable) a descriptive name. Minitab has a first row reserved for this name so that a column with 50 rows ends in row 50 and not 51 (the row for the name does not count).

15.1.2   Overview of some Minitab options

Minitab enables a great variety of statistical analyses, from classic basic statistics (Figure 15.2, top) to more sophisticated ones, such as modeling with regression equations, ANOVA, generalized linear models, factorial designs, surface response methodology, multivariate data analysis, or time series analysis. There is an emphasis on tools oriented to industrial statistics and quality management, some very general (Figure 15.2, bottom), others more specific, such as robust designs (Taguchi method) or statistical process control (SPC). For a comprehensive treatment of industrial statistics with Minitab see Kenett and Zacks (2014).

15.1.2.1   Minitab help

Another strong point in Minitab is its help system, which is very thorough and carefully written. The help files not only explain how to conduct statistical analysis in Minitab, but also the conceptual ideas behind the tool are clearly explained. Reading Minitab help can be a quick way to learn statistics! It is possible to access Minitab help through the main menu but also using the help button in each dialog box. This leads to a detailed explanation of the tool being used at the moment, with examples and interpretation of the results (Figure 15.3).

Minitab also has a wizard (called Assistant) that gives advice and guidance on the techniques to use depending on the characteristics of the data and the purpose of the analysis.

15.1.2.2   Minitab macros

In addition to using Minitab in a point-and-click manner through the menus, it is possible to write commands in the session window (when the enable commands option from the editor menu is activated). For example, if we write

```
MTB > random 10000 c1
MTB > histogram c1
```

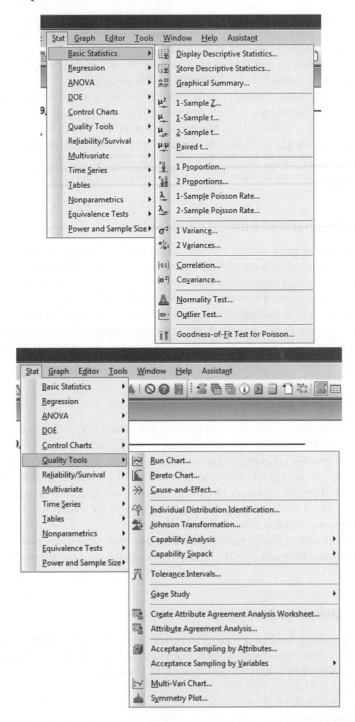

Figure 15.2   Some menu options for basic statistical analysis and quality tools.

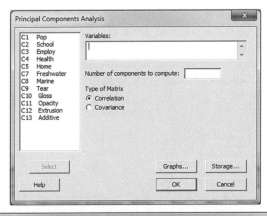

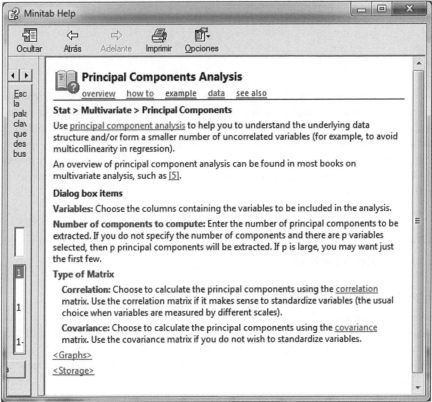

Figure 15.3   A screenshot of Minitab help.

10 000 random numbers are generated (by default, they follow a normal distribution with mean 0 and standard deviation 1) and stored in column C1. After that, a histogram with the data is drawn. To generate random numbers from a normal distribution with mean 10 and standard deviation 2 and to change the appearance of the histogram (e.g., horizontal scale, tick marks, bins, font size), the commands needed are the following:

```
SUBC> normal 10 2.
MTB > Histogram C1;
SUBC>   Scale 1;
SUBC>     PSize 12;
SUBC>     MODEL 1;
SUBC>       Tick 0 2 4 6 8 10 12 14 16 18 20;
SUBC>       Min 0;
SUBC>       Max 20;
SUBC>     EndMODEL;
SUBC>   Midpoint;
SUBC>   NInterval 50.
```

There is no need to memorize these commands because they appear printed on the session window while using the menus. They can later be copied and pasted (and modified as desired), and the macro is ready to be executed. These commands can be used together with usual flow programing commands: if, do, while, etc., so that programs similar to functions in R statistical scripting language can be created. However, it is not a good idea to use Minitab macros for intensive simulations, as its performance is slow.

15.2   Components and dimensions of InfoQ and Minitab

15.2.1   Components of InfoQ and Minitab

When studying how Minitab develops the four components of InfoQ (goal, data, analysis, and utility), it is obvious that the specific characteristics of a statistical software such as Minitab are especially relevant when dealing with data, analysis, and utility components of InfoQ. The following subsections deal with InfoQ components in Minitab.

15.2.1.1   Goal

Like other packages, Minitab statistical software offers a wide range of possibilities both for exploratory data analysis and for confirmatory analysis, following the classification of Tukey (1977). In particular, classical graphics (such as histograms, dotplots, scatterplots, and boxplots) are done in a breeze with Minitab and can be used as part of an exploratory analysis (Figure 15.2). If the objective of the study is more in the confirmatory analysis side, a wide variety of tests can be conducted.

Quality Companion, another product from Minitab Inc., can also be used to clarify the goal of a study. Quality Companion provides tools for organizing projects, emphasizing in a step-by-step roadmap achieved results. It also offers a set of "soft tools," such as process mapping, templates for brainstorming, and reporting. Although it can be used under any improvement framework, it is especially suited for developing projects following the Six Sigma methodology (see Kenett and Zacks, 2014).

15.2.1.2 Data

As mentioned in the introduction, Minitab is able to manage large volumes of data in a structure similar to Excel sheets (although with the suitable restriction of having the same kind of data in a column).

It is easy to manage data in Excel and then use it in Minitab for statistical analysis. The most straightforward method is simply copying data from Excel and pasting it in Minitab. It is also possible to open Excel data files (and files in other formats) and establish more complex connections with Excel (such as establishing a dynamic data exchange (DDE) between Excel and Minitab).

15.2.1.3 Analysis

Minitab includes the most common methods of data analysis, both parametric and nonparametric. It does not include Bayesian methods, although it is possible to write macros to use Bayesian methods, as done, for example, in Albert (1993) or Marriott and Naylor (1993).

15.2.1.4 Utility

Minitab provides measures of utility of the performed analysis. For instance, when modeling with regression equations, goodness-of-fit measures are presented. In hypothesis testing, it is possible to compute the power of the test; when performing time series analysis, confidence intervals for the forecasts are offered (being more or less wide, that is, more or less "useful," depending on the amount and variability of the data). It is also possible to perform fitting tests for a wide range of distributions.

15.2.2   Dimensions of InfoQ and Minitab

15.2.2.1   Data resolution

Working with data at an adequate level of precision and aggregation is a feature that lies more in the field of data collection planning and the application of appropriate measurement instruments than in the statistical software package used. Minitab uses 64 bits of memory to represent a numeric value; this allows working with up to 15 or 16 digits without rounding error. This accuracy is enough in the vast majority of cases.

One can get valuable information and identify the source of a problem when stratifying data by origin. For example, imagine that in the manufacturing of a product defective parts are produced very often. We have data about the machine, operator, shift, provider of the raw material, environmental conditions, etc. under which each unit is manufactured. Looking at all these data, we can probably identify the cause of the excessive number of defective parts. However, if only the number of defective units is known, identifying the cause will be much harder. Minitab can stratify the data according to their origin very easily (but if this information is not available, this is something impossible to perform, regardless of the software used).

Data can be aggregated in the most convenient form. Sometimes this aggregation is done automatically, as when a histogram is drawn and the number of data points in each interval is decided by the program. But it can also be done manually, encoding the data with a value depending on the interval to which they belong.

Minitab includes ways to represent data in aggregate form without losing the original values, such as stem-and-leaf plots. The histogram in Figure 15.4 (left) represents the heartbeats per minute for each of 92 students. We can see one student with heartbeats between 45 and 50 per minute, two students with heartbeats between 50 and 55, and so on. However, it is not possible to recover the exact number of heartbeats for each student. From the original data, you can build the histogram, but from the histogram you cannot reproduce the original data. The right panel of Figure 15.4 shows the stem-and-leaf plot with the same data. Each value is divided into two parts: the most significant is the stem, and the least significant (in this case the units) correspond to the leaves. It can be seen that the smallest value is 48, then 54 comes twice, etc. The profile of this diagram is identical to the histogram, so it contains the same information, but the original values of the dataset are not lost.

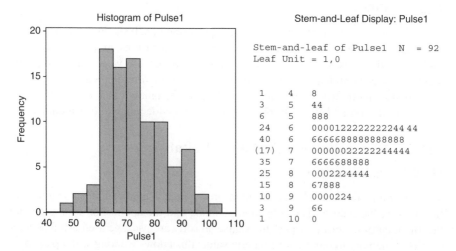

Figure 15.4   *A histogram (left) and its corresponding stem-and-leaf graph (right), of heartbeats per minute of students in a class.*

Being able to look at the specific values can sometimes be useful to correctly interpret the data. For example, in this case, the students were instructed to measure their heartbeats during one minute. However, we can see that, from the 92 values, only two are odd and the rest are even; this suggests that the measurements were done during only 30 seconds, and then multiplied by two, or during only 15 seconds, and then multiplied by 4! As an error at the beginning or end of the measurement period is common, it is not the same having an inaccuracy of ±2 (when measuring one minute) and an inaccuracy of ±8 (when measuring 15 seconds and multiplying by 4).

15.2.2.2   Data structure

Minitab can work with numerical data (called Numeric), text (called Text), or dates (called Date/Time). When the format is of type date, it is possible to perform common operations for this type of data, such as computing the number of days between two dates.

There are no tools for working with textual data. In general, text variables have the only purpose of identifying the source of the data or properties later used to stratify. The most common text variable in Minitab is a categorical variable with different levels. The values of a numerical variable can then be compared depending on the levels of this text variable (for instance, comparing sales (numerical variable) by region (textual variable)).

Contrary to what happens in Excel, each column in the datasheet must contain a single type of data. So if a column contains numeric values, no cell in that column can contain text or any content that is not numeric (except an asterisk, *, which represents a missing value in Minitab). This asterisk can be entered manually to represent a missing value (or when we want to ignore the existing value in an analysis); it can be placed automatically when a cell in a numeric column is left blank; or when the data obtained by some calculations do not result in a value which is a real number (such as when trying to compute the square root of a negative number). Minitab pays attention to the presence of missing values, and when computing statistical summaries of each variable (column), we get the information on the number of missing values.

It is possible to change the data type of a column by using the Data > Change Data Type menu option in Minitab. This is especially useful when the data type of the column is clearly not correct, something that sometimes happens when pasting data from other programs.

15.2.2.3   Data integration

There are different ways to input data into Minitab. An easy and straightforward system, already mentioned, is copying and pasting from another program. Copying and pasting data works very well and much better than in other statistical packages (such as SPSS). Minitab is able to open Excel files, by choosing Excel format in the Open Worksheet dialog box, and text files. However, it cannot directly open files from other statistical packages. If there are problems opening a text file, a safer option is using the File > Other Files > Import Special Text... menu option; this gives the possibility to custom define the format of the file. Minitab can also import data from a database using the Open Database Connectivity (ODBC) protocol. This allows opening data stored in database applications such as Access or SQL (having the ODBC driver for the desired database is a requirement for this functionality to work).

With an ODBC data connection, the link is not dynamic. However, it is possible to use DDE (dynamic data exchange) to dynamically exchange data between older versions of Minitab and other applications. A common use of this functionality was linking a Minitab datasheet with Excel: data then remains synchronized in both programs. So you can, for example, manage the data using Excel facilities and perform statistical analysis with Minitab without being continually moving from one program to another. This functionality has been discontinued in recent versions of Minitab. Figure 15.5 shows an example, where data from Excel is linked into Minitab, and results from Minitab are also linked in Excel.

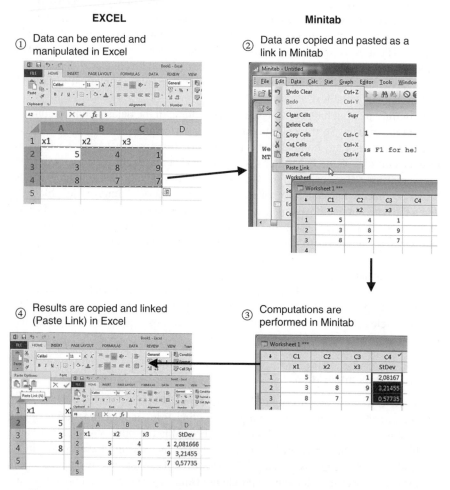

Figure 15.5   An example of a DDE connection between Excel and Minitab.

15.2.2.4   Temporal relevance

Analyzing data by taking into account the date and time when the data was collected is easy and offers many presentation possibilities using time series plots. For example, Figure 15.6 shows the evolution of the number of defects detected in the final inspection of a product during a month. The horizontal axis contains both the days of the week and the days of the month: the chart clearly shows the difference between weeks.

Sometimes the relevance of the temporal evolution is not so obvious. A company conducted a study on the complaints received in the last nine months. When drawing a Pareto chart of the causes of complaints, the most frequent was the one coded as B, representing almost half of the total (Figure 15.7, top). Without more analyses, it seems reasonable focusing on cause B. But when we look at the month in which the complaint was done, we realize that almost all type D complaints appeared in the last studied month (September), so probably the priority is solving the problem with type D complaints as quickly as possible (Figure 15.7, bottom). Minitab can stratify Pareto charts very easily, also with regard to time-related variables or the order of the data collection.

Another area in which the temporal evolution of the data is of great importance is statistical process control. Minitab can build a wide range of control charts, both univariate and multivariate (Figure 15.8), with many possibilities of format and presentation. However, keep in mind that these are really useful when graphics are built and analyzed in real time, and this requires solving the problem of capturing data and incorporating them into Minitab continuously, at specific moments in time. Minitab is useful for computing control limits in control charts based on already collected data, but it is probably not the best software for implementing real-time statistical process control in a production line: there are many software packages focused on SPC that can perform this better, with more useful functionality.

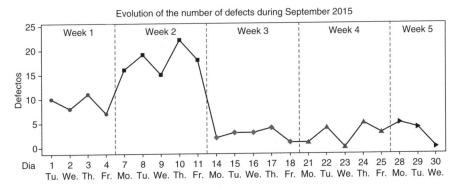

Figure 15.6   An example of the number of defects during a month, showed in a time series plot.

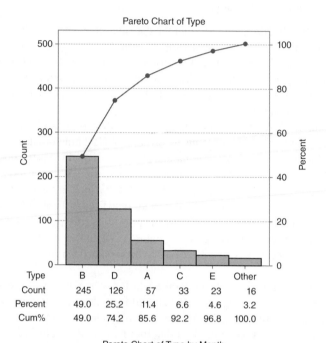

Type	B	D	A	C	E	Other
Count	245	126	57	33	23	16
Percent	49.0	25.2	11.4	6.6	4.6	3.2
Cum%	49.0	74.2	85.6	92.2	96.8	100.0

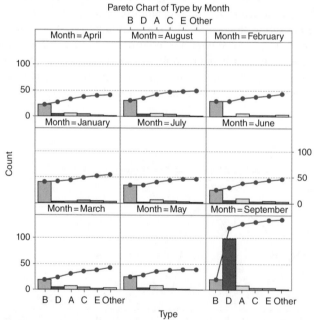

Figure 15.7   An example of a Pareto chart with all the data together (top) and stratifying by month (bottom).

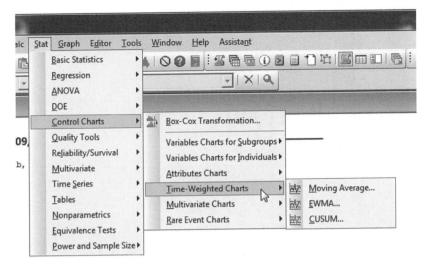

Figure 15.8   A screenshot showing different types of control charts in Minitab.

15.2.2.5   Chronology of data and goal

Besides forecasting models based on time series, Minitab offers extensive possibilities in modeling regression equations (Figure 15.9): from just adding the fitted line to a bivariate diagram to using sophisticated modeling techniques, including the calculation of all possible models with up to 31 independent variables. This involves calculating and comparing more than two billion equations, of which the best are presented according to the most common goodness-of-fit criteria: R^2, adjusted R^2, Mallows' Cp, or standard deviation of the residuals.

If the aim is an explanatory model, that is, a model to explain the behavior of the response by identifying the most influential variables and how they behave, we will be interested in selecting a (usually small) subset of variables that are basically independent and that lead to a model with a reasonable physical interpretation.

However, if the aim is a predictive model, that is, what matters is making good predictions of the value of the response, without the priority of discovering which variables are influencing the response, the independence of the regressors is not that important, although surely the ease to measure them is.

Minitab offers a variety of graphical and quantitative support for the construction and analysis of models. But of course, it cannot replace the expertise of the analyst, who must be also guided by the intended use of the model.

15.2.2.6   Generalizability

Generalizing to the whole population of findings obtained from a sample (statistical generalization) gives good or bad results depending on the quality (representativeness) of the data. Generalizing from one population to another (scientific generalization) has more to do with experience and scientific knowledge than with the data or the statistical analysis. A well-designed and conducted experiment with

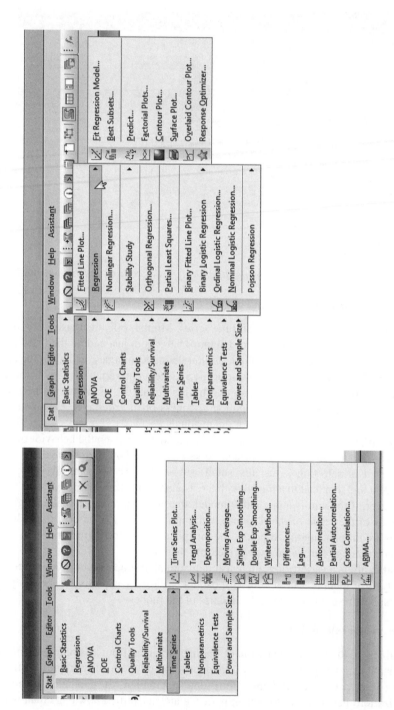

Figure 15.9 A screenshot with different modeling possibilities in Minitab.

laboratory rats may reveal that a certain type of food produces a greater resistance to fatigue in rats, but inferring from this result that the effects in humans will be the same is a leap not justified without further analysis.

Regarding the statistical generalization, Minitab includes techniques for population parameter estimation, comparison of treatments—parametric and non-parametric—and goodness-of-fit tests. With Minitab, it is easy to answer the typical question of "what sample size do I need so my conclusions are valid?" For example, in a comparison of means with independent samples (2-Sample-t test), a dialog box appears when using the Stat > Power and Sample Size dialog box, where one must introduce values for three of the four variables that appear, and Minitab calculates the value of the fourth variable. So, if you want to detect a difference of three units with a power of 90% and the standard deviation of the population is five units, 60 observations in each sample are required. The significance level and the alternative hypothesis (0.05 and "other than" type, respectively) are included by default but can be changed via the options button (Figure 15.10).

15.2.2.7 Operationalization

The selection of variables for explanatory models can be performed using modeling techniques with regression equations. In the output from a regression analysis Minitab offers, in addition to measures concerning the statistical significance of each coefficient, the variance inflation factor (VIF) value associated with each variable. The VIF measures the degree of relationship of a variable with the other explanatory variables. Ideally, the variables are independent of each other so that the role of each one in the model can be assessed properly (this implies having VIFs close to 1).

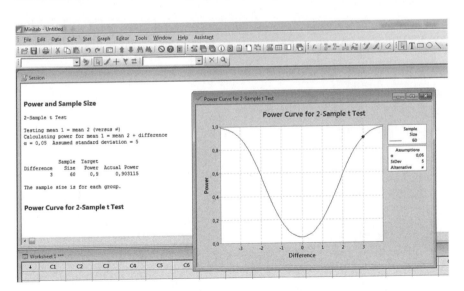

Figure 15.10   Output from the power and sample size procedure for the comparison of means test.

When what really matters is not which are the variables involved in the model but the measurement accuracy of each one of them, as in predictive models, Minitab includes a wide range of techniques to assess the quality of the measurement system (Figure 15.11).

In industrial environments, the variability of critical to quality variables is an issue of concern. This variability in the data is due to the inherent variability of the parts and also to the variability introduced by the measuring system (sometimes even more important than the real variability among parts). The variability of the measuring system can be decomposed into variability caused by the measuring device (repeatability) and variability introduced by the way the operators use the machine, by environmental conditions or by other factors (reproducibility). For more on Gage R&R using the InfoQ lens, see Chapter 11.

Following an orderly process of data collection, Minitab includes analysis techniques that break down the total observed variability in each of its sources: part to part, repeatability, and reproducibility, the latter with each of its components. Of course, the variability introduced by the measurement system must be consistent with tolerances of the measured magnitude. If these tolerances are $\tau \pm \delta$, it is not possible to have a measurement system with an accuracy of $\pm \varepsilon$, ε being of the same order of magnitude as δ. Typically, the measuring system is considered suitable if $\varepsilon < 0.1 \delta$, and Minitab provides this information directly.

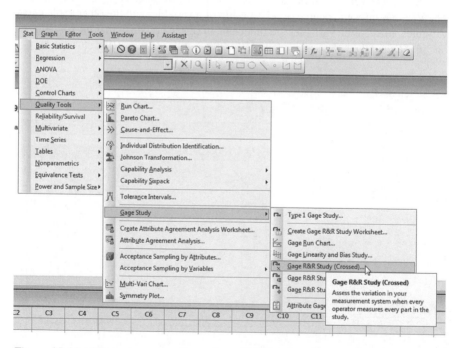

Figure 15.11   The menu option for a Gage R&R study to validate the measurement system in Minitab.

15.2.2.8   Communication

Minitab has a structure that facilitates the management of the obtained results and makes the preparation of a final report fast and convenient. If Minitab is used in conjunction with Quality Companion, the preparation of a report or a presentation can be even faster (basically "dropping" results into a presentation template).

A notable feature in Minitab is the ease of presenting graphics in a clear and clean way in both written reports and presentations. Imagine the following situation: the manager of a bakery was concerned because he had detected loaves of bread below the minimum allowed weight. To study the origin of the problem, 80 units made with each machine were taken. Figure 15.12 shows the histograms of the weights depending on the machine in which they were produced. The nominal value is 220 grams and the tolerances are ±10 grams. A mere look at the histograms clearly reveals that the problem lies in machine 2, which is not centered.

In the field of the analysis of variability, Minitab also includes not so classic graphs that are useful to identify the causes that affect the quality characteristics of a product. The following is an example based on data from a real case study. In a production process of glass bottles, the manager was concerned about the excessive variability in the weight of the bottles (what really matters is the resistance to internal pressure, but the weight is closely related to it and is much easier to measure). The molds are grouped in boxes; each box comprises ten sections with two cavities in which a molten glass drop (coming from two different nozzles) is introduced (Figure 15.13). It was not known if the variability was produced by the different weights of the drops, the characteristics of the boxes, or its sections (each section carries a separate cooling system).

To analyze the origin of the variability, the bottles of five consecutive boxes were weighed each hour during one day. The graph in Figure 15.14, called multivari chart, shows how the weights evolved throughout the day (seven samples were taken) depending on the box, the section, and the drop of each bottle. The graph clearly shows that the variability is essentially related to time.

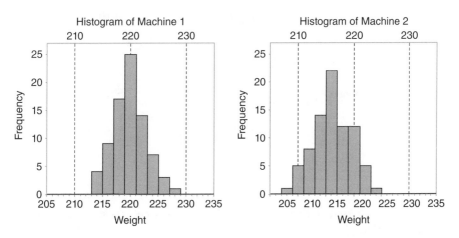

Figure 15.12   Representation of results (using histograms) in the case study of the bakery.

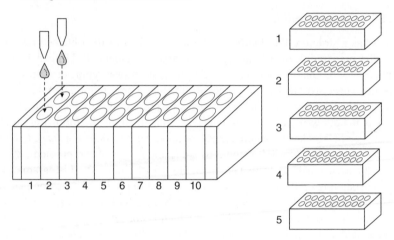

Figure 15.13   Schematic representation of the data collection procedure for the glass bottles case study.

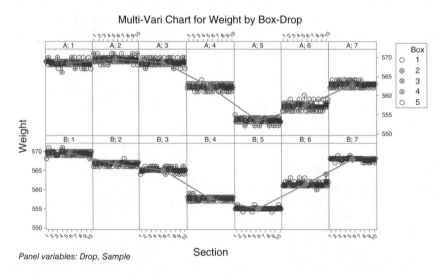

Figure 15.14   Representation of results (using a multivari chart) in the case study of the glass bottles.

15.3   Examples of InfoQ with Minitab

Following Hand's (2008) definition of statistics as "the technology of extracting meaning from data", we present two examples of applying f to a resource X for a given purpose g using Minitab. After each one of the case studies, the InfoQ assessment is offered. The examples are based on situations shown in Grima, Marco and Tort-Martorell (2012).

15.3.1   Example 1: Explaining power plant yield

Technicians in a thermal power plant have collected data on several variables during 50 days of operation. The available data are the following:

Name	Content
Yield	Yield of the thermal power plant
Power	Average power
Fuel	Combustible used (0: fuel, 1: gas)
FF	Form factor of the power curve. Measures the variation of the power throughout the day
Steam temp	Live steam temperature
Air temp	Air temperature (environment)
Seawater temp	Seawater temperature (cold source)
Day	Weekdays (1: Monday, 2: Tuesday, …)

Our aim is building a model that explains the plant's yield as a function of the variables available.

The chosen analysis technique f will be a regression analysis, with Yield as response. In preparation for the analysis, we realize that the variable Day, as such (qualitative variable with more than two categories), cannot be directly included in the model. However, we can encode it as 0 = working day (Monday to Friday) and 1 = weekend (Saturday and Sunday). This distinction seems reasonable to the technicians, because on weekends there is less demand and that could affect the performance of the power plant. Using Minitab coding capabilities (Data>Code>Numeric to Numeric), we create the new variable.

A first exploratory data analysis scatterplot of all pairs of variables (called matrix plot in Minitab) is shown in Figure 15.15. As there are many variables, the graph is not too clear. However, the scatterplot of yield versus power (Figure 15.16) shows a single point far from the others (a day with both high yield and high power).

The technicians know that a yield of 0.44 is an impossible value. Hence, this value must be due to an error and therefore this day is excluded from the study. Redoing the scatterplot without that point (Figure 15.17), we see the point marked by

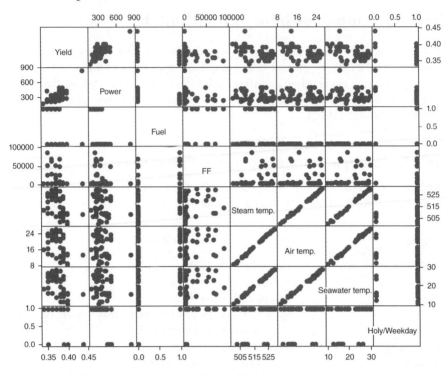

Figure 15.15   Matrix plot of all variables in the power plant case study.

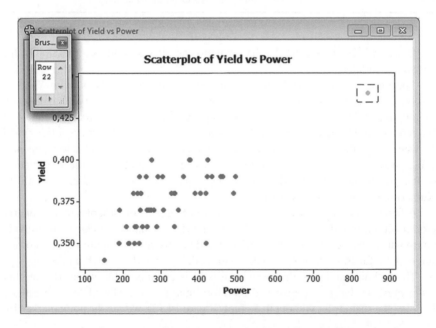

Figure 15.16   Scatterplot of yield versus power (with outlier) in the power plant case study.

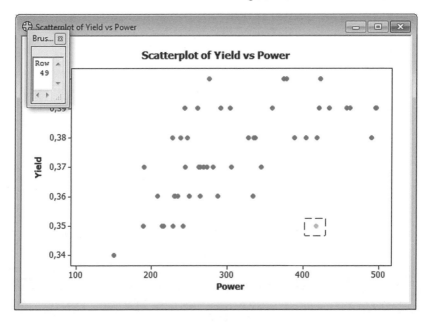

Figure 15.17   Scatterplot of yield versus power (without outlier) in the power plant case study.

a square (row 49) appears isolated and perhaps it would be better not to include it in the study. Looking at the reports from that day, we discover that the power plant was restarted. The technicians consider that this day should not be included in the study because there are very few days with a restart of the power plant and because the performance during these days is exceptionally low. The model should not be based on such days. Hence, this data point is also deleted.

Before starting the search for the best model, we add some variables, transformations of the original ones, which may improve the final model. The transformations considered reasonable are as follows:

- Squared term for power: The scatterplot of yield versus power shows a relation that can be better described by means of a quadratic curve than by a straight line.

- The inverse of the power: This type of nonlinear relationship can also be modeled through the inverse of the power. In addition, our knowledge of the yield's formula makes us think that the inverse of the power is a reasonable variable.

- The logarithm of the form factor: The form factor shows some values grouped near zero (see dotplot in Figure 15.18).

  After a logarithmic transformation is applied, the new dotplot shows better behavior (Figure 15.19).

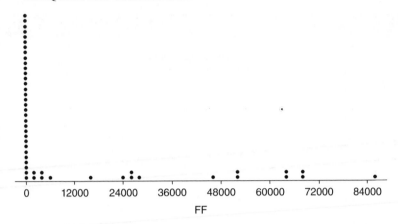

Figure 15.18   Dotplot of factor form in the power plant case study.

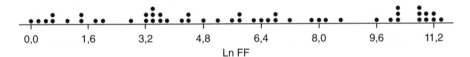

Figure 15.19   Dotplot of the logarithm of factor form in the power plant case study.

Using the best subsets command in Minitab (Stat>Regression>Regression> Best Subsets), we can generate all possible models. The models are ordered based on R^2. Other goodness-of-fit indicators (U) are also presented as follows:

```
Response is Yield
49 cases used, 1 cases contain missing values

                                            S H
                                      S     e o
                                      t     a l
                                      e A   t y
                                      a i e / 1
                                      m r r W P /
                                P     t t t e o P L
                                o F   e e e k w o n
                                w u   m m m d e w
                      Mallows         e e F p p p a r e F
Vars  R-Sq  R-Sq(adj)   Cp        S   r l F . . . y 2 r F
  1   60.4    59.5     97.8  0.0099946  X
  1   51.1    50.0    131.0  0.011103                   X
  2   75.1    74.0     47.1  0.0080112  X                X
  2   72.6    71.4     56.1  0.0084063           X       X
  3   83.0    81.8     20.8  0.0066924  X         X      X
  3   82.9    81.7     21.4  0.0067234  X X              X
  4   87.1    85.9      8.1  0.0058957  X         X      X X
```

4	87.1	85.9	8.2	0.0059020	X	X		X X
5	89.0	87.7	3.2	0.0055029	X X		X	X X←
5	89.0	87.7	3.4	0.0055123	X X		X	X X
6	89.4	87.8	4.0	0.0054796	X X		X	X X X
6	89.3	87.8	4.2	0.0054952	X X		X	X X X
7	89.5	87.7	5.4	0.0055060	X X	X X		X X X
7	89.4	87.5	5.9	0.0055430	X X X	X		X X X
8	89.6	87.5	7.2	0.0055632	X X	X X X		X X X
8	89.6	87.4	7.3	0.0055672	X X X X X			X X X
9	89.6	87.2	9.1	0.0056234	X X X X X X			X X X
9	89.6	87.2	9.1	0.0056291	X X X X X		X	X X X
10	89.7	86.9	11.0	0.0056933	X X X X X X X		X	X X

The most interesting model is indicated with an arrow (low values of Cp and S, high values of adjusted R^2). This model can be fitted with the following result:

```
The regression equation is
Yield = 0.476 - 0.000078 Power - 0.0122 Fuel - 0.000808 Seawater
   temp.
        - 14.1 1/Power - 0.000959 Ln FF
48 cases used, 2 cases contain missing values
Predictor              Coef      SE Coef        T      P
Constant            0.47636      0.01806    26.38  0.000
Power           -0.00007802  0.00002876    -2.71  0.010
Fuel              -0.012194     0.002103    -5.80  0.000
Seawater temp.    -0.0008076    0.0001559   -5.18  0.000
1/Power             -14.114        2.554    -5.53  0.000
Ln FF            -0.0009585    0.0002325    -4.12  0.000
S = 0.00550286    R-Sq = 89.0%    R-Sq(adj) = 87.7%
Analysis of Variance
Source          DF           SS          MS        F      P
Regression       5    0.0103261   0.0020652    68.20  0.000
Residual Error  42    0.0012718   0.0000303
Total           47    0.0115979
Unusual Observations
Obs   Power     Yield        Fit    SE Fit    Residual  St Resid
  4     150  0.340000   0.342651  0.004282   -0.002651     -0.77 X
 15     263  0.370000   0.356510  0.001672    0.013490      2.57R
 24     335  0.380000   0.397380  0.001903   -0.017380     -3.37R
R denotes an observation with a large standardized residual.
X denotes an observation whose X value gives it large leverage.
```

Nothing alarming appears when checking the residuals with a normal probability plot and a scatterplot of residuals versus fitted values, and the obtained model can be accepted without problems. The model can be used to predict yield based on power, fuel used, form factor, and seawater temperature.

We conclude this example with Table 15.1, the InfoQ assessment. The final InfoQ score obtained is 78%.

Table 15.1   InfoQ assessment for Example 1.

InfoQ dimension	Score	Comments
Data resolution	3	Data at the day level is probably sufficient, although it could be beneficial having them at a lower aggregation level
Data structure	5	Outliers removed from the study are confirmed by technical expertise, number of decimal places in variables is sufficient
Data integration	5	Data are automatically registered from the power plant information systems, in a single database
Temporal relevance	3	Once the model is built, new predictions can be produced immediately. It is not clear if the model will be valid in case of profound changes in the process
Chronology of data and goal	5	Getting new data points for the model poses no difficulties
Generalizability	4	Although data from only 50 days were collected and in a single power plant, yield is based on physical principles, so it can probably be generalized
Operationalization	5	The model can be used for predicting yield in an easy way
Communication	4	Although the model can be interpreted, some kind of graphical representation (a dashboard, for instance) could facilitate understanding
Total score	78%	

15.3.2   Example 2: Optimizing hardness of steering wheels

A manufacturer of steering wheels for cars had problems with the hardness of its product. The manufacturing process consists of injecting polyurethane into a mold.

To discover which variables affect the breakage index, a 2^3 factorial experiment is carried out with variables P (injection pressure), R (ratio of the two components of the polyurethane), and T (injection temperature). After properly choosing the levels and given the large variability detected in the hardness, a decision was made to replicate the experiment. The obtained results are shown in Table 15.2 (column Hardness1 is the first replicate, Hardness2 the second replicate).

The aim is to analyze how each of the factors affects the hardness.

Minitab can be used both for creating the design matrix for a factorial design of experiments (thus getting the set of experimental conditions) and for analyzing the results.

Table 15.2   Results of the factorial experimental design of the steering wheels.

P	R	T	Hardness1	Hardness2
−1	−1	−1	35	18
1	−1	−1	62	47
−1	1	−1	28	31
1	1	−1	55	56
−1	−1	1	49	26
1	−1	1	48	31
−1	1	1	34	39
1	1	1	45	44

As we have replicates, it is possible to estimate the variance of the response, and Minitab hence carries out significance tests for each of the coefficients in the model. The results are the following:

```
Estimated Effects and Coefficients for Hardness (coded units)
Term       Effect    Coef   SE Coef      T        P
Constant            40,500    2,312   17,52   0,000
P          16.000    8.000    2.312    3.46   0.009
R           2.000    1.000    2.312    0.43   0.677
T          -2.000   -1.000    2.312   -0.43   0.677
P*R         1.000    0.500    2.312    0.22   0.834
P*T       -11.000   -5.500    2.312   -2.38   0.045
R*T        -0.000   -0.000    2.312   -0.00   1.000
P*R*T       2.000    1.000    2.312    0.43   0.677
S = 9.24662      PRESS = 2736
R-Sq = 69.52%    R-Sq(pred) = 0.00%   R-Sq(adj) = 42.85%
```

The ratio of both components of the polyurethane (factor R) is inert within the range of values used in the experiment. Only pressure (P) and injection temperature (T) are active factors. We now have a look at the interaction between these two active factors (Figure 15.20).

The maximum resistance is obtained with pressure level + (denoted 1) and temperature level − (denoted −1). If at any time it is necessary to work with pressure level −, the temperature should be set at level +.

Although company technicians were satisfied with the information extracted from the experiment, a later study on the hardness of the steering wheels revealed that the previous experiment was not properly randomized. Initially, the first eight replicates were done, and then the other eight, with the aggravating circumstance that two weeks passed between both sets of experiments. Environmental conditions such as temperature and humidity may affect the characteristics of the components of polyurethane.

Therefore it was decided to reanalyze the experiment considering it as a 2^4 design, where a new factor W (level − for the eight runs of the first replication and level + for

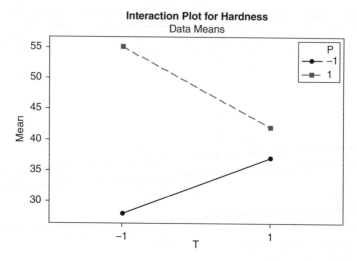

Figure 15.20   Interaction plot for pressure and temperature in the steering wheels case study.

the eight runs of the second one) represents the differences (environmental or of other type) occurring during the two weeks that passed between the first and the second round. We'll assume that these differences are due to weather changes (and thus we call the factor W "weather").

As there is no replication now and no estimation of response variability is available, a common procedure to identify significant factors in a factorial design is representing the effects in a normal probability plot (Q–Q plot). Figure 15.21 shows the results given by Minitab.

It is clear that the significant effects are A (P: pressure) and interaction AC (PT: pressure–temperature). Up to this point, the result is the same as in the previous analysis. However, also D (W: weather) and the interaction BD (RW: ratio–weather) are significant. Analyzing interaction PT, conclusions are the same as before. However, analyzing interaction RW, we obtain the interaction plot shown in Figure 15.22.

If R is at level –, weather clearly affects the hardness of the steering wheels; however, if it is at level +, this influence is much less. Previously, when the environmental conditions were ignored, R appeared as inert; so its value was insignificant. Now, in view of its interaction with weather, it can be used to neutralize the influence of environmental conditions and get a robust product independent of its production conditions.

We conclude this example with Table 15.3, the InfoQ assessment. The final InfoQ obtained is 87%.

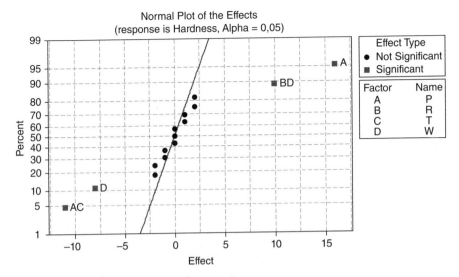

Figure 15.21   Normal probability plot of the effects in the steering wheels case study.

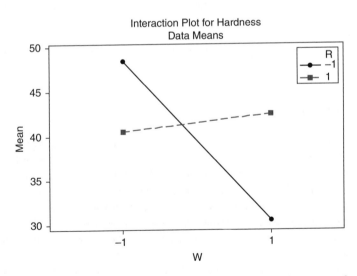

Figure 15.22   Interaction plot for ratio and weather in the steering wheels case study.

Table 15.3   InfoQ assessment for Example 2.

InfoQ dimension	Score	Comments
Data resolution	5	Data are collected in an experimental setting, thus assuring appropriateness
Data structure	4	The assumption of having real replicates appears incorrect but has been corrected with the reanalysis of the data
Data integration	5	Data are collected from the experimental setting, in a controlled environment
Temporal relevance	4	The model obtained taking into account machine parameters poses no difficulties, although changing environmental conditions could decrease model validity
Chronology of data and goal	4	Machine conditions coming from control factors can be set without difficulties, although this is not true for environmental conditions (could be considered a noise factor)
Generalizability	4	Discoveries made from the experiment will be useful for other similar machines producing wheels
Operationalization	5	The knowledge gained from the experiment can be used to define appropriate technical conditions
Communication	5	Graphical output showing main effects and interaction plots clearly communicate the effect of each factor
Total score	87%	

15.4   Summary

Minitab is characterized by its ease of use. You can start using it almost without any prior training and autonomously progress in the understanding of its possibilities. From the point of view of the dimensions of InfoQ, Minitab lets you work with data on the desired resolution and easily organize it in a familiar structure of rows and columns that is very practical. Data can be easily integrated from a spreadsheet (although direct integration of data captured by external sensors is not provided). It has plenty of opportunities for graphical analysis and use of statistical methods. Its output is provided in a manner that is easy to use for fast and effective communication.

Besides developing aspects of Minitab that can support each of the eight dimensions of InfoQ, the two case studies offered in this chapter are examples of the simplicity and efficiency of Minitab as a statistical software for quality improvement and data-driven decisions.

References

Albert, J.H. (1993) Teaching Bayesian statistics using sampling methods and Minitab. *The American Statistician*, 47, 3, pp. 182–191.

Grima, P., Marco, L. and Tort-Martorell, X. (2012) *Industrial Statistics with Minitab*. John Wiley & Sons, Ltd, Chichester, UK.

Hand, D.J. (2008) *Statistics: A Very Short Introduction*. Oxford University Press, Oxford.

Kenett, R.S. and Zacks, S. (2014) *Modern Industrial Statistics: With Applications Using R, Minitab and JMP*, 2nd edition. John Wiley & Sons, Ltd, Chichester, UK.

Marriott, J.M. and Naylor, J.C. (1993) Teaching Bayes on Minitab. *Journal of the Royal Statistical Society, Series C (Applied Statistics)*, 42, 1, pp. 223–232.

Tukey, J.W. (1977) *Exploratory Data Analysis*. Addison-Wesley, Reading, MA.

16

InfoQ support with JMP

Ian Cox

JMP Division, SAS Institute, Cary, NC, USA

16.1 Introduction

JMP is a desktop product from SAS that is designed to allow researchers, engineers, and scientists to get the most value from data generated by measurements. Like all software, it is *enabling* in the sense that it has the potential to allow users to do things that would otherwise be difficult or impossible for them to do through another way. The InfoQ framework allows consideration and discussion of what "value" actually means in specific situations, a more rational appraisal of possible alternative approaches or, in the worst case, a clearer indication of the limitations of an approach demanded by practical considerations.

Clearly the skills, aptitudes, and knowledge of the user are a vital consideration in the design and development of software that aspires to enable, and, as with everything, there is variation in these three aspects. To tame this complexity, modern software development is usually couched in terms of a repertoire of use cases (Cockburn, 2001), often refined by an agile approach for delivery (Shore and Warden, 2007). Such use cases can generally be grouped into two types, which we will label as *planned* and *unplanned*. A planned use case is characterized by the fact that the pathway from data to information and action is known (or stipulated) in advance, whereas for an unplanned use case, this is not so. For a given application area, unplanned use cases tend to come before planned ones, and the latter often emerge from what is learned through the former. The considerations of InfoQ can and should

Information Quality: The Potential of Data and Analytics to Generate Knowledge,
First Edition. Ron S. Kenett and Galit Shmueli.
© 2017 John Wiley & Sons, Ltd. Published 2017 by John Wiley & Sons, Ltd.
Companion website: www.wiley.com/go/information_quality

be applied to both types. Note also that, generally, planned use cases often imply or demand some level of automation to make them practically feasible and often have to support multiple users (with their attendant variation). Associated with this is the idea of encapsulating the best practice so as to democratize the use of data to do something useful and valuable.

Every software product has strengths and weaknesses. Vendors or open-source communities attempt to serve markets and users to the best of their ability, and many users present a proliferation of use cases that they might be interested in. Users are inventive and demanding, so, from the point of view of the vendor at least, this tension is best managed by some overarching principles that guide product concept and development. Additionally, software products have to continue development from where they currently are, and older products may be at risk from a choice of technology, architecture, or platform that originally made sense but now puts constraints on what is possible.

In the case of JMP, we aim to support the process of statistical discovery, shown in Figure 16.1. Many of our users work in companies and environments in which innovation is a prerequisite for survival, and innovation demands learning something new about the way the relevant system of value production functions or can be made to function through appropriate redesign. New insights are serendipitous, and JMP aims to reduce the barriers to encountering such insights by offering a synergistic blend of visualization and statistical modeling in one environment. In other words, one of the guiding principles of JMP is to offer the best support possible for unplanned use cases that are in scope and can be reasonably delivered by desktop software.

JMP was first brought to market in 1989 and at the time of writing is at version 12 on Windows and OS X. It uses an in-memory architecture, so that the local RAM

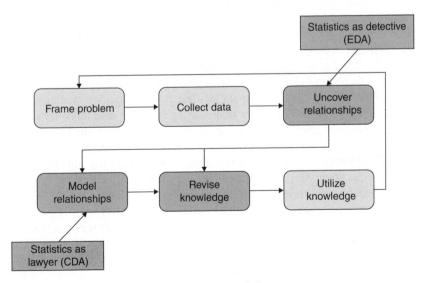

Figure 16.1   Statistical discovery.

determines how much data you can work with. The user interface of JMP was inspired by some key ideas which were new at the time but which continue to serve users well today:

- Hot spots for optional analyses

- Hypertext for organizing, folding, and unfolding

- Boxes and glue for layout

- Smart scrolling to keep titles sticky

- Color coding for preattentive cognition

- Dynamic linking

Like every well-established software product, raw functionality has grown over the years.

Although support for unplanned use cases has been a particular focus, another guiding principle of JMP is that, when functionality is added, it is always made scriptable, that is, accessible via the JMP Scripting Language (JSL). This allows JMP to fully support the automated, planned use cases mentioned earlier. JSL is an interpreted, full-fledged scripting language that supports a variety of programming styles. JMP also provides an integrated development environment and the associated tools for building and distributing statistical discovery applications for workgroups with common requirements. Interactive use of JMP generates JSL behind the scenes, and this code can be easily reused to reduce the burden of application development. JSL also allows JMP to interoperate with other software (such as SAS, R, and MATLAB) when this is needed or beneficial for moving from data to information and action.

Use cases are always indicative rather than exhaustive or comprehensive. This chapter contains two examples to illustrate how JMP supports InfoQ. Each example starts with a tabular synopsis of the scenario to orient the reader. As mentioned earlier, in so far as InfoQ is a way of thinking that guides our approach to exploiting data and preventing its abuse, it should be clear that the practice of statistical discovery could benefit considerably from this discipline. Moreover, in so far as InfoQ can be systematized, one can also envisage that its repeatable application can be facilitated through software. To illustrate the application-building capabilities of JMP, we briefly show a prototype of this after the two examples. Finally, we close with some commentary on how various JMP capabilities align with the four InfoQ components (Chapter 2) and the eight InfoQ dimensions (Chapter 3).

We encourage the reader to download a trial copy of JMP from www.jmp.com/trial to see the materials in this chapter in a "live" setting. Download the *InfoQ.jmpaddin* file from the book's website, and if JMP is installed, double-click on this file to create an "InfoQ" menu under the "Add-Ins" main menu. All the materials here are available as submenus from the "InfoQ" menu, and tooltips explain the intended purpose of each. For a comprehensive treatment of modern industrial statistics with JMP see Kenett and Zacks (2014).

16.2   Example 1: Controlling a film deposition process

Table 16.1   Synopsis of Example 1.

Application area	Quality engineering, reliability, and six sigma
Industry niche	Semiconductor fabrication
System under study	Low-pressure chemical vapour decomposition (LPCVD) diffusion step growing silicon nitride on semiconductor wafers
Practical goal (operations manager)	From the operations manager to the diffusion engineer: "Implement statistical process control (SPC) at the LPCVD step so my technicians only chase real problems"
Analytical goal (diffusion engineer)	Using available process data, construct, deploy, and maintain appropriate control charts to balance risks appropriately
Major steps	Assess data quality and take remedial actions Visualize and estimate patterns of variation Construct appropriate charts and calculate limits Implement control methodology
Analytical tools	Multivariate missing maps Linked histograms and bar charts Spatial maps Mixed models Control charts

Manufacturing working semiconductor devices is nontrivial. Many steps are involved and one of the key aspects is that although it is the final individual chips that count, in the early stages these have to be fabricated on wafers. When each wafer is cut up, it forms many chips (ideally all identical). For operational reasons wafers are often arranged into groups (lots or batches) and can be processed at a particular manufacturing step as a run consisting of either part of a lot or of several lots. This example is the latter situation: low-pressure chemical vapour deposition (LPCVD) is conducted in a furnace tube, which holds four lots (each of 24 wafers). The engineering objective is to grow a new layer of silicon nitride on the surface of each wafer that has properties such that the final device works as designed. The furnace tube is heated, and gases flow from one end of the tube to the other, passing over and between the wafers. The process is executed according to a recipe that controls things like the rate of heating, temperature, and gas flows and pressures. After each run of the furnace tube, measurements of responses are made on four nonproduct wafers, each one adjacent to a lot in the tube. The measurements made on each monitor wafer are assumed to be representative of the properties of the adjacent lot. A summary of the example and the data analysis methods it uses is presented in Table 16.1.

To meet customer demand, build schedules require that four distinct furnace tubes or entities are available. The cycle time for the LPCVD step is several hours, and manufacturing operations require that each of the four entities can also be used for several other steps. Ideally, as lots arrive from various prior steps in the process, entities are available to process them in such a way that the end-to-end cycle time of each is as short as possible.

Although the LPCVD step is generally well understood, things can and do go wrong, at which point technicians intervene to try to fix the problem, ideally finding the root cause and removing it. The operations manager has issued an apparently simple directive to the diffusion engineer—"Implement statistical process control (SPC) at the LPCVD step so my technicians only chase real problems." From the InfoQ point of view, this is the practical goal, at least from the frame of reference of the operations manager. Although not stated directly, we can assume that the operations manager wants a decision rule that directs whether or not technicians need to intervene each time a furnace has been run. Other practical goals are certainly possible, so this highlights an important issue, namely, that the successful application of InfoQ requires good communication, particularly in the case of goal definition (see Chapter 2).

Note that, as mentioned in Chapter 3, SPC and the related concept of rational subgrouping is a special case of the data resolution dimension of InfoQ. However, note also that, as originally conceived, the idea of rational subgrouping was designed to handle situations in which differences in the product stream can be usefully modeled with just a single source of variation. In such a case the focus of the analysis lies in understanding how this variation shows itself within and between the chosen subgroups. However the batch-oriented LPCVD process here is clearly more complex and, depending on the measurement scheme that is used, may result in data that is correspondingly less simple. Depending on exactly how it is handled, this increased data complexity may or may not increase the InfoQ analysis quality and how this contributes to the InfoQ goal quality.

For reasons of expediency, we assume that the diffusion engineer is presented with a single JMP table containing data already extracted from the manufacturing execution system. Of course, accessing data and getting it "in" to any analytical environment is a big and important topic and one that is crucial to InfoQ. Considerations of data structure, resolution, and temporal relevance start with this step, and, in extreme cases, it may be that the analytical environment is overwhelmed to the point that no analysis can proceed. Systems that work well with textbook data may or may not perform so well in the real world. In some cases, knowledge of how to coax the best from any given environment may make marginal cases feasible, but often such knowledge is in short supply.

In so far as we sidestep this key issue, we only give a partial picture of JMP capabilities aligned with the requirements of InfoQ. However, most of the analyses in JMP, whether graphical or numerical or a combination thereof, consume a single table, so this is a sensible starting point.

Like most statistical software, in JMP tables, rows (columns) represent the units that were measured (values of a measurement made on those units). JMP allows each column to be one of the following four data types:

1. Numeric

2. Character

3. Row state

4. Expression

Values within a column have to be of the same type. Data types 1 and 2 are obvious enough; 3 is reflective of the fact that rows can be designated with colors, markers, and so on, whereas 4 allows a table cell to contain any JSL expression. A special case of an expression is an image, which can be very useful in providing additional context and meaning, not easily quantifiable, to a given row. Any column with data type 1 or 2 is also assigned a modeling type (continuous, ordinal or nominal) that affects the details of how JMP handles this column (both graphically and numerically). As well as containing data, a JMP table can also contain metadata, relating either to specific columns or the entire table. Such metadata is persisted when the JMP table is saved to the operating system and, for instance, can consist of JSL scripts that direct JMP as to how the data in the table should be treated.

A partial view of the diffusion engineer's table is shown in Figure 16.2. Note that the table under the "InfoQ" menu also contains some saved scripts to allow you to easily follow the steps described here. These scripts are in the tables panel of the JMP table, below the little red triangle (LRT) hot spot at the top left of the figure. To run one of these scripts, click on the LRT and select Run Script.

The first challenge the diffusion engineer faces is to make an assessment of data quality. By inspection of the data grid and the columns and rows panels, you can see that the data appears to have come from 200 furnace runs and that for each run, nine process conditions were logged ("Initial Pump and Purge Cycles" to "Final Pump and Purge Cycles"). In addition, a film thickness was measured 49 times for each unit or wafer. Note that the recorded data is ambiguous, in the sense that knowledge of exactly how the measurements were made and the subsequent values stored and extracted is required to fully understand its meaning. For example, each run number seems to be associated with four rows, and although it is tempting to assume that each row relates to one of the monitor wafers and that the wafer that occupied a particular position in the tube always has the same row-wise placement, this is nevertheless an assumption that should be checked. The script "Add Wafer ID" adds a new column to the table assuming this is indeed true. Furthermore, although the 49 measurements of film thickness could be repeated measurements made at a particular spot on the monitor wafer (e.g., the center), it is much more likely that they are measurements made at different points over the wafer to try to understand within-wafer variation in thickness. We pursue this point later.

Given that we have identified two groups of columns (one for inputs and one for outputs), it is useful to explicitly form these groups since it will simplify subsequent JMP dialogs. The script "Group Columns" does this.

The structure of the data can be further verified by using "Analyze > Distribution" (or the third saved script)—the equal heights of the bars in the Run Number bar chart is a visual confirmation that each run is indeed associated with four rows.

The script "Add Spec Limits" adds the upper, lower, and target values to each of the Film Thickness columns (this can be done interactively via the "Column Information" dialog and the "Cols > Standardize Attributes" menu). This is an example of adding metadata as a column property and, although not of primary interest in the consideration of process control, would be used automatically by JMP if we attempt to assess capability.

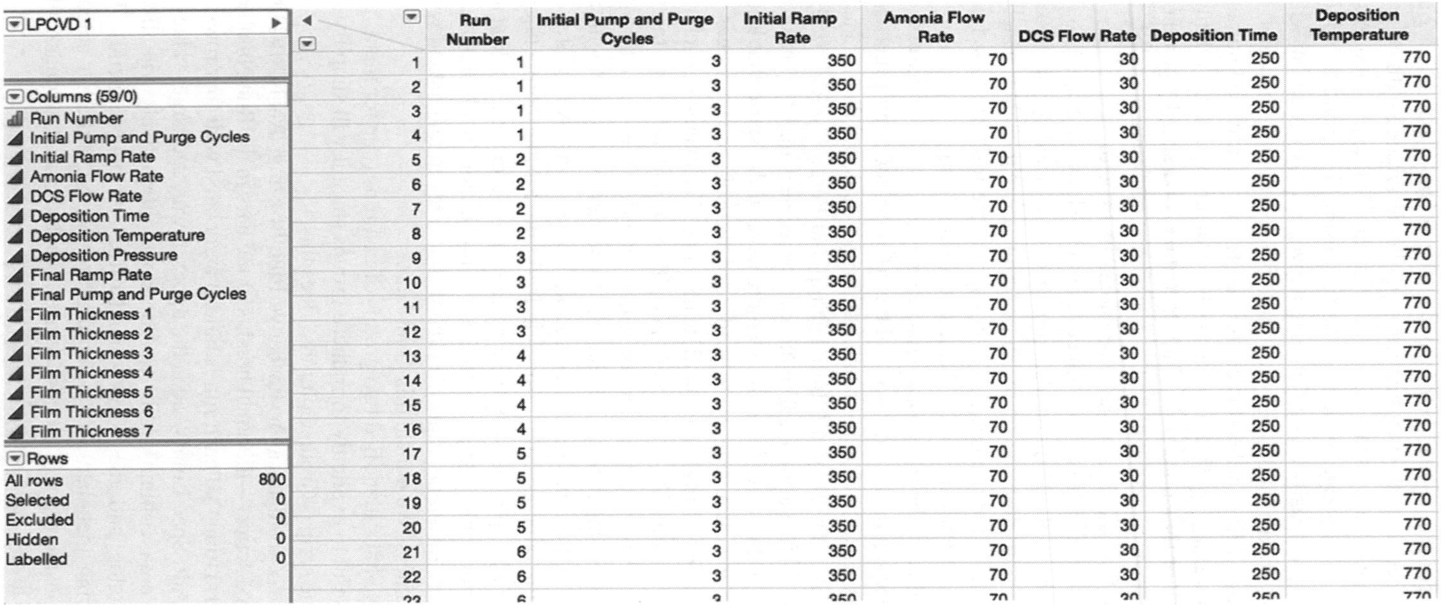

LPCVD 1

Columns (59/0)
- Run Number
- Initial Pump and Purge Cycles
- Initial Ramp Rate
- Amonia Flow Rate
- DCS Flow Rate
- Deposition Time
- Deposition Pressure
- Deposition Temperature
- Final Ramp Rate
- Final Pump and Purge Cycles
- Film Thickness 1
- Film Thickness 2
- Film Thickness 3
- Film Thickness 4
- Film Thickness 5
- Film Thickness 6
- Film Thickness 7

Rows
All rows	800
Selected	0
Excluded	0
Hidden	0
Labelled	0

	Run Number	Initial Pump and Purge Cycles	Initial Ramp Rate	Amonia Flow Rate	DCS Flow Rate	Deposition Time	Deposition Temperature
1	1		3	350	70	30	250	770
2	1		3	350	70	30	250	770
3	1		3	350	70	30	250	770
4	1		3	350	70	30	250	770
5	2		3	350	70	30	250	770
6	2		3	350	70	30	250	770
7	2		3	350	70	30	250	770
8	2		3	350	70	30	250	770
9	3		3	350	70	30	250	770
10	3		3	350	70	30	250	770
11	3		3	350	70	30	250	770
12	3		3	350	70	30	250	770
13	4		3	350	70	30	250	770
14	4		3	350	70	30	250	770
15	4		3	350	70	30	250	770
16	4		3	350	70	30	250	770
17	5		3	350	70	30	250	770
18	5		3	350	70	30	250	770
19	5		3	350	70	30	250	770
20	5		3	350	70	30	250	770
21	6		3	350	70	30	250	770
22	6		3	350	70	30	250	770
23	6		3	350	70	30	250	770

Figure 16.2   The LPCVD data (partial view).

One of the key considerations of data quality is the extent of missingness in the data. The script "Missing Response Values" produces a multivariate view of missingness in a summary table (see Figure 16.3).

Every row in this summary table is linked to a row in the LPCVD 1 table that has at least one missing thickness measurement. We see immediately that only 747 of the 800 rows are complete and also the counts and structure of the rows that are incomplete. Highlighting rows two to eight of the summary table and exploiting the fact that this is linked to the detail-level data, which in turn is linked to the distributions made earlier, we arrive at an interesting insight, namely, that missing values of thickness are always associated with Wafer 2. Scrolling the bar chart vertically also reveals which runs have missing values within them.

Although seen here in a simple setting, this interactive linking of graphical displays with data, possibly also within data hierarchies, is a very powerful capability for uncovering relationships and modeling relationships (see Figure 16.1).

The question of how to best handle missing data is subtle. Clearly the "Wafer 2" issue needs to be followed up. Given, though, that the analytical objective is to construct some kind of SPC chart, we lose little in deleting every run that has one or more missing thickness values to form a final table that is complete and balanced. Note here that, generally, there is an interaction between the missingness of the data and the implementation of the desired analysis method. For example, because JMP implements restricted estimation maximum likelihood (REML, 2015) in the case of linear regression models, it does not require you to decimate your data so that it is balanced. Additionally, when data is more scarce, it may be that some form of data imputation will help. JMP provides a variety of methods (JMP Missing Values, 2015) and, in the case of predictive modeling, can automatically use auxiliary indicator variables related to the missingness of predictors to improve prediction accuracy (JMP Missing Values, 2015). For some further discussion, see Cox et al. (2016).

To select all rows for each run that has a missing value, select the Run Number column in the LPCVD 1 table, and use Rows > Row Selection > Select Matching Cells. Then select Rows > Delete Rows. Alternatively use the script "Delete Incomplete Runs." This produces a table with data for 147 lots and $4 \times 147 = 588$ rows.

Looking at the columns in the process conditions group, it is easy to demonstrate using similar approaches that there are no missing values and each column has only one value, corresponding to the set point used in production. These columns contain no useful information for the analysis goal, and they can be safely deleted using the Cols > Delete Columns menu or the "Delete Process Conditions" script.

We now return to the question of how the film thickness measurements are made on a monitor wafer. As mentioned earlier, it is probable that they represent values from specific wafer locations or sites, and the diffusion engineer confirms this with the engineer responsible for the metrology tool. The engineer also gets a map of where the sites are. Given that we want to look at the measurements en masse, it makes sense to stack the data so that they all appear in a single column. In the stacked data, each row corresponds to a site on a wafer in a run. If it later turns out that we want to focus on measurements from a single site, we can use JMP's Local Data Filter to show just the sites we are interested in. You can use Tables > Stack or the

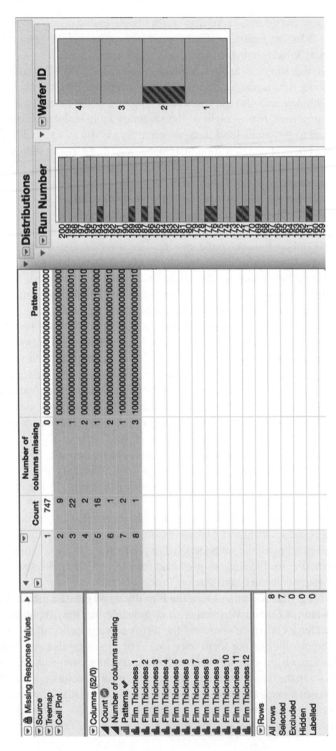

Figure 16.3   Pattern of missing thickness data.

script "Stack Measurements" to rearrange the data into a new table, LPCVD 2. Note that the values inside the Site column were inherited from the column names in LCCVD 1. For what follows, it is more convenient to truncate them, and this can be done using Cols > Utilities > Recode or via the "Recode Site" script. To see the scripts mentioned in the following text, use the LPCVD 2 table accessible via the add-in.

Note that LPCVD 2 contains no obvious spatial information, yet because JMP can use ESRI shapefiles and name files (Shapelife, 2015), it is still possible to map such data from a table like this. Running the saved script "Make Map Files" will set things up correctly. Normally such files would be provided and managed by an administrator and provided "unseen" to a group of users who need no knowledge of them. In this case, though, the requisite files are placed on the Desktop so that you can more easily locate and delete them. Using this mechanism maintains the interactivity of JMP, since you can click on a shape or shapes to select the associated rows.

Figure 16.4 shows a Graph Builder report (use Graphs > Graph Builder or the script "Map 1"). Note that the Site column in LPCVD 2 is assigned to the Map Shapes role to automatically generate the map. Note also the clear spatial variation of thickness between the measured locations.

Graph Builder allows you to build many graphical reports simply by dragging and dropping a column to one of the many different drop zones and by selecting the appropriate graph type from the palette. You can create trellis plots with nested x and/or y grouping variables and customize the properties of many of the graphical elements.

The script "Map 2" shows overall thicknesses for each wafer, while "Map 3" is similar, but uses the Local Data Filter to show results for each run. The Animation option under the LRT allows you to loop through each run and see the corresponding values in succession. The script "Graph Builder" does something similar, but looking at the thickness variation along the furnace tube rather than within the wafer. The script "Variability Chart" gives a conventional, if perhaps more unwieldy, view of the same thing.

The overall message from the graphical displays used is that there are several sources of variation in film thickness and that the source of variation between sites and between wafers has both a fixed and a random part. It should be clear that the more commonly used SPC charts will probably not have the desired statistical properties (because they are based on a model that is too simple), so they are unlikely to be of much practical use.

You can generate an XBar-R chart of film thickness using Analyze > Quality and Process > Control Chart Builder or via the "Control Chart Builder 1" script (Figure 16.5). Using Run Number in the x role means that each run will be considered a rational subgroup, and JMP will automatically aggregate the data in LPCVD 2 without the need for further data management steps. More than 10 of the runs are flagged as being out of control.

You can obtain a more appropriate chart by right-clicking and asking to Add Dispersion Chart (or running the script "Control Chart Builder 2") to give Figure 16.6. Such "three-way" charts are commonly used when attempting to introduce SPC for batch processes (Threeway chart, 2015), but may still not fit every case. Note that the limits of the topmost chart are now wider than in Figure 16.5, to the point that none of the runs are considered to be out of control.

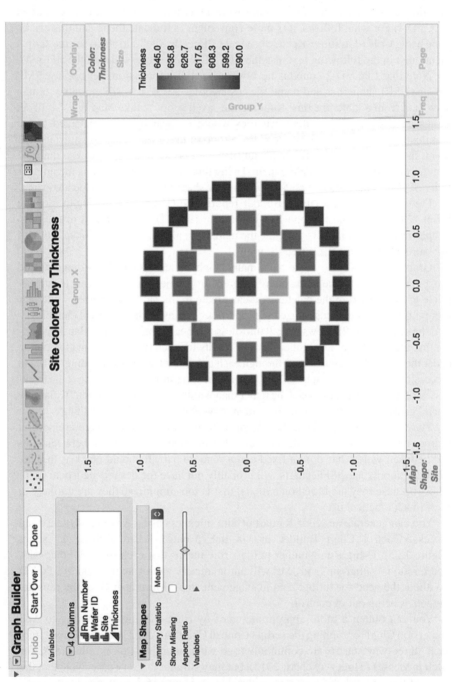

Figure 16.4   Map of all the thickness values.

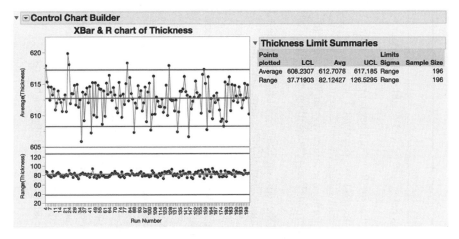

Figure 16.5   XBar-R chart of film thickness.

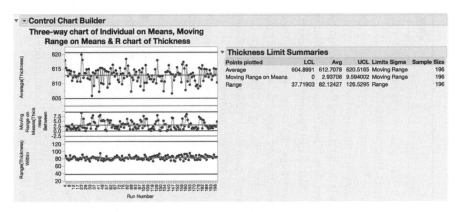

Figure 16.6   Three-way chart of film thickness.

The topic of SPC for complex processes is a broad and interesting one. For example, CUSUM charts of variance components have been used to signal a change in the process (Wetherill and Brown, 1991, Yaschin, 1994, and Kenett and Zacks, 2014). The saved script "Fit Model" estimates the variance components associated with the within-wafer, between-wafer (within a run), and between-run random effects but taking into account the systematic variation within and between wafers. By combining inbuilt functionality with JSL, control limits for such control schemes can easily be calculated with JMP. Note, however, that JMP is not the best vehicle for implementing such schemes in real time, in which case consideration must be given to how limits are passed to, or updated within, an online system.

Finally, note that an adequate SPC model is only part of a control methodology. The XBar-R chart and its variants are general-purpose tools. Specific knowledge of how the system has failed in the past can often be exploited to prompt more timely

Table 16.2 InfoQ assessment for Example 1.

InfoQ dimension	Score	Comments
Data resolution	Very high	Contingent on the measurement process for film thickness being adequate. Possible redundant data in relation to the practical goal
Data structure	High	Some plausible assumptions were required, for instance, that the Run Number represented the processing sequence through the furnace tube. Although the data was decimated (at the Run Number level) due to missing values within runs, the remaining data was sufficient
Data integration	Very high	Film thickness data is readily available from the manufacturing execution system. Map data (in the form of ESRI shapefiles and name files) enrich the film thickness data without requiring specialist knowledge on the part of the diffusion engineer
Temporal relevance	Very high	Film thickness data and map data are readily extracted, and the requisite chart construction can be automated in JMP, including the deployment of associated limits to an online system. Once the required analysis pathway is defined, data volumes are such that its runtime is very small compared to the time taken for a furnace run
Chronology of data and goal	Very high	Contingent on updating the associated limits in the online system with an appropriate frequency. SPC charts can be considered as graphical statistical models, and statistical models always degrade over time
Generalizability	Acceptable	Only data from one of the four furnace tubes was used. With 200 runs, it is almost certain that the furnace tube was used for other steps and/or subject to maintenance during this time, and these may or may not have had an effect on the subsequent runs
Operationalization	Low	Action operationalization is unclear since the analysis here is incomplete. Deciding to use the topmost chart in Figure 16.6 as the decision rule to prompt operator intervention would increase this rating. Statistical considerations aside, one important test for a control methodology is "with what frequency are special causes actually found after interventions?" This information is only available retrospectively after the control scheme is implemented
Communication	High	Even though manufacturing communities tend to be close-knit, statistical approaches can sometimes be viewed with distrust

interventions, for example, by using Cuscore charts on specific metrics (Ramirez, 1998). Finally, we need to recognize that the practical goal (in this case, set by the operations manager) is what drives the analysis goal. Although it may be tempting to wring every last drop of information from the data by using more sophisticated analyses, this may not be necessary. Indeed, simpler approaches requiring fewer measurements may have a hidden value simply because they do require fewer measurements. Even though the sampling scheme described here makes sense when the objective is to gather information on lots as they move through the process (in support of yield improvement or problem-solving initiatives), they may not be well aligned with the stated practical goal. Indeed, it might be reasonable to expect that we could exploit the correlations in the data to reduce the number of measurements taken but not compromise the final decision rule that dictates when operators intervene.

We conclude this example with Table 16.2, an InfoQ assessment. As well as the score for each dimension, there is a comments column with some notes as to why the score was chosen. The final InfoQ is calculated as 72%.

16.3   Example 2: Predicting water quality in the Savannah River Basin

Table 16.3   Synopsis of Example 2.

Application area	Statistics, predictive modeling and data mining
Industry niche	Ecology
System under study	Water quality in the Savannah River Basin
Practical goal	Make assessments of water quality in remote regions without the need for expensive field trips to take water samples
Analytical goal	Build a model that makes good predictions of aquatic biota properties from remote sensing data from satellites
Major steps	Assess data quality and take remedial actions
	Visualize and estimate patterns of variation
	Partition data for predictive modeling
	Build predictive model using partial least squares
	Refine predictive model
	Profile the model and assess prediction accuracy
Analytical tools	Multivariate missing maps
	Linked histograms and bar charts
	Spatial maps
	Scatter plot matrix and correlation
	Partial least squares
	Prediction profiler
	Actual versus predicted plots

The second example is taken from the book *Discovering Partial Least Squares with JMP* (Cox and Gaudard, 2013), which contains a more extensive treatment of this case. The data was originally provided by Nash and Chaloud (2011).

The practical goal is to make assessments of water quality in remote regions without the need for expensive field trips to take water samples. The term *water quality* refers to the biological, chemical, and physical conditions of a body of water and is a measure of its ability to support beneficial uses. A summary of example 2 is presented in Table 16.3.

Of particular interest to landscape ecologists is the relationship between landscape conditions (predictors, or Xs) and indicators of water quality (responses, or Ys). In their attempts to develop statistically valid predictive models that relate the two (the analytic goal), they often find themselves with a small number of observations and a large number of highly correlated Xs. An additional challenge is the low level of signal relative to noise in the relationship between Xs and Ys. In the application of standard multiple or multivariate regression, these conditions usually compromise the modeling process in one way or another, often requiring the selection and use of a subset of the potential predictors. To meet the analytic goal, we will use partial least squares (PLS) (or projection on latent structures) as the analytic technique, since it may be expected to require fewer modeling compromises and predict well.

The data is in a JMP table WaterQuality2 and shown partially in Figure 16.7. As with the first example, we do not consider how this table was built. As before, we use saved scripts to help you walk along the chosen analysis path, though in the interests of brevity we do not highlight every step.

Each row (unit) describes a water sample taken at a particular field station in the Savannah River Basin. There are 86 samples. The field station is described by the columns in the station descriptors group (seven of them), and there are four Ys and

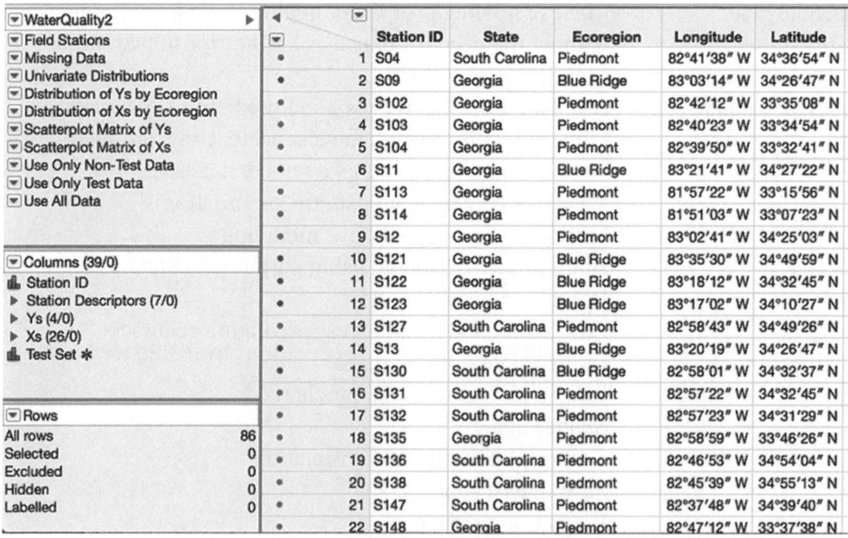

	Station ID	State	Ecoregion	Longitude	Latitude
1	S04	South Carolina	Piedmont	82°41′38″ W	34°36′54″ N
2	S09	Georgia	Blue Ridge	83°03′14″ W	34°26′47″ N
3	S102	Georgia	Piedmont	82°42′12″ W	33°35′08″ N
4	S103	Georgia	Piedmont	82°40′23″ W	33°34′54″ N
5	S104	Georgia	Piedmont	82°39′50″ W	33°32′41″ N
6	S11	Georgia	Blue Ridge	83°21′41″ W	34°27′22″ N
7	S113	Georgia	Piedmont	81°57′22″ W	33°15′56″ N
8	S114	Georgia	Piedmont	81°51′03″ W	33°07′23″ N
9	S12	Georgia	Piedmont	83°02′41″ W	34°25′03″ N
10	S121	Georgia	Blue Ridge	83°35′30″ W	34°49′59″ N
11	S122	Georgia	Blue Ridge	83°18′12″ W	34°32′45″ N
12	S123	Georgia	Blue Ridge	83°17′02″ W	34°10′27″ N
13	S127	South Carolina	Piedmont	82°58′43″ W	34°49′26″ N
14	S13	Georgia	Blue Ridge	83°20′19″ W	34°26′47″ N
15	S130	South Carolina	Blue Ridge	82°58′01″ W	34°32′37″ N
16	S131	South Carolina	Piedmont	82°57′22″ W	34°32′45″ N
17	S132	South Carolina	Piedmont	82°57′23″ W	34°31′29″ N
18	S135	Georgia	Piedmont	82°58′59″ W	33°46′26″ N
19	S136	South Carolina	Piedmont	82°46′53″ W	34°54′04″ N
20	S138	South Carolina	Piedmont	82°45′39″ W	34°55′13″ N
21	S147	South Carolina	Piedmont	82°37′48″ W	34°39′40″ N
22	S148	Georgia	Piedmont	82°47′12″ W	33°37′38″ N

WaterQuality2
Field Stations
Missing Data
Univariate Distributions
Distribution of Ys by Ecoregion
Distribution of Xs by Ecoregion
Scatterplot Matrix of Ys
Scatterplot Matrix of Xs
Use Only Non-Test Data
Use Only Test Data
Use All Data

Columns (39/0)
Station ID
Station Descriptors (7/0)
Ys (4/0)
Xs (26/0)
Test Set

Rows
All rows   86
Selected   0
Excluded   0
Hidden   0
Labelled   0

Figure 16.7   The water quality data.

26 Xs in the respective column groups. The last column, Test Set, will be described later. The row states were obtained by marking and coloring rows by Ecoregion (Rows > Color or Mark by Column).

The location of the field stations can be seen by running the script with the same name (Figure 16.8). Note that if the requisite columns are given the appropriate format, you can ask JMP to show background maps that provide additional context to the point cloud that is plotted. Maps can be stored locally or obtained on demand from a designated map server over the Internet.

The Ys are described in Table 16.4. Note that AGPT has been transformed logarithmically.

We refer to Cox and Gaudard (2013) for a description of the Xs—they are values derived from satellite images in the neighborhood of the field station from where the sample was taken. For example, column f is the percentage of forest cover, and column x is the mean slope of the terrain.

Running the "Missing Data" script reveals that five rows have missing values only for HAB, while one row has missing values for both RICH and EPT. Note that it is typical to have missing values when data are collected in the field. There are no missing values in the Xs. Dropping six of the 86 available rows is not desirable if it can be avoided, and the PLS platform in JMP Pro provides both mean and iterative expectation-maximization (EM) imputation. Using JMP Pro will allow us to make best use of the data, and using JMP will simply drop any incomplete cases.

Exploratory data analysis (EDA) (or "Uncovering Relationships" in Figure 16.1) can be conducted in the normal way. Given the heuristic nature of EDA, defining a rigid process can be self-defeating. The user interface of JMP, coupled with its inter-activity, linking, and filtering capabilities, encourages you to take a free-form approach. Nonetheless, we would recommend you to look at the suggestions in Cox et al. (2016). In this case, the correlation between Xs and Ys is of specific interest. Running the script "Scatterplot Matrix of Ys" shows the bivariate correlations of the Ys (Figure 16.9). Note that there is some suggestion of difference by ecoregion.

More extensive EDA confirms the fact that ecoregion has an important impact on the patterns of variation in the data. This suggests that we should consider two modeling approaches:

1. Produce a model for the Savannah River Basin as a whole, but include ecoregion as an additional **X**.

2. Produce three models, one for each ecoregion.

Approach (2) necessarily restricts the number of observations available for building and testing each model to the number of samples taken in that ecoregion. But, even so, it might provide better predictions than (1) for similar ecoregion located elsewhere. Approach (1) gives us more data to work with and might produce an omnibus model that has more practical utility, assuming it can predict well. In the interests of brevity, we only consider (1) here.

One of the hallmarks of predictive modeling is that we split the available data into groups for different phases of the model building process. As the name implies, our

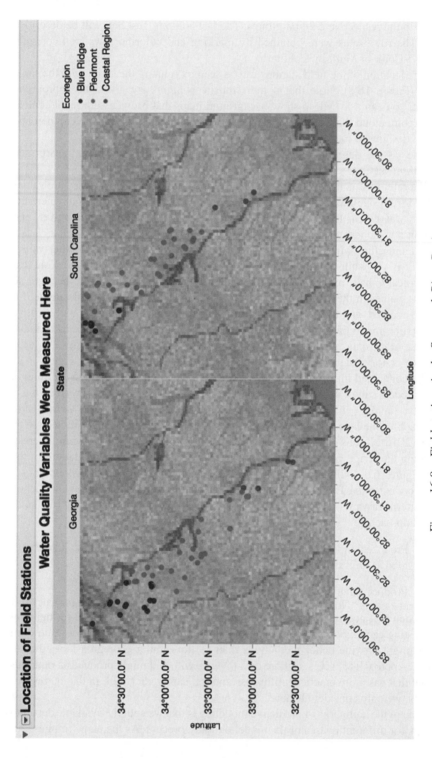

Figure 16.8   Field stations in the Savannah River Basin.

Table 16.4   Ys for the PLS model.

Column name	Full name	Description
AGPT	Algal growth potential test	An indicator of the level of nutrients that are biologically available to support algal growth. Higher levels of nutrients are indicated by higher values
HAB	Macroinvertebrate habitat	A weighted composite score derived from visual observations of stream habitat characteristics. Higher scores indicate better habitat conditions for macroinvertebrate populations
RICH	Macroinvertebrate species richness	A count of the number of taxa observed in a sample collected from a 100 meters stream segment. Higher numbers indicate greater diversity. For this study, counts exceeding 26 indicated nonimpaired conditions, while counts below 11 indicated severely impaired conditions
EPT	Ephemeroptera/ Plecoptera/ Trichoptera index	An index derived by assessing the density of three orders of macroinvertebrate that are known to be sensitive to environmental conditions. The orders are Ephemeroptera (mayflies), Plecoptera (stoneflies), and Trichoptera (caddisflies). For this study, values exceeding 10% indicate nonimpaired conditions, while values of 1% or below indicate severely impaired conditions

fundamental objective is to make statements about data not yet acquired, so it is advisable to hold back some of the data to try to assess the likely prediction accuracy. The data that we do not hold back is usually split further in some way, in an attempt to assure that we model signal rather than noise.

The last column in WaterQuality2.jmp is called Test Set. It contains the values "0" (labeled as "No") and "1" (labeled as "Yes"). This column was constructed by selecting a stratified random sample of the full table, taking a proportion of 0.3 of the rows for each level of ecoregion. You can review the outcome of this random selection by selecting Analyze>Distribution to look at the distributions of test set and ecoregion. Clicking on the Yes bar shows that there is an appropriate number of rows highlighted for each level of ecoregion.

Although we could use hide and exclude attributes in the row state, it can sometimes be less confusing to make a subset table to work with. Tables>Subset will respect any row selection currently in force, and you can make such a selection from

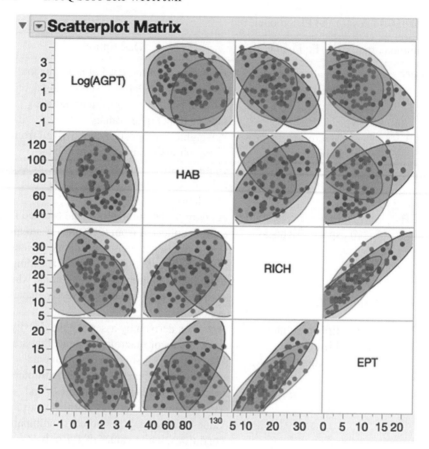

Figure 16.9   Bivariate correlation of Ys.

a distribution (Analyze > Distribution) or tabulation (Analyze > Tabulate) of Test Set. The table WaterQuality2_Train contains the required rows, plus various saved scripts.

Running the "Fit Model Launch" script gives Figure 16.10. PLS, along with many other regression-based modeling methods, are available as different personalities in the Fit Model platform (Analyze > Fit Model).

Having assigned columns to the necessary roles, you can fit and compare a variety of models (see Figure 16.11). The Model Launch outline node allows you to specify the Validation Method the first time it appears. This is the specific mechanism used to avoid overfitting, and the default is k-fold cross-validation with sevenfolds. The output from each model fit is contained under the respective outline node (all closed in Figure 16.11). Additional options for each fit are found under the LRT of each outline node.

One of the key decisions in PLS modeling is how many latent factors to included in the model. JMP will define a best model according to certain criteria, but you can override this if you want to. Each model term plays a dual role, both in dimensionality reduction and in explaining the variation in the Ys. So, aside from the usual diagnostics

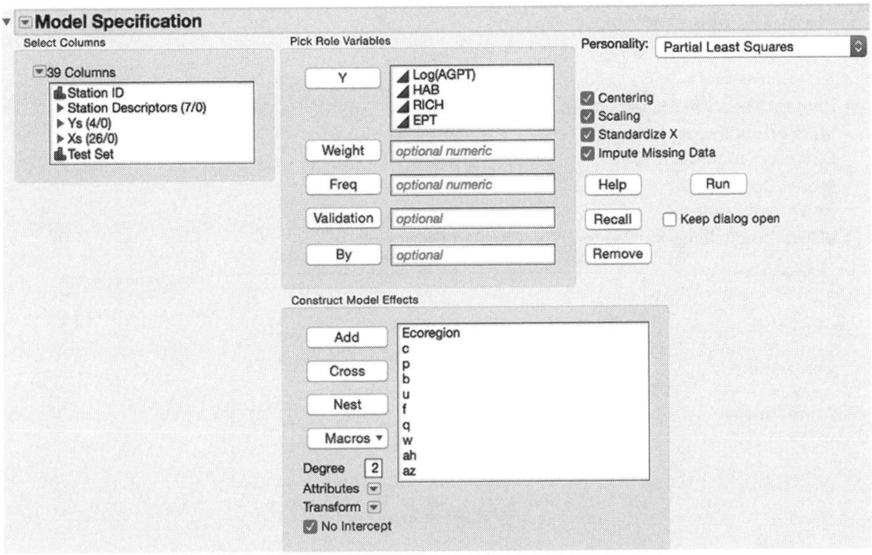

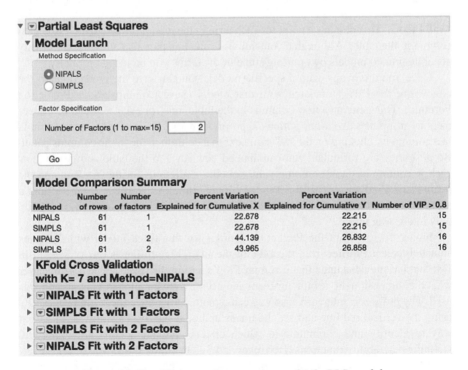

Figure 16.10   The PLS personality of fit model.

Figure 16.11   Fitting and comparing multiple PLS models.

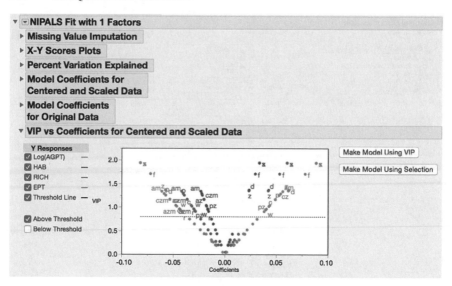

Figure 16.12   The dual role of terms in a PLS model.

relating to residuals (which should have no structure if the model is well specified), the VIP versus Coefficients plot shown in Figure 16.12 is very useful (run the script "VIP Plots"). This plot has a point for each term in the model (linked to the associated column in the table). Although a contentious issue, this plot also allows you to easily generate pruned models by omitting some of the terms you included originally.

Once you are happy with a specific model, you can save the predictions to the table from the LRT associated with that model (Save Columns > Save Prediction Formula). This generates new columns in the table, one for each Y. It can be convenient to group these columns with an appropriate name. You can inspect the formula in a column by clicking on the "+" sign next to a column name in the columns panel. Because they are formulas, you can append new rows to the table and (so long as each X column has a nonmissing value) obtain predictions for the Ys. Running the script "Add Predictions for PLS Pruned Model" mimics the effect of performing a pruned two factor PLS fit (with NIPALS and sevenfold cross-validation) involving only interesting terms.

Figure 16.13 shows the Prediction Profiler for this fitted model with Y scales adjusted (select Profiler from the LRT of the fit). This profiler is one of several in JMP but has the advantage that it can profile Ys within the space of the Xs no matter what the dimensionality of the problem and no matter what modeling approach was used. The profiler is interactive, so you can dynamically adjust the settings of the Xs using the vertical red line and see the traces update instantly. This provides a visual way to identify and communicate which effects and interactions are important. Defining desirability functions (Derringer and Suich, 1980) allows you to find X settings that simultaneously optimize the Ys, and using the integrated simulator allows this optimization to take into account preset variation in the Xs.

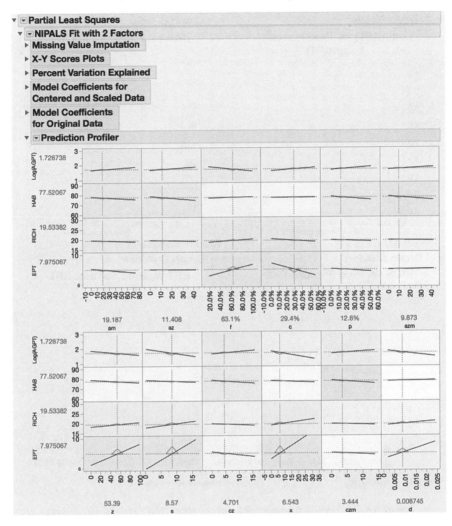

Figure 16.13   Interactively profiling four Ys in the space of 12 Xs.

Opening the table WaterQuality2_Test (containing the test data we set aside ear-lier) and running the script "Add Test Data" in the table WaterQuality2_Train make a new table, WaterQuality3, with all of the rows recombined (you can alternatively use Tables > Concatenate). Running the script "Actual by Predicted" in this new table shows these plots for all the data, and running the script "Use only Test Data" (which just manipulates the row states appropriately) updates the display to give Figure 16.14, showing the performance of the chosen model on the unseen test data.

We conclude this example with Table 16.5, an InfoQ assessment. As well as the score for each dimension, there is a comments column with some notes as to why the score was chosen. The final InfoQ is calculated as 58%.

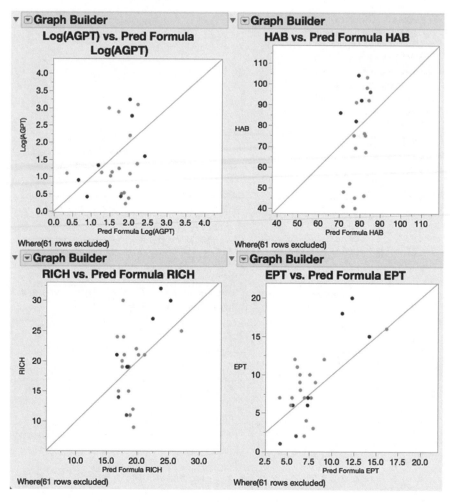

Figure 16.14   Prediction accuracy of the final PLS model for test data.

16.4   A JMP application to score the InfoQ dimensions

As noted already, part of the power of InfoQ is to force active consideration of crucial issues that too often are taken for granted. Although generating a single InfoQ score is attractive, it may sometimes be easier to generate a range of scores reflecting uncertainty in the assessment of the dimensions. Selecting Add-Ins > InfoQ > InfoQ brings up a simple application that facilitates this. You can adjust the range of scores for each dimension and see the resulting score update dynamically. The five-point scoring algorithm set out in Chapter 3 is implemented, but the use of the slider boxes would make a continuous scale possible if needed.

Table 16.5   InfoQ assessment for Example 2.

InfoQ dimension	Score	Comments
Data resolution	Acceptable	Contingent on the sampling and measurement process for the Ys being adequate. Contingent on an appropriate operationalization of the "neighborhood" of a field station
Data structure	High	Using appropriate imputation helps overcome limitations of data collected in the field, so long as the missingness is "reasonable" (as in this case)
Data integration	Very high	Contingent on upstream feature extraction to generate values for the Xs being easy to automate
Temporal relevance	High	The requisite model construction can be automated in JMP, including the deployment of the resulting score code to another system should this be required. Some oversight would be advised, but this would not be a bottleneck. Once the required analysis pathway is defined, data volumes are such that the execution time is very small
Chronology of data and goal	Low	The timing of new satellite images is not under direct control. Gathering new Y values is time consuming. So updating the model to reflect changes in land use over time may not be easy
Generalizability	Low	As noted, constraints on the size of the data and the pattern of variation therein interact with the efficacy of the modeling approach and at what level of granularity this is performed. This determines the quality of the analytic goal and hence the utility of the practical goal. Using PLS1 (in which the Ys are analyzed separately, rather than together as here) may improve prediction accuracy but is unlikely to remove the effect of ecoregion
Operationalization	Very high	Once the modeling is completed to your satisfaction, the outcome is encapsulated in score code which can be deployed without any understanding of how it was generated
Communication	High	PLS is a relatively sophisticated technique, possibly hard to explain should this be needed. The Profiler and related tools in JMP may help

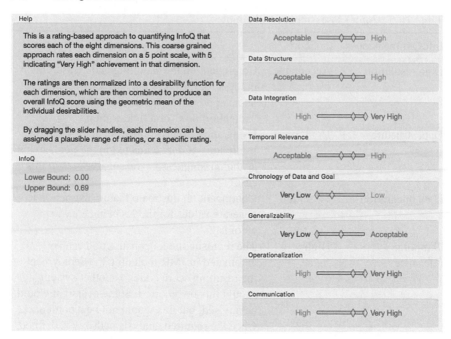

Figure 16.15   InfoQ assessment of Example 2 with uncertainty.

When initialized, the InfoQ score is undefined and the sliders span the whole range of each dimension. Figure 16.15 shows an assessment of Example 2. The InfoQ score lies in the range 0–69%.

16.5   JMP capabilities and InfoQ

Definitively mapping JMP capabilities to InfoQ is difficult, because the InfoQ agenda seeks to illuminate and encourage the productive use of all data and because JMP is very functional and becoming more so. Generally, we would expect enabling software to have most affinity with two of the InfoQ components (data quality and analysis quality), impacting the other two components only indirectly and in an emergent way.

Following through the two examples in this chapter should suggest the kind of user experience that JMP provides, and as implied in the Introduction, *how* functionality is made accessible is very influential in determining its effective use. To this extent, lists of capabilities, features, and functions of software can be misleading, or at best unhelpful, particularly when one is confronted with real-world data.

We therefore restrict ourselves to a few general comments as follows:

1. In Chapter 1, InfoQ is defined in terms of its four components g, X, f, and U as InfoQ $(f, X, g, U) = U\{f(X|g)\}$, where f is the "analysis quality" and U is the "quality of the utility measure." JMP provides the flexibility to define custom

loss functions (Desirability, 2015) which, used appropriately, can be used to increase f and U and hence improve InfoQ for given g and X.

2. The distinction between planned and unplanned use cases is important, and JMP was designed from the outset to support the latter through interactivity, the combined use of graphical and tabular output and a free-form approach to analysis where, as far as possible, the software "gets out of the way" leaving you free to tease out the nuances of the data at hand and model its essential features.

3. The link between data, models, and graphical displays thereof is powerful in both "Uncovering Relationships" and "Modeling Relationships," and JMP attempts to provide a comprehensive and unified environment to exploit this.

4. JSL provides the basis to customize or extend core capabilities and support specific users and groups of users in a way they find helpful. In relative terms, the barriers to doing this are small.

5. JSL also provides the means to automate planned use cases from "end to end," interoperating with other software as needed. Such workflows are easily deployed to others and can encapsulate best practice and help codify the knowledge that results.

6. Although not shown here, JMP has a particular strength in the field of statistically designed experiments (DOE for short). As digital marketing really takes hold, DOE is likely to find increasing application outside its agricultural and manufacturing roots and in areas where InfoQ has much relevance.

Finally, we remark again that JMP is a desktop product. Its memory-based architecture and multithreaded code base make very good use of such hardware, but there are many problems that will remain outside its grasp. In such "big data" scenarios, it can still serve as a useful prototyping environment.

16.6   Summary

Increasing data availability and computing power, coupled with the continuing necessity to improve and innovate, make it an interesting time for anyone inclined to take a rational view of how systems of value production operate or can be made to operate. InfoQ provides a much needed language and framework within which vital issues can be surfaced and discussed. Seasoned practitioners have always appreciated that the useful application of statistical thinking is largely contextual and that, while it can be mastered by an individual, it is often difficult to codify this knowledge and expertise in such a way it can be transferred to others. The relatively recent intertwining of statistics with other disciplines such as machine learning and data science only compounds this issue, which, ironically, is further exacerbated by the ready availability of software like JMP. Nonetheless, it is clear that we definitely need software for working with data. In so far as JMP has always aimed to support researchers, scientists, and engineers with statistical discovery, it is reasonable to expect that JMP can also help such users pursue the InfoQ agenda as it takes root.

References

Cockburn, A. (2001) *Writing Effective Use Cases*. Addison-Wesley, Boston.

Cox, I. and Gaudard, M. (2013) *Discovering Partial Least Squares with JMP*. SAS Institute Press, Cary, NC.

Cox, I., Gaudard, M. and Stephens, M. (2016) *Visual Six Sigma: Making Data Analysis Lean*, 2nd edition. John Wiley & Sons, Inc., Hoboken, NJ.

Derringer, G. and Suich, R. (1980) Simultaneous optimization of several response variables. *Journal of Quality Technology*, 12(4), pp. 214–219.

Desirability. http://www.jmp.com/support/help/Additional_Examples_2.shtml#289104 (accessed December 7, 2015).

JMP Missing Values. http://www.jmp.com/support/help/Explore_Missing_Values_Utility. shtml (accessed December 7, 2015).

Kenett, R.S. and Zacks, S. (2014) *Modern Industrial Statistics: With Applications in R, MINITAB and JMP*, 2nd edition. John Wiley & Sons, Ltd, Chichester, UK.

Nash, M.S. and Chaloud, D.J. (2011) Partial Least Squares Analysis of Landscape and Surface Water Biota Associations in the Savannah River Basin, International Scholarly Research Network, ISRN Ecology, Vol. 2011, Article ID 571749, 11 pages.

Ramirez, J.G. (1998) Monitoring clean room air using cuscore charts. *Quality and Reliability Engineering International*, 14(4), pp. 281–289.

REML. https://en.wikipedia.org/wiki/Restricted_maximum_likelihood (accessed December 7, 2015).

Shapelife. https://en.wikipedia.org/wiki/Shapefile (accessed December 7, 2015).

Shore, J. and Warden, D. (2007) *The Art of Agile Development*. O'Reilly Media, Sebastopol, CA.

Threeway Chart. http://blogs.sas.com/content/jmp/2012/04/18/whats-a-three-way-chart-and-why-would-i-need-one/ (accessed December 7, 2015).

Wetherill, G.B. and Brown, D.W. (1991) *Statistical Process Control: Theory and Practice*, 3rd edition. Chapman and Hall, London.

Yaschin, E. (1994) Monitoring variance components. *Technometrics*, 36(4), pp. 379–393.

Index

Page numbers in **bold** indicate tables; page numbers in *italics* indicate figures

Information Quality: The Potential of Data and Analytics to Generate Knowledge,
First Edition. Ron S. Kenett and Galit Shmueli.
© 2017 John Wiley & Sons, Ltd. Published 2017 by John Wiley & Sons, Ltd.
Companion website: www.wiley.com/go/information_quality